Electromagnetics

Mehdi Rahmani-Andebili

Electromagnetics

Practice Problems, Methods, and Solutions

Mehdi Rahmani-Andebili
Electrical and Computer Engineering
The University of Alabama
Tuscaloosa, AL, USA

ISBN 978-3-031-95399-6   ISBN 978-3-031-95400-9   (eBook)
https://doi.org/10.1007/978-3-031-95400-9

© The Editor(s) (if applicable) and The Author(s), under exclusive license to Springer Nature Switzerland AG 2025

This work is subject to copyright. All rights are solely and exclusively licensed by the Publisher, whether the whole or part of the material is concerned, specifically the rights of translation, reprinting, reuse of illustrations, recitation, broadcasting, reproduction on microfilms or in any other physical way, and transmission or information storage and retrieval, electronic adaptation, computer software, or by similar or dissimilar methodology now known or hereafter developed.

The use of general descriptive names, registered names, trademarks, service marks, etc. in this publication does not imply, even in the absence of a specific statement, that such names are exempt from the relevant protective laws and regulations and therefore free for general use.

The publisher, the authors and the editors are safe to assume that the advice and information in this book are believed to be true and accurate at the date of publication. Neither the publisher nor the authors or the editors give a warranty, expressed or implied, with respect to the material contained herein or for any errors or omissions that may have been made. The publisher remains neutral with regard to jurisdictional claims in published maps and institutional affiliations.

This Springer imprint is published by the registered company Springer Nature Switzerland AG
The registered company address is: Gewerbestrasse 11, 6330 Cham, Switzerland

If disposing of this product, please recycle the paper.

Preface

Electromagnetics is one of the most fundamental courses of Electrical Engineering and Physics majors that is taught at universities and colleges worldwide for junior students. This textbook has been prepared for instructors and students. In each chapter of the textbook, different types of problems and exercises have been presented that are categorized as follows:

- *Problems with detailed solution*: They have been designed to teach students the subjects in detail. Moreover, they have been categorized into different levels based on their difficulty levels (easy, normal, and hard) and calculation amounts (small, normal, and large). This classification helps students study the book in the most efficient way.
- *Partially solved exercises*: They have been designed to encourage students to practice problems while guiding them through the problem-solving procedure and hinting the required formulas.
- *Exercises with final answer*: They have been designed to encourage students to practice more by themselves while hinting them by the final answer as well as to help instructors to give tests or quizzes.

In the following, the description of each chapter of the textbook is briefly presented.

Chapters 1 and 2 cover the Cartesian, cylindrical, and spherical coordinate systems along with many examples and exercises. In these chapters, the conversions between the coordinate systems are investigated, and unit vector, dot product (inner product), cross product (vector product), position vector, and distance vector are introduced.

Chapters 3 and 4 teach the gradient, divergence, curl, and Laplacian operators in Cartesian, cylindrical, and spherical coordinate systems and solve many examples and exercises.

Chapters 5 and 6 study the electric flux and electric field in Cartesian, cylindrical, and spherical coordinate systems. Herein, the examples and exercises are various and comprehensive.

Chapters 7 and 8 review the electric potential in Cartesian, cylindrical, and spherical coordinate systems along with a variety of examples and exercises.

Chapters 9 and 10 investigate the electric potential energy due to discrete charge distribution and continuous charge distribution in Cartesian, cylindrical, and spherical coordinate systems.

Chapters 11 and 12 are concerned with the polarization and electric field in dielectrics as well as boundary conditions in Cartesian, cylindrical, and spherical coordinate systems.

Chapters 13 and 14 are related to the flat, cylindrical, and spherical capacitors. In this regard, various and comprehensive examples and exercises are solved.

Chapters 15 and 16 cover the subjects, including method of image charge for grounded conductors and isolated conductors.

Chapters 17 and 18 are concerned with the flat, cylindrical, and spherical resistors and boundary conditions for electric current. Herein, a variety of examples and exercises are solved.

Chapters 19 and 20 investigate the magnetic flux and magnetic field due to linear, surface, and volume currents with some interesting examples and exercises.

Chapters 21 and 22 review the electromagnetic force and electromagnetic torque with some useful examples and exercises.

Chapters 23 and 24 study the Ampere's circuital law, magnetic energy, and magnetic energy density.

Chapters 25 and 26 teach the magnetic vector potential along with several examples and exercises.

Chapters 27 and 28 cover the magnetization in magnetic materials with some useful examples.

Chapters 29 and 30 are concerned with the boundary conditions in magnetic field and method of image current in magnetostatics.

Chapters 31 and 32 investigate electromagnetic induction with some useful and interesting examples.

It should be mentioned that General Physics II is a perquisite for Electromagnetics. Hence, the underprepared students are suggested to review the related parts of it before studying Electromagnetics. The subjects covered in the textbook of "M. Rahmani-Andebili, *General Physics II: Practice Problems, Methods, and Solutions*, Springer Nature, 2025" are as follows:

- *Electrostatics*
- *Electrical Charge, Capacitance, Capacitor, Current, Resistance, and Resistor*
- *Magnetic Field*
- *Electromagnetic Induction*
- *Fluid Dynamics*
- *Thermodynamics*
- *Transverse and Longitudinal Waves*
- *Light, Mirrors, and Lenses*

Since the textbook includes the basic and advanced problems with very detailed problem solutions, it can be used as a practicing study guide by students and as a supplementary teaching source by instructors. Moreover, since the problems and exercises have very detailed solutions, the textbook is helpful for underprepared students. In addition, it is beneficial for knowledgeable students because it includes advanced problems and exercises.

In preparing the problems and solutions, care has been taken to use methods typically found in the primary instructor-recommended textbooks. By considering this key point, the textbook is in the direction of instructors' lectures, and the instructors will not see any untaught and unusual problem solutions in their students' answer sheets.

Students and instructors are welcome to email their comments to me at mehdi.rahmani.andebili@gmail.com if they find any misprint or any other possible mistakes in the textbook. Their name and contribution will be mentioned in the next version of the textbook.

Tuscaloosa, AL   Mehdi Rahmani-Andebili

The Other Books Published by the Author

AC Electric Machines: Practice Problems, Methods, and Solutions

AC Electrical Circuits Analysis: Practice Problems, Methods, and Solutions, Second Edition

Advanced Electrical Circuits Analysis: Practice Problems, Methods, and Solutions, Second Edition

Applications of Artificial Intelligence in Planning and Operation of Smart Grid

Applications of Fuzzy Logic in Planning and Operation of Smart Grids

Calculus I: Practice Problems, Methods, and Solutions, Second Edition

Calculus II: Practice Problems, Methods, and Solutions

Calculus III: Practice Problems, Methods, and Solutions

DC Electric Machines, Electromechanical Energy Conversion Principles, and Magnetic Circuit Analysis: Practice Problems, Methods, and Solutions

DC Electrical Circuits Analysis: Practice Problems, Methods, and Solutions, Second Edition

Design, Control, and Operation of Microgrids in Smart Grids

Differential Equations: Practice Problems, Methods, and Solutions

Feedback Control Systems Analysis and Design: Practice Problems, Methods, and Solutions

General Physics I: Practice Problems, Methods, and Solutions

General Physics II: Practice Problems, Methods, and Solutions

Mathematics of Engineering and Science: Practice Problems, Methods, and Solutions

MATLAB Lessons, Examples, and Exercises: A Tutorial for Beginners and Experts

Planning and Operation of Electric Vehicles in Smart Grid

Planning and Operation of Plug-in Electric Vehicles: Technical, Geographical, and Social Aspects

Power System Analysis: Comprehensive Lessons

Power System Analysis: Practice Problems, Methods, and Solutions

Precalculus: Practice Problems, Methods, and Solutions, Second Edition

Operation of Smart Homes

Contents

1 Cartesian, Cylindrical, and Spherical Coordinate Systems: Part A 1
- 1.1 Unit Vector 1
- 1.2 Dot Product (Inner Product) 2
- 1.3 Cross Product (Vector Product) 4
- 1.4 Cartesian Coordinate System 6
- 1.5 Cylindrical Coordinate System 7
- 1.6 Spherical Coordinate System 13
- 1.7 Cartesian to Cylindrical Coordinate Conversion 19
- 1.8 Cartesian to Spherical Coordinate Conversion 20
- 1.9 Cylindrical to Cartesian Coordinate Conversion 21
- 1.10 Spherical to Cartesian Coordinate Conversion 22
- 1.11 Spherical to Cylindrical Coordinate Conversion 23
- 1.12 Cylindrical to Spherical Coordinate Conversion 24
- 1.13 Position Vector 25
- 1.14 Distance Vector 26
- References 27

2 Cartesian, Cylindrical, and Spherical Coordinate Systems: Part B 29
- 2.1 Unit Vector 29
- 2.2 Dot Product (Inner Product) 30
- 2.3 Cross Product (Vector Product) 32
- 2.4 Cartesian Coordinate System 33
- 2.5 Cylindrical Coordinate System 36
- 2.6 Spherical Coordinate System 45
- 2.7 Cartesian to Cylindrical Coordinate Conversion 54
- 2.8 Cartesian to Spherical Coordinate Conversion 55
- 2.9 Cylindrical to Cartesian Coordinates Conversion 55
- 2.10 Spherical to Cartesian Coordinates Conversion 56
- 2.11 Spherical to Cylindrical Coordinates Conversion 56
- 2.12 Cylindrical to Spherical Coordinates Conversion 57
- 2.13 Position Vector 58
- 2.14 Distance Vector 59
- References 59

3 Gradient, Divergence, Curl, and Laplacian: Part A 61
- 3.1 Gradient 61
- 3.2 Gradient of a Scalar in the Cartesian Coordinate System 61
- 3.3 Gradient of a Scalar in Cylindrical Coordinate System 63
- 3.4 Gradient of a Scalar in Spherical Coordinate System 64
- 3.5 Divergence 66
- 3.6 Divergence of a Vector in the Cartesian Coordinate System 67
- 3.7 Divergence of a Vector in Cylindrical Coordinate System 69

3.8	Divergence of a Vector in Spherical Coordinate System	70
3.9	Curl	72
3.10	Curl of a Vector in the Cartesian Coordinate System	73
3.11	Curl of a Vector in Cylindrical Coordinate System	75
3.12	Curl of a Vector in Spherical Coordinate System	76
3.13	Laplacian	78
3.14	Laplacian of a Scalar in the Cartesian Coordinate System	78
3.15	Laplacian of a Scalar in Cylindrical Coordinate System	79
3.16	Laplacian of a Scalar in Spherical Coordinate System	80
	References	82

4 Gradient, Divergence, Curl, and Laplacian: Part B 83

4.1	Gradient	83
4.2	Gradient of a Scalar in the Cartesian Coordinate System	83
4.3	Gradient of a Scalar in Cylindrical Coordinate System	84
4.4	Gradient of a Scalar in Spherical Coordinate System	86
4.5	Divergence	87
4.6	Divergence of a Vector in the Cartesian Coordinate System	88
4.7	Divergence of a Vector in Cylindrical Coordinate System	90
4.8	Divergence of a Vector in Spherical Coordinate System	91
4.9	Curl	92
4.10	Curl of a Vector in the Cartesian Coordinate System	94
4.11	Curl of a Vector in Cylindrical Coordinate System	95
4.12	Curl of a Vector in Spherical Coordinate System	96
4.13	Laplacian	98
4.14	Laplacian of a Scalar in the Cartesian Coordinate System	98
4.15	Laplacian of a Scalar in Cylindrical Coordinate System	100
4.16	Laplacian of a Scalar in Spherical Coordinate System	100
	References	102

5 Electric Field and Electric Flux: Part A 103

5.1	Electric Flux	103
5.2	Electric Field in the Cartesian Coordinate System	106
5.3	Electric Field in Cylindrical Coordinate System	111
5.4	Electric Field in Spherical Coordinate System	121
	References	130

6 Electric Field and Electric Flux: Part B 131

6.1	Electric Flux	131
6.2	Electric Field in Cartesian Coordinate System	135
6.3	Electric Field in Cylindrical Coordinate System	139
6.4	Electric Field in Spherical Coordinate System	163
	References	187

7 Electric Potential: Part A 189

7.1	Electric Potential in the Cartesian Coordinate System	189
7.2	Electric Potential in Cylindrical Coordinate System	191
7.3	Electric Potential in Spherical Coordinate System	195
	References	203

8 Electric Potential: Part B 205

8.1	Electric Potential in the Cartesian Coordinate System	205
8.2	Electric Potential in Cylindrical Coordinate System	207
8.3	Electric Potential in Spherical Coordinate System	216
	References	231

9	**Electric Potential Energy: Part A**	233
	9.1 Electric Potential Energy Due to Discrete Charge Distribution	233
	9.2 Electric Potential Energy Due to Continuous Charge Distribution	238
	References	242
10	**Electric Potential Energy: Part B**	243
	10.1 Electric Potential Energy Due to Discrete Charge Distribution	243
	10.2 Electric Potential Energy Due to Continuous Charge Distribution	247
	References	251
11	**Polarization and Electric Field in Dielectrics and Boundary Conditions: Part A**	253
	11.1 Polarization and Electric Field in Dielectrics and Boundary Conditions in the Cartesian Coordinate System	253
	11.2 Polarization and Electric Field in Dielectrics in Cylindrical Coordinate System	256
	11.3 Polarization and Electric Field in Dielectrics in Spherical Coordinate System	260
	References	264
12	**Polarization and Electric Field in Dielectrics and Boundary Conditions: Part B**	265
	12.1 Polarization and Electric Field in Dielectrics and Boundary Conditions in the Cartesian Coordinate System	265
	12.2 Polarization and Electric Field in Dielectrics in Cylindrical Coordinate System	270
	12.3 Polarization and Electric Field in Dielectrics in Spherical Coordinate System	276
	References	282
13	**Flat, Cylindrical, and Spherical Capacitors: Part A**	283
	13.1 Flat Capacitors	283
	13.2 Cylindrical Capacitors	287
	13.3 Spherical Capacitors	290
	References	294
14	**Flat, Cylindrical, and Spherical Capacitors: Part B**	295
	14.1 Flat Capacitors	295
	14.2 Cylindrical Capacitors	302
	14.3 Spherical Capacitors	311
	References	323
15	**Method of Image Charge in Electrostatics: Part A**	325
	15.1 Method of Image Charge for Grounded Conductors	325
	15.2 Method of Image Charge for Isolated Conductors	328
	References	330
16	**Method of Image Charge in Electrostatics: Part B**	331
	16.1 Method of Image Charge for Grounded Conductors	331
	16.2 Method of Image Charge for Isolated Conductors	335
	References	339
17	**Flat, Cylindrical, and Spherical Resistors and Boundary Conditions for Electric Current: Part A**	341
	17.1 Boundary Conditions for Electric Current	341
	17.2 Flat Resistors	343

	17.3	Cylindrical Resistors	347
	17.4	Spherical Resistors	351
	References		354

18 Flat, Cylindrical, and Spherical Resistors and Boundary Conditions for Electric Current: Part B ... 355
- 18.1 Boundary Conditions for Electric Current ... 355
- 18.2 Flat Resistors ... 357
- 18.3 Cylindrical Resistors ... 364
- 18.4 Spherical Resistors ... 372
- References ... 381

19 Magnetic Field and Magnetic Flux: Part A ... 383
- 19.1 Magnetic Field Due to a Linear Current ... 383
- 19.2 Magnetic Field Due to a Surface Current ... 390
- 19.3 Magnetic Field Due to a Volume Current ... 392
- 19.4 Magnetic Flux ... 392
- References ... 393

20 Magnetic Field and Magnetic Flux: Part B ... 395
- 20.1 Magnetic Field Due to a Linear Current ... 395
- 20.2 Magnetic Field Due to a Surface Current ... 403
- 20.3 Magnetic Field Due to a Volume Current ... 406
- 20.4 Magnetic Flux ... 407
- References ... 408

21 Electromagnetic Force and Torque: Part A ... 409
- 21.1 Electromagnetic Force ... 409
- 21.2 Electromagnetic Torque ... 413
- References ... 414

22 Electromagnetic Force and Torque: Part B ... 415
- 22.1 Electromagnetic Force ... 415
- 22.2 Electromagnetic Torque ... 421
- References ... 422

23 Ampere's Circuital Law and Magnetic Energy: Part A ... 423
- 23.1 Ampere's Circuital Law ... 423
- 23.2 Magnetic Energy and Magnetic Energy Density ... 426
- References ... 427

24 Ampere's Circuital Law and Magnetic Energy: Part B ... 429
- 24.1 Ampere's Circuital Law ... 429
- 24.2 Magnetic Energy and Magnetic Energy Density ... 436
- References ... 438

25 Magnetic Vector Potential: Part A ... 439
- References ... 441

26 Magnetic Vector Potential: Part B ... 443
- References ... 445

27 Magnetization: Part A ... 447
- References ... 451

28 Magnetization: Part B ... 453
- References ... 458

29 Boundary Conditions and Method of Image Current in Magnetostatics: Part A ... 459
- 29.1 Boundary Conditions in Magnetic Field ... 459
- 29.2 Method of Image Current in Magnetostatics ... 462
- References ... 463

30 Boundary Conditions and Method of Image Current in Magnetostatics: Part B ... 465
- 30.1 Boundary Conditions in Magnetic Field ... 465
- 30.2 Method of Image Current in Magnetostatics ... 469
- References ... 470

31 Electromagnetic Induction: Part A ... 471
- References ... 474

32 Electromagnetic induction: Part B ... 475
- References ... 481

Index ... 483

About the Author

Mehdi Rahmani-Andebili is an Assistant Professor with the Department of Electrical and Computer Engineering at the University of Alabama. He received his first M.Sc. and Ph.D. degrees in Electrical Engineering from Tarbiat Modares University and Clemson University in 2011 and 2016, respectively, and his second Ph.D. (2nd M.Sc.) degree in Physics and Astronomy from the University of Alabama in Huntsville in 2019. Moreover, he was a Postdoctoral Fellow at the Sharif University of Technology from 2016 to 2017. As a Professor, he has taught a wide selection of courses and labs, including Power System Analysis, Electric Machines, Feedback Control Systems Analysis and Design, Signals and Systems, Electromagnetics, Electronics, Digital Logic, Industrial Electronics, Renewable Distributed Generation and Storage, AC Electrical Circuits Analysis, DC Electrical Circuits Analysis, Electrical Circuits and Devices, Fundamentals of Electrical Engineering, Essentials of Electrical Engineering Technology, and Algebra and Calculus-Based Physics. Dr. Rahmani-Andebili has over 300 single-author or first-author publications, including journal papers, conference papers, textbooks, books, and book chapters, in the fields of Electrical Engineering, Power Engineering, Physics, Mathematics, and Computer Programming. He is an IEEE Senior Member and a permanent reviewer of many reputable journals. His research areas include Smart Grid, Applications of Artificial Intelligence in Planning and Operation of Power Systems, Integration of Renewables and Energy Storages into Power Systems, Energy Scheduling and Demand-Side Management, Electric Vehicles and Distributed Generation, and Advanced Optimization Techniques in Power System Studies.

1. Cartesian, Cylindrical, and Spherical Coordinate Systems: Part A

Abstract

In this chapter, the basic and advanced problems of cartesian, cylindrical, and spherical coordinate systems are studied. The subjects include unit vector; dot product (inner product); cross product (vector product); differential length vectors, differential area vectors, and differential volume in cartesian, cylindrical, and spherical coordinate systems; conversions between the coordinate systems; position vector; and distance vector. Herein, different types of problems and exercises are presented that are categorized as follows.

- **Problems with detailed solution**: They have been designed to teach students the subjects in detail. Moreover, they have been categorized in different levels based on their difficulty levels (easy, normal, and hard) and calculation amounts (small, normal, and large).
- **Partially solved exercises**: They have been designed to encourage students to practice problems while guiding them through the problem-solving procedure and hinting the required formulas.
- **Exercises with final answer**: They have been designed to encourage students to practice more by themselves while hinting them by the final answer as well as to help instructors to give tests or quizzes.

1.1 Unit Vector

Problem

1.1. Calculate the unit vector of the vector $\vec{R} = \hat{x} + 2\hat{y} + 3\hat{z}$ [1–8].
Difficulty level • Easy ○ Normal ○ Hard
Calculation amount • Small ○ Normal ○ Large

1) $\hat{R} = \dfrac{1}{\sqrt{14}}$
2) $\hat{R} = \sqrt{14}(\hat{x} + 2\hat{y} + 3\hat{z})$
3) $\hat{R} = \sqrt{14}$
4) $\hat{R} = \dfrac{1}{\sqrt{14}}(\hat{x} + 2\hat{y} + 3\hat{z})$

Partially Solved Exercise

1.2. Calculate the unit vector of the vector below.

$$\vec{R} = \sqrt{3}\hat{x} + \sqrt{3}\hat{y} + \sqrt{3}\hat{z}$$

Solution

As we know, the unit vector ($\hat{R}$) of the vector $\vec{R} = R_1\hat{x} + R_2\hat{y} + R_3\hat{z}$ can be calculated as follows.

$$\hat{R} = \frac{\vec{R}}{R} = \frac{\vec{R}}{\sqrt{(R_1)^2 + (R_2)^2 + (R_3)^2}}$$

where, R is the magnitude of the vector.

Hence:

$$\hat{R} = \frac{(\quad\quad\quad)}{\sqrt{(\;)^2 + (\;)^2 + (\;)^2}} = \frac{(\quad\quad\quad)}{(\;)}$$

$$\Rightarrow \hat{R} = \frac{\sqrt{3}}{3}\hat{x} + \frac{\sqrt{3}}{3}\hat{y} + \frac{\sqrt{3}}{3}\hat{z}$$

Exercise

1.3. Calculate the unit vector of the following vector.

$$\vec{R} = 5\hat{z}$$

Final Answer

$\hat{R} = \hat{z}$

Exercise

1.4. Identify a vector that its magnitude and unit vector are $R = 8$ and $\hat{R} = \left(\frac{\sqrt{6}}{4}\hat{x} + \frac{1}{4}\hat{y} + \frac{3}{4}\hat{z}\right)$, respectively.

Final Answer

$\vec{R} = 2\sqrt{6}\hat{x} + 2\hat{y} + 6\hat{z}$

1.2 Dot Product (Inner Product)

Problem

1.5. Which one of the following choices is wrong about the dot product of the vectors below?

$$\vec{a} = a_1\hat{x} + a_2\hat{y} + a_3\hat{z}$$

$$\vec{b} = b_1\hat{x} + b_2\hat{y} + b_3\hat{z}$$

1.2 Dot Product (Inner Product)

Difficulty level ○ Easy ● Normal ○ Hard
Calculation amount ● Small ○ Normal ○ Large

1) $\vec{a}\cdot\vec{b} = a_1 b_1 + a_2 b_2 + a_3 b_3$
2) $\vec{a}\cdot\vec{b} = ab\cos\theta$
3) $\vec{a}\cdot\vec{b} = -\vec{b}\cdot\vec{a}$
4) $\vec{a}\cdot\vec{a} = a^2$

Problem

1.6. Calculate the angle between the vectors below.

$$\vec{a} = -3\hat{x} - 3\hat{y} + 3\hat{z}$$

$$\vec{b} = 2\hat{x} - 2\hat{y} + 2\hat{z}$$

Difficulty level ● Easy ○ Normal ○ Hard
Calculation amount ● Small ○ Normal ○ Large

1) $\theta = \arccos\frac{1}{3}$
2) $\theta = \arccos\frac{1}{2}$
3) $\theta = \arccos\frac{1}{5}$
4) $\theta = \arccos\frac{1}{4}$

Partially Solved Exercise

1.7. Calculate the angle between the following vectors.

$$\vec{a} = \hat{x} + \hat{y} + \hat{z}$$

$$\vec{b} = 2\hat{y} - \hat{z}$$

Solution

The problem can be solved as follows.

$$\cos\theta = \frac{\vec{a}\cdot\vec{b}}{|\vec{a}||\vec{b}|} = \frac{(\qquad)}{(\sqrt{\qquad})(\sqrt{\qquad})}$$

$$\Rightarrow \cos\theta = \frac{1}{(\quad)\times(\quad)} = (\quad)$$

$$\Rightarrow \theta = \arccos\frac{1}{\sqrt{15}}$$

Notes

In this problem, the relations below have been used.

The angle between two vectors can be calculated as follows.

$$\cos\theta = \frac{\vec{a}.\vec{b}}{ab}$$

The inner product (dot product) of two vectors can be calculated as follows.

$$\vec{a}.\vec{b} = (a_1\hat{x} + a_2\hat{y} + a_3\hat{z}).(b_1\hat{x} + b_2\hat{y} + b_3\hat{z}) = a_1b_1 + a_2b_2 + a_3b_3$$

The magnitude of a vector can be calculated as follows.

$$a = |\vec{a}| = |a_1\hat{x} + a_2\hat{y} + a_3\hat{z}| = \sqrt{(a_1)^2 + (a_2)^2 + (a_3)^2}$$

Problem

1.8. Determine the values of the parameter "a" so that the vector $\vec{A}$ is perpendicular to the vector $\vec{B}$.

$$\vec{A} = a\hat{x} + 3\hat{y} + \hat{z}$$

$$\vec{B} = a\hat{x} - a\hat{y} + 2\hat{z}$$

Difficulty level   ○ Easy   ● Normal   ○ Hard
Calculation amount ● Small ○ Normal   ○ Large
1) $a = 1, 0$
2) $a = 2, 1$
3) $a = 3, 1$
4) $a = 3, 2$

1.3 Cross Product (Vector Product)

Problem

1.9. Which one of the following choices is wrong about the cross product of the vectors below?

$$\vec{a} = a_1\hat{x} + a_2\hat{y} + a_3\hat{z}$$

$$\vec{b} = b_1\hat{x} + b_2\hat{y} + b_3\hat{z}$$

1.3 Cross Product (Vector Product)

Difficulty level ○ Easy ● Normal ○ Hard
Calculation amount ● Small ○ Normal ○ Large

1) $\vec{a} \times \vec{b} = (a_2 b_3 - a_3 b_2)\hat{x} + (a_3 b_1 - a_1 b_3)\hat{y} + (a_1 b_2 - a_2 b_1)\hat{z}$
2) $\vec{a} \times \vec{b} = ab \sin\theta$
3) $\vec{a} \times \vec{b} = -\vec{b} \times \vec{a}$
4) $\vec{a} \times \vec{a} = a^2$

Problem

1.10. Determine the vector product of the following vectors.

$$\vec{a} = 2\hat{x} - 4\hat{y} + \hat{z}$$

$$\vec{b} = 6\hat{x} + 8\hat{y} - 15\hat{z}$$

Difficulty level ○ Easy ● Normal ○ Hard
Calculation amount ● Small ○ Normal ○ Large

1) $\vec{a} \times \vec{b} = 52\hat{x} + 36\hat{y} + 40\hat{z}$
2) $\vec{a} \times \vec{b} = 5\hat{x} + 36\hat{y} + 40\hat{z}$
3) $\vec{a} \times \vec{b} = 52\hat{x} + 6\hat{y} + 40\hat{z}$
4) $\vec{a} \times \vec{b} = 52\hat{x} + 36\hat{y} + 4\hat{z}$

Partially Solved Exercise

1.11. Calculate the cross product of the vectors below.

$$\vec{a} = \hat{x} - \hat{y} + \hat{z}$$

$$\vec{b} - \hat{x} \mid \hat{y} + \hat{z}$$

Solution

The problem can be solved as follows.

$$\vec{a} \times \vec{b} = \begin{vmatrix} \hat{x} & \hat{y} & \hat{z} \\ 1 & -1 & 1 \\ 1 & 1 & 1 \end{vmatrix} = (\quad)\hat{x} + (\quad)\hat{y} + (\quad)\hat{z}$$

$$\Rightarrow \vec{a} \times \vec{b} = (\quad)\hat{x} + (\quad)\hat{y} + (\quad)\hat{z}$$

Notes

In this problem, the relation below has been used.

The vector product (cross product) of two vectors can be calculated as follows.

$$\vec{a} \times \vec{b} = \begin{vmatrix} \hat{x} & \hat{y} & \hat{z} \\ a_1 & a_2 & a_3 \\ b_1 & b_2 & b_3 \end{vmatrix} = (a_2 b_3 - a_3 b_2)\hat{x} + (a_3 b_1 - a_1 b_3)\hat{y} + (a_1 b_2 - a_2 b_1)\hat{z}$$

1.4 Cartesian Coordinate System

Exercise

1.12. What are the variables of the Cartesian (rectangular) coordinate system?

Final Answer

(x, y, z)

Problem

1.13. What are the range of variables in the Cartesian coordinate system?
Difficulty level   ● Easy   ○ Normal   ○ Hard
Calculation amount   ● Small   ○ Normal   ○ Large
1) $-\infty < x, y < \infty, 0 \leq z < \infty$
2) $0 \leq x, y, z < \infty$
3) $0 < x, y, z < \infty$
4) $-\infty < x, y, z < \infty$

Problem

1.14. Which one of the following choices is wrong about the cross product of the unit vectors in the Cartesian coordinate system?
Difficulty level   ● Easy   ○ Normal   ○ Hard
Calculation amount   ● Small   ○ Normal   ○ Large
1) $\hat{x} \times \hat{y} = \hat{z}$
2) $\hat{z} \times \hat{y} = -\hat{x}$
3) $\hat{z} \times \hat{x} = \hat{y}$
4) $\hat{x} \times \hat{z} = \hat{y}$

Problem

1.15. What is the overall differential length vector in the Cartesian coordinate system?
Difficulty level   ● Easy   ○ Normal   ○ Hard
Calculation amount   ● Small   ○ Normal   ○ Large

1) $\vec{dl} = dx\hat{x} + dx\hat{y} + dx\hat{z}$
2) $\vec{dl} = dx\hat{x} + dy\hat{y} + dx\hat{z}$
3) $\vec{dl} = (dx + dy + dz)\hat{x}$
4) $\vec{dl} = dx\hat{x} + dy\hat{y} + dz\hat{z}$

Problem

1.16. Which one of the following choices correctly presents one of the differential area vectors in Cartesian coordinate system?
Difficulty level   ● Easy   ○ Normal   ○ Hard
Calculation amount   ● Small   ○ Normal   ○ Large
1) $\vec{dS_x} = dxdz\hat{x}$
2) $\vec{dS_y} = dxdy\hat{y}$
3) $\vec{dS_z} = dxdy\hat{z}$
4) $\vec{dS_z} = dydz\hat{z}$

Problem

1.17. What is the differential volume in the Cartesian coordinate system?
Difficulty level   ● Easy   ○ Normal   ○ Hard
Calculation amount   ● Small   ○ Normal   ○ Large
1) $dV = dxdy$
2) $dV = (dx)^3$
3) $dV = dxdydz$
4) $dV = dx + dz$

1.5 Cylindrical Coordinate System

Exercise

1.18. What are the variables of the cylindrical coordinate system?

Final Answer

(r, φ, z)

Problem

1.19. What is the range of variables in the cylindrical coordinate system?
Difficulty level   ○ Easy   ● Normal   ○ Hard
Calculation amount   ● Small   ○ Normal   ○ Large
1) $-\infty < r < \infty$, $0 \leq \varphi < 2\pi$, $-\infty < z < \infty$
2) $0 \leq r < \infty$, $0 \leq \varphi < \pi$, $-\infty < z < \infty$
3) $0 \leq r < \infty$, $0 \leq \varphi < 2\pi$, $0 \leq z < \infty$
4) $0 \leq r < \infty$, $0 \leq \varphi < 2\pi$, $-\infty < z < \infty$

Problem

1.20. Which one of the following choices is wrong about the cross product of the unit vectors in the cylindrical coordinate system?
Difficulty level   ○ Easy   ● Normal   ○ Hard
Calculation amount   ● Small   ○ Normal   ○ Large
1) $\hat{\varphi} \times \hat{r} = -\hat{z}$
2) $\hat{z} \times \hat{\varphi} = \hat{r}$
3) $\hat{\varphi} \times \hat{z} = \hat{r}$
4) $\hat{z} \times \hat{r} = \hat{\varphi}$

Problem

1.21. What is the overall differential length vector in the cylindrical coordinate system?
Difficulty level   ○ Easy   ● Normal   ○ Hard
Calculation amount   ● Small   ○ Normal   ○ Large
1) $\vec{dl} = dr\hat{r} + rd\varphi\hat{\varphi} + dz\hat{z}$
2) $\vec{dl} = rdr\hat{r} + d\varphi\hat{\varphi} + dz\hat{z}$
3) $\vec{dl} = rdr\hat{r} + rd\varphi\hat{\varphi} + dz\hat{z}$
4) $\vec{dl} = dr\hat{r} + d\varphi\hat{\varphi} + dz\hat{z}$

Problem

1.22. Which one of the following choices is wrong about the differential area vectors in cylindrical coordinate system?
Difficulty level   ○ Easy   ● Normal   ○ Hard
Calculation amount   ● Small   ○ Normal   ○ Large
1) $\vec{dS_r} = rd\varphi dz\hat{r}$
2) $\vec{dS_\varphi} = drdz\hat{\varphi}$
3) $\vec{dS_z} = rdrd\varphi\hat{z}$
4) $\vec{dS_\varphi} = drd\varphi\hat{\varphi}$

Problem

1.23. Calculate the area of a cylindrical surface described by $r = 1$, $30° \leq \varphi \leq 60°$, $0 \leq z \leq 1$. The area is graphically shown in Fig. 1.1.
Difficulty level   ○ Easy   ● Normal   ○ Hard
Calculation amount   ● Small   ○ Normal   ○ Large
1) $\frac{1}{6}\pi$
2) $\frac{1}{10}\pi$
3) $\frac{1}{30}\pi$
4) $\frac{1}{18}\pi$

1.5 Cylindrical Coordinate System

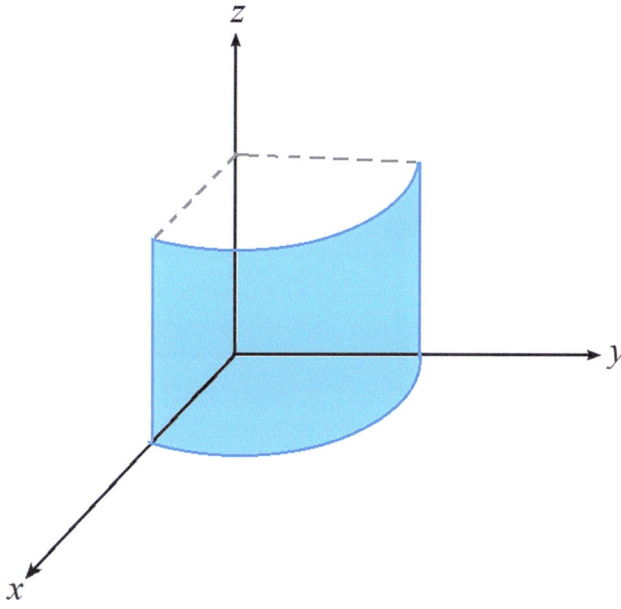

Fig. 1.1 A cylindrical surface

Partially Solved Exercise

1.24. Calculate the area of a cylindrical surface described by $r = 10$, $0 \leq \varphi \leq \frac{\pi}{2}$, $1 \leq z \leq 3$.

Solution

In this problem, r is constant for the surface. Thus, we need to use the relation below.

$$S_r = \iint dS_r$$

Therefore:

$$S_r = \int_{z=1}^{z=3} \int_{\varphi=0}^{\varphi=\frac{\pi}{2}} r\, d\varphi dz = (\quad) \times \int_{z=1}^{z=3} \int_{\varphi=0}^{\varphi=\frac{\pi}{2}} d\varphi dz$$

$$\Rightarrow S_r = (\quad) \Big|_1^3 (\quad) \Big|_0 = (\quad - \quad)(\quad - \quad) = (\quad) \times (\quad)$$

$$\Rightarrow S_r = 10\pi$$

Problem

1.25. Calculate the area of a cylindrical surface described by $0.1 \leq r \leq 0.5$, $\varphi = 90°$, $0 \leq z \leq 2$. The area is graphically shown in Fig. 1.2.

Difficulty level ○ Easy ● Normal ○ Hard
Calculation amount ● Small ○ Normal ○ Large
1) 2
2) 0.4
3) 0.8
4) 1.6

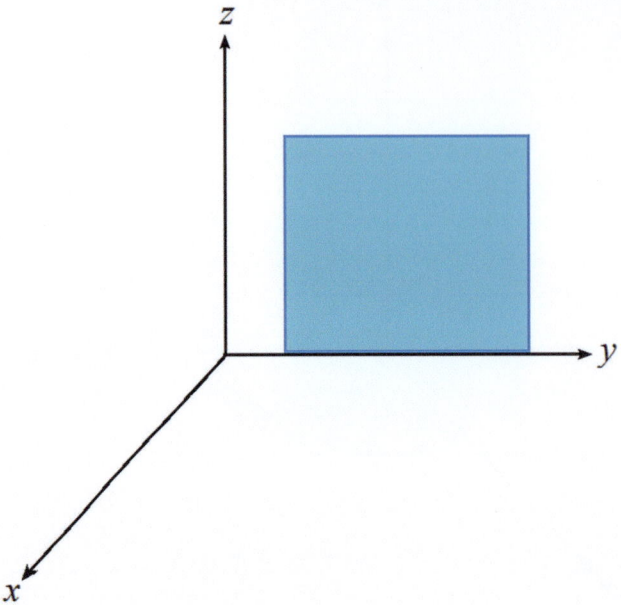

Fig. 1.2 A cylindrical surface

Partially Solved Exercise

1.26. Calculate the area of a cylindrical surface described by $0 \leq r \leq 2$, $\varphi = \dfrac{\pi}{6}$, $0 \leq z \leq 2$.

Solution

Since φ is constant for the surface, we need to use the relation below.

$$S_\varphi = \iint dS_\varphi$$

Therefore:

$$S_\varphi = \int_{z=0}^{z=2} \int_{r=0}^{r=2} dr\,dz$$

$$\Rightarrow S_\varphi = (\quad)\bigg|_0^2 (\quad)\bigg|_0^2 = (\quad - \quad)(\quad - \quad) = (\quad)(\quad)$$

$$\Rightarrow S_\varphi = 4$$

Exercise

1.27. Calculate the area of a cylindrical surface described by $2 \leq r \leq 5$, $\varphi = 45°$, $-2 \leq z \leq 2$.

Final Answer

$S = 12$

1.5 Cylindrical Coordinate System

Problem

1.28. Calculate the area of a cylindrical surface described by $1 \leq r \leq 2$, $30° \leq \varphi \leq 60°$, $z = 1$. The area is graphically shown in Fig. 1.3.

Difficulty level ○ Easy ● Normal ○ Hard
Calculation amount ● Small ○ Normal ○ Large

1) $\frac{1}{4}\pi$
2) $\frac{1}{3}\pi$
3) $\frac{1}{2}\pi$
4) π

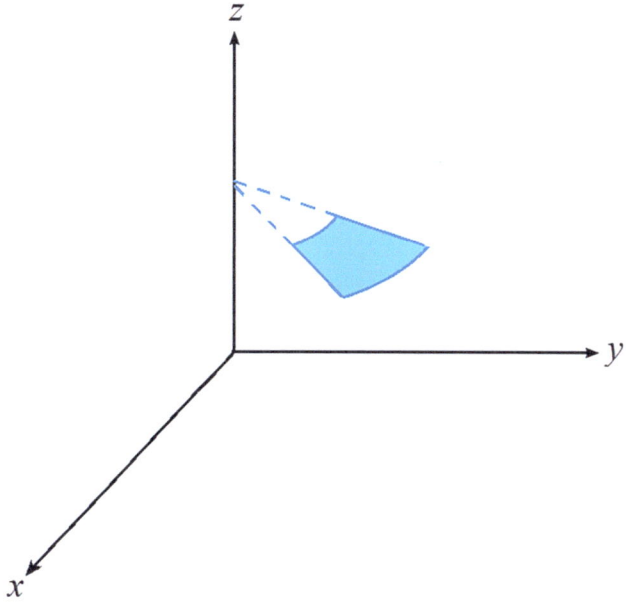

Fig. 1.3 A cylindrical surface

Partially Solved Exercise

1.29. Calculate the area of a cylindrical surface described by $2 \leq r \leq 5$, $90° \leq \varphi \leq 180°$, $z = 0$.

Solution

In this problem, z is constant; therefore, we must use the relation below.

$$S_z = \iint dS_z$$

Hence:

$$S_z = \int_{\varphi=\frac{\pi}{2}}^{\varphi=\pi} \int_{r=2}^{r=5} r\, dr\, d\varphi$$

$$\Rightarrow S_z = (\quad)\Big|_{\frac{\pi}{2}}^{\pi} (\quad)\Big|_{2}^{5} = (\quad - \quad)(\quad - \quad) = (\quad)(\quad)$$

$$\Rightarrow S_z = \frac{21\pi}{4}$$

Problem

1.30. What is the differential volume in the cylindrical coordinate system?

Difficulty level   ○ Easy   ● Normal   ○ Hard
Calculation amount ● Small ○ Normal   ○ Large

1) $dV = dr d\varphi dz$
2) $dV = r dr d\varphi dz$
3) $dV = r^2 dr d\varphi dz$
4) $dV = r dr dz$

Problem

1.31. Calculate the cylindrical volume described by $2 \leq r \leq 5$, $90° \leq \varphi \leq 180°$, $0 \leq z \leq 2$. The volume is graphically shown in Fig. 1.4.

Difficulty level   ○ Easy   ● Normal   ○ Hard
Calculation amount ○ Small ● Normal   ○ Large

1) $V = \dfrac{21}{2}$
2) $V = \dfrac{21}{4}$
3) $V = \dfrac{21\pi}{4}$
4) $V = \dfrac{21\pi}{2}$

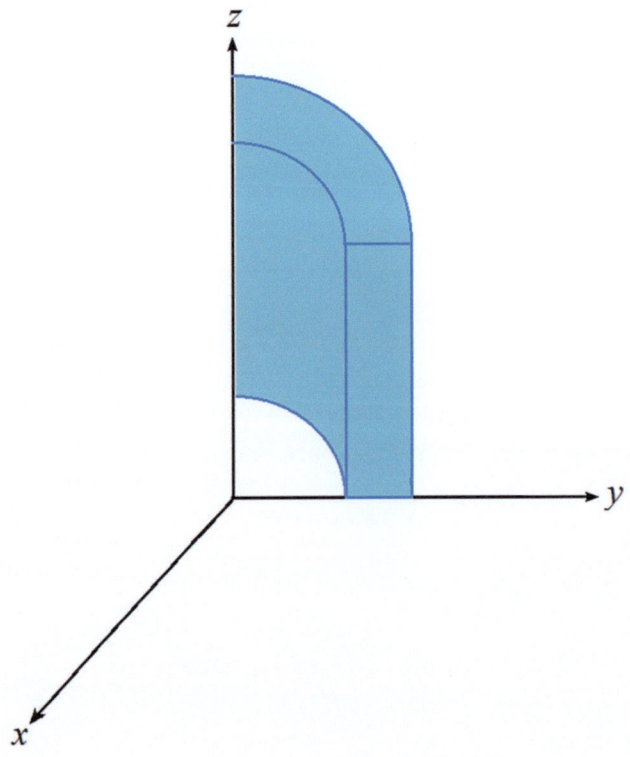

Fig. 1.4 A cylindrical volume

Partially Solved Exercise

1.32. Calculate the cylindrical volume described by $0 \leq r \leq \sqrt{2}$, $0 \leq \varphi \leq \frac{\pi}{6}$, $-1 \leq z \leq 1$.

Solution
The problem can be solved as follows.

$$V = \iiint dV$$

$$V = \int_{z=()}^{z=()} \int_{\varphi=()}^{\varphi=()} \int_{r=()}^{r=()} r\, dr\, d\varphi\, dz$$

$$\Rightarrow V = () \Big|_{()}^{()} \cdot () \Big|_{()}^{()} \cdot () \Big|_{()}^{()}$$

$$\Rightarrow V = (-)(-)(-) = ()()()$$

$$\Rightarrow V = \frac{\pi}{3}$$

1.6 Spherical Coordinate System

Exercise
1.33. What are the variables of the spherical coordinate system?

Final Answer
(R, θ, φ)

Problem
1.34. What is the range of variables in the spherical coordinate system?
Difficulty level ○ Easy ● Normal ○ Hard
Calculation amount ● Small ○ Normal ○ Large
1) $0 \leq R < \infty$, $0 \leq \theta < 2\pi$, $0 \leq \varphi < 2\pi$
2) $-\infty < R < \infty$, $0 \leq \theta < 2\pi$, $0 \leq \varphi < 2\pi$
3) $0 \leq R < \infty$, $0 \leq \theta < \pi$, $0 \leq \varphi < \pi$
4) $0 \leq R < \infty$, $0 \leq \theta < \pi$, $0 \leq \varphi < 2\pi$

Problem

1.35. Which one of the following choices is wrong about the cross product of the unit vectors in the spherical coordinate system?

Difficulty level ○ Easy ● Normal ○ Hard
Calculation amount ● Small ○ Normal ○ Large

1) $\hat{R} \times \hat{\theta} = -\hat{\varphi}$
2) $\hat{\theta} \times \hat{R} = -\hat{\varphi}$
3) $\hat{\theta} \times \hat{\varphi} = \hat{R}$
4) $\hat{\varphi} \times \hat{R} = \hat{\theta}$

Problem

1.36. What is the overall differential length vector in the spherical coordinate system?

Difficulty level ○ Easy ● Normal ○ Hard
Calculation amount ● Small ○ Normal ○ Large

1) $\vec{dl} = dR\hat{R} + Rd\theta\hat{\theta} + R\sin\theta\, d\varphi\hat{\varphi}$
2) $\vec{dl} = dR\hat{R} + d\theta\hat{\theta} + R\sin\theta\, d\varphi\hat{\varphi}$
3) $\vec{dl} = dR\hat{R} + Rd\theta\hat{\theta} + R\cos\theta\, d\varphi\hat{\varphi}$
4) $\vec{dl} = dR\hat{R} + d\theta\hat{\theta} + \sin\theta\, d\varphi\hat{\varphi}$

Problem

1.37. Which one of the following choices is wrong about the differential area vectors in spherical coordinate system?

Difficulty level ○ Easy ● Normal ○ Hard
Calculation amount ● Small ○ Normal ○ Large

1) $\vec{dS_R} = R^2 \sin\theta\, d\theta d\varphi \hat{R}$
2) $\vec{dS_\theta} = R\sin\theta\, dRd\varphi \hat{\theta}$
3) $\vec{dS_\varphi} = RdRd\theta \hat{\varphi}$
4) $\vec{dS_\varphi} = RdRd\varphi \hat{\varphi}$

Problem

1.38. Calculate the area of a spherical surface described by $R = 3$, $0 \leq \theta \leq \frac{\pi}{4}$, $0 \leq \varphi \leq 2\pi$. The area is graphically shown in Fig. 1.5.

Difficulty level ○ Easy ● Normal ○ Hard
Calculation amount ○ Small ● Normal ○ Large

1) $3\pi(2 - \sqrt{2})$
2) $9\pi(2 - \sqrt{2})$
3) 3π
4) 9π

1.6 Spherical Coordinate System

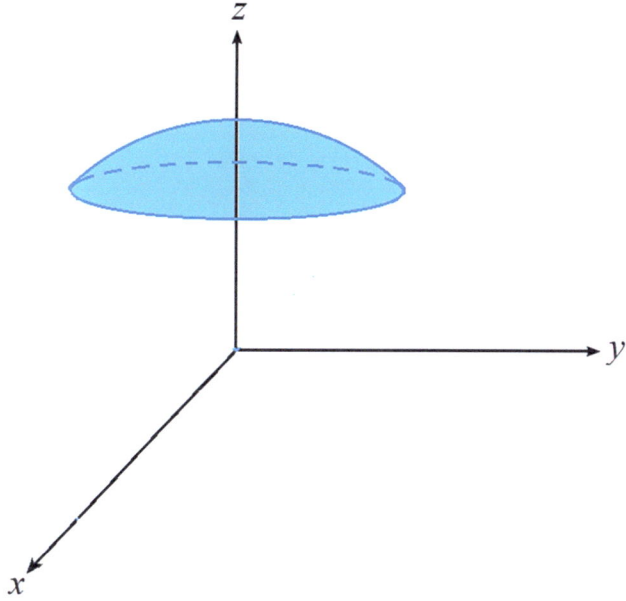

Fig. 1.5 A spherical surface

Partially Solved Exercise

1.39. Calculate the area of a spherical surface described by $R = 1, 0° \leq \theta \leq 60°, 0° \leq \varphi \leq 180°$.

Solution

Since R is constant for the surface, we need to use the relation below.

$$S_R = \iint dS_R$$

Therefore:

$$S_R = \int_{\theta=(\ \ )}^{\theta=(\ \ )} \int_{\varphi=(\ \ )}^{\varphi=(\ \ )} R^2 \sin\theta \, d\theta d\varphi = (\ \ )^2 \int_{\theta=(\ \ )}^{\theta=(\ \ )} \int_{\varphi=(\ \ )}^{\varphi=(\ \ )} \sin\theta \, d\theta d\varphi$$

$$\Rightarrow S_R = (\ \ ) \Big|_{(\ \ )}^{(\ \ )} (\ \ ) \Big|_{(\ \ )}^{(\ \ )} = (\ \ -\ \ )(\ \ -\ \ ) = (\ \ )(\ \ )$$

$$\Rightarrow S_R = \frac{\pi}{2}$$

Problem

1.40. Calculate the area of a spherical surface described by $1 \leq R \leq 5, \theta = 60°, 0° \leq \varphi \leq 360°$. The area is graphically shown in Fig. 1.6.

Difficulty level   ○ Easy   ● Normal   ○ Hard
Calculation amount   ○ Small   ● Normal   ○ Large

1) $12\sqrt{3}\pi$
2) $12\sqrt{3}$
3) 12π
4) $\dfrac{25\sqrt{3}\pi}{2}$

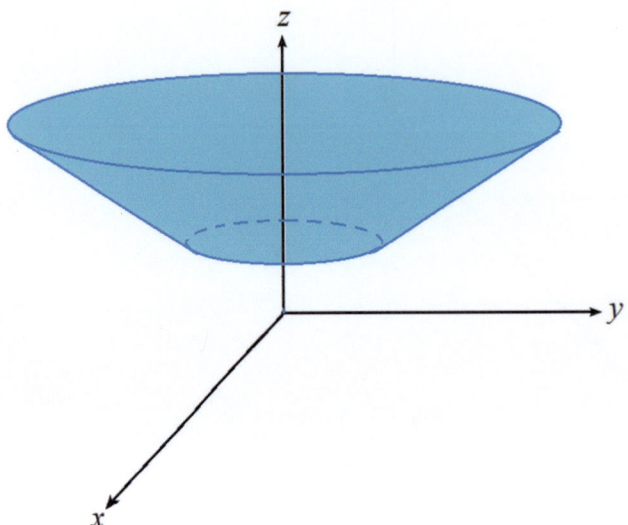

Fig. 1.6 A spherical surface

Partially Solved Exercise

1.41. Calculate the area of a spherical surface described by $0 \leq R \leq 1$, $\theta = 90°$, $0° \leq \varphi \leq 180°$.

Solution

Since θ is constant for the surface, we need to use the relation below.

$$S_\theta = \iint dS_\theta$$

Therefore:

$$S_\theta = \int_{R=(\ )}^{R=(\ )} \int_{\varphi=(\ )}^{\varphi=(\ )} R\sin\theta \, dR d\varphi = \sin(\ ) \int_{R=(\ )}^{R=(\ )} \int_{\varphi=(\ )}^{\varphi=(\ )} R \, dR d\varphi$$

$$\Rightarrow S_\theta = (\ ) \times (\ ) \Big|_{(\ )}^{(\ )} \, (\ ) \Big|_{(\ )}^{(\ )}$$

$$\Rightarrow S_\theta = (\ ) \times (\ \ - \ )(\ \ - \ ) = (\ )(\ )$$

$$\Rightarrow S_\theta = \dfrac{\pi}{2}$$

1.6 Spherical Coordinate System

Problem

1.42. Calculate the area of a spherical surface described by $1 \leq R \leq 2, 0° \leq \theta \leq 60°, \varphi = 90°$. The area is graphically shown in Fig. 1.7

Difficulty level   ○ Easy   ● Normal   ○ Hard
Calculation amount   ○ Small   ● Normal   ○ Large

1) $\dfrac{\pi}{3}$
2) $\dfrac{\pi}{6}$
3) $\dfrac{\pi}{2}$
4) $\dfrac{\pi}{12}$

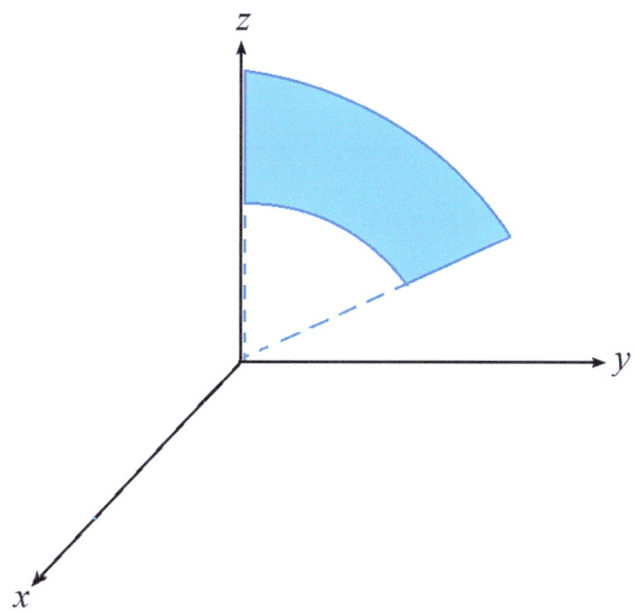

Fig. 1.7 A spherical surface

Partially Solved Exercise

1.43. Calculate the area of a spherical surface described by $0 \leq R \leq 2, 0° \leq \theta \leq 90°, \varphi = 45°$.

Solution

Since φ is constant for the surface, we need to use the relation below.

$$S_\varphi = \iint dS_\varphi$$

Therefore:

$$S_\varphi = \int_{R=(\quad)}^{R=(\quad)} \int_{\theta=(\quad)}^{\theta=(\quad)} R\,dR\,d\theta$$

$$\Rightarrow S_\varphi = \left.\left(\right)\right|_{\binom{}{}}^{\binom{}{}} \left.\left(\right)\right|_{\binom{}{}}^{\binom{}{}} = \left(- \right)\left(- \right) = \left(\right)\left(\right)$$

$$\Rightarrow S_\varphi = \pi$$

Problem

1.44. What is the differential volume in the spherical coordinate system?
Difficulty level   ○ Easy   ● Normal   ○ Hard
Calculation amount   ● Small   ○ Normal   ○ Large
1) $dV = R \sin\theta \, dR d\theta d\varphi$
2) $dV = R^2 \sin\theta \, dR d\theta d\varphi$
3) $dV = R^2 \cos\theta \, dR d\theta d\varphi$
4) $dV = R \sin\theta \, d\theta d\varphi$

Problem

1.45. Calculate the spherical volume described by $0 \leq R \leq 5$, $30° \leq \theta \leq 60°$, $0° \leq \varphi \leq 360°$. The volume is graphically shown in Fig. 1.8
Difficulty level   ○ Easy   ● Normal   ○ Hard
Calculation amount   ○ Small   ● Normal   ○ Large
1) $V = \dfrac{125(\sqrt{3}-1)}{3}\pi$
2) $V = \dfrac{125(\sqrt{3}-1)}{3}$
3) $V = \dfrac{125(\sqrt{2}-1)}{3}\pi$
4) $V = \dfrac{125(\sqrt{2}-1)}{3}$

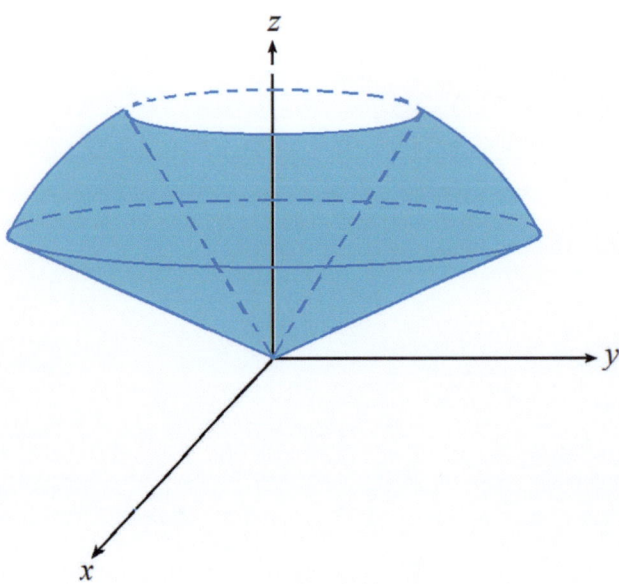

Fig. 1.8 A spherical volume

1.7 Cartesian to Cylindrical Coordinate Conversion

Partially Solved Exercise

1.46. Calculate the spherical volume described by $1 \leq R \leq 2$, $0° \leq \theta \leq 90°$, $30° \leq \varphi \leq 90°$.

Solution

The problem can be solved as follows.

$$V = \iiint dV$$

$$V = \int_{\varphi=()}^{\varphi=()} \int_{\theta=()}^{\theta=()} \int_{R=()}^{R=()} R^2 \sin\theta \, dR \, d\theta \, d\varphi$$

$$\Rightarrow V = () \left|_{()}^{()}\right. () \left|_{()}^{()}\right. () \left|_{()}^{()}\right.$$

$$\Rightarrow V = (-)(-)(-) = ()()()$$

$$\Rightarrow V = \frac{7}{9}\pi$$

1.7 Cartesian to Cylindrical Coordinate Conversion

Problem

1.47. Convert the coordinates of the point $(x, y, z) = (1, 2, 0)$ from Cartesian to cylindrical.
Difficulty level   ○ Easy   ● Normal   ○ Hard
Calculation amount ● Small ○ Normal   ○ Large
1) $(r, \varphi, z) = (\sqrt{5}, 63.4°, 0)$
2) $(r, \varphi, z) = (\sqrt{5}, 90°, 0)$
3) $(r, \varphi, z) = (3, 90°, 0)$
4) $(r, \varphi, z) = (1, 63.4°, 0)$

Partially Solved Exercise

1.48. Convert the coordinates of the point $(x, y, z) = (1, 1, 1)$ from Cartesian to cylindrical.

Solution

The relations below are used to convert the coordinates of a point from Cartesian (x, y, z) to cylindrical (r, φ, z).

$$r = \sqrt{x^2 + y^2}$$

$$\varphi = \tan^{-1}\left(\frac{y}{x}\right)$$

$$z = z$$

Therefore:

$$r = \sqrt{(\ )^2 + (\ )^2} = (\ )$$

$$\varphi = \tan^{-1}\left(\frac{(\ )}{(\ )}\right) = \tan^{-1}(\ ) = (\ )$$

$$z = (\ )$$

$$\Rightarrow (r, \varphi, z) = \left(\sqrt{2}, 45°, 1\right)$$

Exercise

1.49. Convert the coordinates of the point $(x, y, z) = (0, 0, 2)$ from Cartesian to cylindrical.

Final Answer

$(r, \varphi, z) = (0, 0°, 2)$

1.8 Cartesian to Spherical Coordinate Conversion

Problem

1.50. Convert the coordinates of the point $(x, y, z) = (1, 2, 0)$ from Cartesian to spherical.
Difficulty level   ○ Easy   ● Normal   ○ Hard
Calculation amount   ● Small   ○ Normal   ○ Large
1) $(R, \theta, \varphi) = \left(\sqrt{5}, 90°, 0°\right)$
2) $(R, \theta, \varphi) = \left(\sqrt{5}, 90°, 63.4°\right)$
3) $(R, \theta, \varphi) = (3, 90°, 63.4°)$
4) $(R, \theta, \varphi) = (1, 0°, 63.4°)$

Partially Solved Exercise

1.51. Convert the coordinates of the point $(x, y, z) = (1, 1, 1)$ from Cartesian to spherical.

Solution

The relations below are used to convert the coordinates of a point from Cartesian (x, y, z) to cylindrical (R, θ, φ).

$$R = \sqrt{x^2 + y^2 + z^2}$$

$$\theta = \tan^{-1}\left(\frac{\sqrt{x^2 + y^2}}{z}\right)$$

$$\varphi = \tan^{-1}\left(\frac{y}{x}\right)$$

1.9 Cylindrical to Cartesian Coordinate Conversion

Therefore:

$$R = \sqrt{()^2 + ()^2 + ()^2} = \sqrt{()}$$

$$\theta = \tan^{-1}\left(\frac{\sqrt{()^2 + ()^2}}{()}\right) = \tan^{-1}() = ()$$

$$\varphi = \tan^{-1}\left(\frac{()}{()}\right) = ()$$

$$\Rightarrow (R, \theta, \varphi) = \left(\sqrt{3}, 54.7°, 45°\right)$$

Exercise

1.52. Convert the coordinates of the point $(x, y, z) = (0, 0, 2)$ from Cartesian to spherical.

Final Answer

$(R, \theta, \varphi) = (2, 0°, 0°)$

1.9 Cylindrical to Cartesian Coordinate Conversion

Problem

1.53. Convert the coordinates of the point $(r, \varphi, z) = (3, 0°, -2)$ from cylindrical to Cartesian.
Difficulty level   ○ Easy   ● Normal   ○ Hard
Calculation amount   ● Small   ○ Normal   ○ Large
1) $(x, y, z) = (3, 0, -2)$
2) $(x, y, z) = (3, 3, -2)$
3) $(x, y, z) = (0, 0, -2)$
4) $(x, y, z) = (3, 0, 0)$

Partially Solved Exercise

1.54. Convert the coordinates of the point $(r, \varphi, z) = (2, 45°, -3)$ from cylindrical to Cartesian.

Solution

The relations below are used to convert the coordinates of a point from cylindrical (r, φ, z) to Cartesian (x, y, z).

$$x = r \cos \varphi$$

$$y = r \sin \varphi$$

$$z = z$$

Thus:

$$x = (\quad)\cos(\quad) = (\quad)$$

$$y = (\quad)\sin(\quad) = (\quad)$$

$$z = (\quad)$$

$$\Rightarrow (x, y, z) = \left(\sqrt{2}, \sqrt{2}, -3\right)$$

Exercise

1.55. Convert the coordinates of the point $(r, \varphi, z) = (1, 90°, 90)$ from cylindrical to Cartesian.

Final Answer

$(x, y, z) = (0, 1, 90)$

1.10 Spherical to Cartesian Coordinate Conversion

Problem

1.56. Convert the coordinates of the point $(R, \theta, \varphi) = \left(\sqrt{3}, 90°, 90°\right)$ from spherical to Cartesian.
Difficulty level   ○ Easy   ● Normal   ○ Hard
Calculation amount ● Small   ○ Normal   ○ Large
1) $(x, y, z) = (1, 1, 1)$
2) $(x, y, z) = \left(\sqrt{3}, \sqrt{3}, \sqrt{3}\right)$
3) $(x, y, z) = \left(1, \sqrt{3}, 1\right)$
4) $(x, y, z) = \left(0, \sqrt{3}, 0\right)$

Partially Solved Exercise

1.57. Convert the coordinates of the point $(R, \theta, \varphi) = (1, 30°, 60°)$ from spherical to Cartesian.

Solution

The relations below are used to convert the coordinates of a point from spherical (R, θ, φ) to Cartesian (x, y, z).

$$x = R \sin\theta \cos\varphi$$

$$y = R \sin\theta \sin\varphi$$

$$z = R \cos\theta$$

Thus:

$$x = (\quad)\sin(\quad)\cos(\quad) = (\quad)$$

$$y = (\quad)\sin(\quad)\sin(\quad) = (\quad)$$

$$z = (\ )\cos(\ ) = (\ )$$

$$\Rightarrow (x, y, z) = \left(\frac{1}{4}, \frac{\sqrt{3}}{4}, \frac{\sqrt{3}}{2}\right)$$

Exercise

1.58. Convert the coordinates of the point $(R, \theta, \varphi) = (0, 45°, 45°)$ from spherical to Cartesian.

Final Answer

$(x, y, z) = (0, 0, 0)$

1.11 Spherical to Cylindrical Coordinate Conversion

Problem

1.59. Convert the coordinates of the point $(R, \theta, \varphi) = (\sqrt{3}, 90°, 90°)$ from spherical to cylindrical.
Difficulty level ○ Easy ● Normal ○ Hard
Calculation amount ● Small ○ Normal ○ Large
1) $(r, \varphi, z) = (0, 0°, \sqrt{3})$
2) $(r, \varphi, z) = (1, 90°, 0)$
3) $(r, \varphi, z) = (\sqrt{3}, 90°, 0)$
4) $(r, \varphi, z) = (\sqrt{3}, 0°, \sqrt{3})$

Partially Solved Exercise

1.60. Convert the coordinates of the point $(R, \theta, \varphi) = (1, 0°, 180°)$ from spherical to cylindrical.

Solution

The relations below are used to convert the coordinates of a point from spherical (R, θ, φ) to cylindrical (r, φ, z).

$$r = R\sin\theta$$

$$\varphi = \varphi$$

$$z = R\cos\theta$$

Therefore:

$$r = (\ )\sin(\ ) = (\ )$$

$$\varphi = (\ )$$

$$z = (\ )\cos(\ ) = (\ )$$

$$\Rightarrow (r, \varphi, z) = (0, 180°, 1)$$

Exercise

1.61. Convert the coordinates of the point $(R, \theta, \varphi) = (10, 90°, 0°)$ from spherical to cylindrical.

Final Answer

$(r, \varphi, z) = (10, 0°, 0)$

1.12 Cylindrical to Spherical Coordinate Conversion

Problem

1.62. Convert the coordinates of the point $(r, \varphi, z) = (10, 90°, 10)$ from cylindrical to spherical.
Difficulty level   ○ Easy   ● Normal   ○ Hard
Calculation amount   ● Small ○ Normal   ○ Large
1) $(R, \theta, \varphi) = \left(10\sqrt{2}, 45°, 90°\right)$
2) $(R, \theta, \varphi) = (10, 90°, 90°)$
3) $(R, \theta, \varphi) = (10, 45°, 45°)$
4) $(R, \theta, \varphi) = \left(10\sqrt{2}, 45°, 0°\right)$

Partially Solved Exercise

1.63. Convert the coordinates of the point $(r, \varphi, z) = \left(1, 30°, \sqrt{3}\right)$ from cylindrical to spherical.

Solution

The relations below are used to convert the coordinates of a point from cylindrical (r, φ, z) to spherical (R, θ, φ).

$$R = \sqrt{r^2 + z^2}$$

$$\theta = \tan^{-1}\left(\frac{r}{z}\right)$$

$$\varphi = \varphi$$

Therefore:

$$R = \sqrt{()^2 + ()^2} = ()$$

$$\theta = \tan^{-1}\left(\frac{()}{()}\right) = \tan^{-1}() = ()$$

$$\varphi = ()$$

$$\Rightarrow (R, \theta, \varphi) = (2, 30°, 30°)$$

Exercise

1.64. Convert the coordinates of the point $(r, \varphi, z) = (1.5, 0°, 0)$ from cylindrical to spherical.

Final Answer

$(R, \theta, \varphi) = (1.5, 90°, 0°)$

1.13 Position Vector

Problem

1.65. Which one of the following choices correctly presents the position vector of the point $P(x_0, y_0, z_0)$ in Cartesian coordinate system?
Difficulty level   ○ Easy   ● Normal   ○ Hard
Calculation amount   ● Small ○ Normal   ○ Large
1) $\overrightarrow{OP} = x_0 \hat{x}$
2) $\overrightarrow{OP} = y_0 \hat{y}$
3) $\overrightarrow{OP} = z_0 \hat{z}$
4) $\overrightarrow{OP} = x_0 \hat{x} + y_0 \hat{y} + z_0 \hat{z}$

Problem

1.66. Which one of the following choices correctly presents the position vector of the point $P(r_0, \varphi_0, z_0)$ in cylindrical coordinate system?
Difficulty level   ○ Easy   ● Normal   ○ Hard
Calculation amount   ● Small ○ Normal   ○ Large
1) $\overrightarrow{OP} = r_0 \hat{r} + \varphi_0 \hat{\varphi}$
2) $\overrightarrow{OP} = \varphi_0 \hat{r} + z_0 \hat{z}$
3) $\overrightarrow{OP} = r_0 \hat{r} + z_0 \hat{z}$
4) $\overrightarrow{OP} = r_0 \hat{r} + \varphi_0 \hat{\varphi} + z_0 \hat{z}$

Problem

1.67. Which one of the following choices correctly presents the position vector of the point $P(R_0, \theta_0, \varphi_0)$ in spherical coordinate system?
Difficulty level   ○ Easy   ● Normal   ○ Hard
Calculation amount   ● Small ○ Normal   ○ Large
1) $\overrightarrow{OP} = R_0 \hat{R}$
2) $\overrightarrow{OP} = \theta_0 \hat{\theta}$
3) $\overrightarrow{OP} = \varphi_0 \hat{\varphi}$
4) $\overrightarrow{OP} = R_0 \hat{R} + \theta_0 \hat{\theta} + \varphi_0 \hat{\varphi}$

Problem

1.68. Calculate the position vector of the point $P = (1, 2, 3)$ in Cartesian coordinate system.
Difficulty level ○ Easy ● Normal ○ Hard
Calculation amount ● Small ○ Normal ○ Large

1) $\overrightarrow{OP} = 3\hat{x} + 2\hat{y} + \hat{z}$
2) $\overrightarrow{OP} = \hat{x} + 2\hat{y} + 3\hat{z}$
3) $\overrightarrow{OP} = \dfrac{1}{\sqrt{13}}(\hat{x} + 2\hat{y} + 3\hat{z})$
4) $\overrightarrow{OP} = \hat{x} + \hat{y} + \hat{z}$

Exercise

1.69. Determine the position vector of the point $P = (0, -10, 1)$ in Cartesian coordinate system.

Final Answer

$\overrightarrow{OP} = -10\hat{y} + \hat{z}$

Exercise

1.70. Determine the position vector of the point $P = (1, 90°, 1)$ in cylindrical coordinate system.

Final Answer

$\overrightarrow{OP} = \hat{r} + \hat{z}$

Exercise

1.71. Determine the position vector of the point $P = (5, 30°, 60°)$ in spherical coordinate system.

Final Answer

$\overrightarrow{OP} = 5\hat{R}$

1.14 Distance Vector

Problem

1.72. Calculate the distance vector from the point $P_1 = (-2, 2, -5)$ to the point $P_2 = (1, 1, 4)$ in the Cartesian coordinate system.
Difficulty level ○ Easy ● Normal ○ Hard
Calculation amount ● Small ○ Normal ○ Large

1) $\overrightarrow{P_1P_2} = -\hat{x} - \hat{y} + 9\hat{z}$
2) $\overrightarrow{P_1P_2} = 2\hat{x} - \hat{y} + \hat{z}$
3) $\overrightarrow{P_1P_2} = 3\hat{x} + \hat{y} + 9\hat{z}$
4) $\overrightarrow{P_1P_2} = 3\hat{x} - \hat{y} + 9\hat{z}$

Exercise

1.73. Determine the distance vector from the point $P_1 = (18, 18, 0)$ to the point $P_2 = (18, 18, 1)$ in the Cartesian coordinate system.

Final Answer

$\overrightarrow{P_1 P_2} = \hat{z}$

References

1. Rahmani-Andebili, M., General Physics II – Practice Problems, Methods, and Solutions, Springer Nature, 2025.
2. Rahmani-Andebili, M., General Physics I – Practice Problems, Methods, and Solutions, Springer Nature, 2025.
3. Rahmani-Andebili, M., Mathematics of Engineering and Science – Practice Problems, Methods, and Solutions, Springer Nature, 2024.
4. Rahmani-Andebili, M., Differential Equations – Practice Problems, Methods, and Solutions, Springer Nature, 2022.
5. Rahmani-Andebili, M., Calculus III – Practice Problems, Methods, and Solutions, Springer Nature, 2023.
6. Rahmani-Andebili, M., Calculus II – Practice Problems, Methods, and Solutions, Springer Nature, 2023.
7. Rahmani-Andebili, M., Calculus I (2nd Ed.) – Practice Problems, Methods, and Solutions, Springer Nature, 2023.
8. Rahmani-Andebili, M., Precalculus (2nd Ed.) – Practice Problems, Methods, and Solutions, Springer Nature, 2024.

Cartesian, Cylindrical, and Spherical Coordinate Systems: Part B

Abstract

In this chapter, the problems of the first chapter are fully solved, in detail, step-by-step, and with different methods.

2.1 Unit Vector

2.1. Based on the information given in the problem, we have:

$$\vec{R} = \hat{x} + 2\hat{y} + 3\hat{z}$$

The unit vector ($\hat{R}$) of the vector $\vec{R} = R_1\hat{x} + R_2\hat{y} + R_3\hat{z}$ can be calculated as follows [1–8].

$$\hat{R} = \frac{\vec{R}}{R} = \frac{\vec{R}}{\sqrt{(R_1)^2 + (R_2)^2 + (R_3)^2}}$$

where, R is the magnitude of the vector.

Therefore, for this problem, we can write as follows.

$$\hat{R} = \frac{\hat{x} + 2\hat{y} + 3\hat{z}}{\sqrt{(1)^2 + (2)^2 + (3)^2}}$$

$$\Rightarrow \hat{R} = \frac{1}{\sqrt{14}}(\hat{x} + 2\hat{y} + 3\hat{z})$$

Choice (4) is the answer.

2.2 Dot Product (Inner Product)

2.5. The following properties are valid for dot products.

$$\vec{a} \cdot \vec{b} = a_1 b_1 + a_2 b_2 + a_3 b_3$$

$$\vec{a} \cdot \vec{b} = ab \cos \theta$$

$$\vec{a} \cdot \vec{b} = \vec{b} \cdot \vec{a}$$

$$\vec{a} \cdot \vec{a} = a^2$$

Choice (3) is the answer.

2.6. Based on the information given in the problem, we have:

$$\vec{a} = -3\hat{x} - 3\hat{y} + 3\hat{z}$$

$$\vec{b} = 2\hat{x} - 2\hat{y} + 2\hat{z}$$

As we know, the angle between two vectors can be calculated as follows.

$$\cos \theta = \frac{\vec{a} \cdot \vec{b}}{ab}$$

Thus, the problem can be solved as follows.

$$\cos \theta = \frac{(-3) \times (2) + (-3) \times (-2) + (3) \times (2)}{\left(\sqrt{(-3)^2 + (-3)^2 + (3)^2} \right) \left(\sqrt{(2)^2 + (-2)^2 + (2)^2} \right)}$$

$$\Rightarrow \cos \theta = \frac{6}{3\sqrt{3} \times 2\sqrt{3}} = \frac{1}{3}$$

$$\Rightarrow \theta = \arccos \frac{1}{3}$$

Choice (1) is the answer.

2.2 Dot Product (Inner Product)

> **Notes**
>
> In this problem, the relations below have been used.
>
> The inner product (dot product) of two vectors can be calculated as follows.
>
> $$\vec{a}\cdot\vec{b} = \left(a_1\hat{i} + a_2\hat{j} + a_3\hat{k}\right)\cdot\left(b_1\hat{i} + b_2\hat{j} + b_3\hat{k}\right) = a_1b_1 + a_2b_2 + a_3b_3$$
>
> The magnitude of a vector can be calculated as follows.
>
> $$a = |\vec{a}| = |a_1\hat{i} + a_2\hat{j} + a_3\hat{k}| = \sqrt{(a_1)^2 + (a_2)^2 + (a_3)^2}$$

2.8. Based on the information given in the problem, we have:

$$\vec{A} = a\hat{x} + 3\hat{y} + \hat{z}$$

$$\vec{B} = a\hat{x} - a\hat{y} + 2\hat{z}$$

As we know, two vectors are perpendicular to each other if their inner product (dot product) is zero. In other words:

$$\vec{A}\cdot\vec{B} = 0$$

Therefore:

$$(a\hat{x} + 3\hat{y} + \hat{z})\cdot(a\hat{x} - a\hat{y} + 2\hat{z}) = 0$$

$$\Rightarrow a^2 - 3a + 2 = 0 \Rightarrow (a-2)(a-1) = 0$$

$$\Rightarrow a = 1, 2$$

Choice (2) is the answer.

> **Notes**
>
> In this problem, the relation below has been used.
>
> The inner product (dot product) of two vectors can be calculated as follows.
>
> $$\vec{a}\cdot\vec{b} = \left(a_1\hat{i} + a_2\hat{j} + a_3\hat{k}\right)\cdot\left(b_1\hat{i} + b_2\hat{j} + b_3\hat{k}\right) = a_1b_1 + a_2b_2 + a_3b_3$$

2.3 Cross Product (Vector Product)

2.9. The following properties are valid for cross products.

$$\vec{a} \times \vec{b} = (a_2 b_3 - a_3 b_2)\hat{x} + (a_3 b_1 - a_1 b_3)\hat{y} + (a_1 b_2 - a_2 b_1)\hat{z}$$

$$\vec{a} \times \vec{b} = ab \sin \theta$$

$$\vec{a} \times \vec{b} = -\vec{b} \times \vec{a}$$

$$\vec{a} \times \vec{a} = 0$$

Choice (4) is the answer.

2.10. Based on the information given in the problem, we have:

$$\vec{a} = 2\hat{x} - 4\hat{y} + \hat{z}$$

$$\vec{b} = 6\hat{x} + 8\hat{y} - 15\hat{z}$$

The problem can be solved as follows.

$$\vec{a} \times \vec{b} = \begin{vmatrix} \hat{x} & \hat{y} & \hat{z} \\ 2 & -4 & 1 \\ 6 & 8 & -15 \end{vmatrix} = (60 - 8)\hat{x} + (6 + 30)\hat{y} + (16 + 24)\hat{z}$$

$$\Rightarrow \vec{a} \times \vec{b} = 52\hat{x} + 36\hat{y} + 40\hat{z}$$

Choice (1) is the answer.

> **Notes**
>
> In this problem, the relation below has been used.
>
> The vector product (cross product) of two vectors can be calculated as follows.
>
> $$\vec{a} \times \vec{b} = \begin{vmatrix} \hat{x} & \hat{y} & \hat{z} \\ a_1 & a_2 & a_3 \\ b_1 & b_2 & b_3 \end{vmatrix} = (a_2 b_3 - a_3 b_2)\hat{x} + (a_3 b_1 - a_1 b_3)\hat{y} + (a_1 b_2 - a_2 b_1)\hat{z}$$

2.4 Cartesian Coordinate System

2.13. As is shown in Fig. 2.1, the range of variables in the Cartesian coordinate system is as follows.

$$-\infty < x < \infty$$

$$-\infty < y < \infty$$

$$-\infty < z < \infty$$

Choice (4) is the answer.

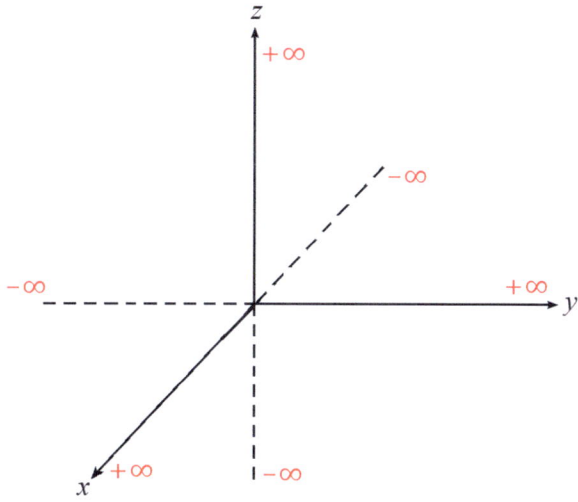

Fig. 2.1 The range of variables in the Cartesian coordinate system

---

2.14. By using Fig. 2.2 and right-hand rule, the cross product of unit vectors in the Cartesian coordinate system can be calculated as follows.

$$\hat{x} \times \hat{y} = \hat{z} \quad \text{or} \quad \hat{y} \times \hat{x} = -\hat{z}$$

$$\hat{y} \times \hat{z} = \hat{x} \quad \text{or} \quad \hat{z} \times \hat{y} = -\hat{x}$$

$$\hat{z} \times \hat{x} = \hat{y} \quad \text{or} \quad \hat{x} \times \hat{z} = -\hat{y}$$

Choice (4) is the answer.

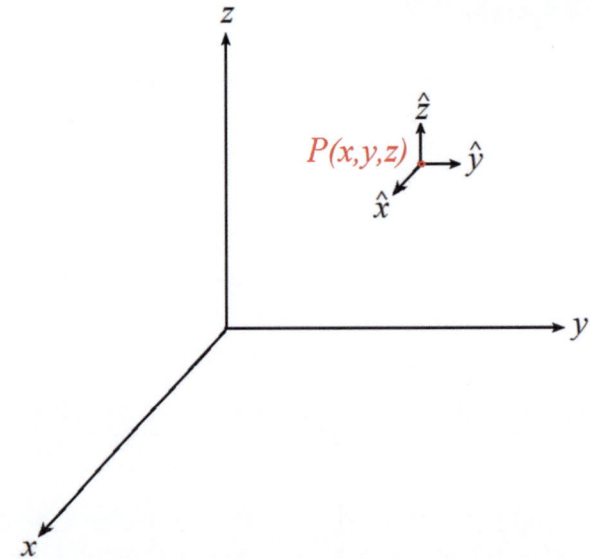

Fig. 2.2 The unit vectors in the Cartesian coordinate system

> **Notes**
> In this problem, the relation below has been used.
> $$\vec{a} \times \vec{b} = -\vec{b} \times \vec{a}$$

2.15. As is illustrated in Fig. 2.3, the overall differential length vector in the Cartesian coordinate system is as follows.

$$\vec{dl} = dx\hat{x} + dy\hat{y} + dz\hat{z}$$

Choice (4) is the answer.

2.4 Cartesian Coordinate System

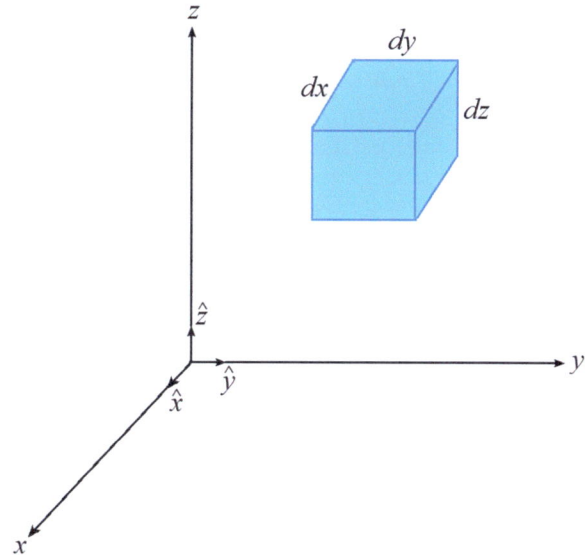

Fig. 2.3 The differential length vectors in the Cartesian coordinate system

2.16. As is illustrated in Fig. 2.4, the three differential area vectors in the Cartesian coordinate system are as follows.

$$\overrightarrow{dS_x} = (dy)(dz)\hat{x} \Rightarrow \overrightarrow{dS_x} = dydz\hat{x}$$

$$\overrightarrow{dS_y} = (dx)(dz)\hat{y} \Rightarrow \overrightarrow{dS_y} = dxdz\hat{y}$$

$$\overrightarrow{dS_z} = (dx)(dy)\hat{z} \Rightarrow \overrightarrow{dS_z} = dxdy\hat{z}$$

Choice (3) is the answer.

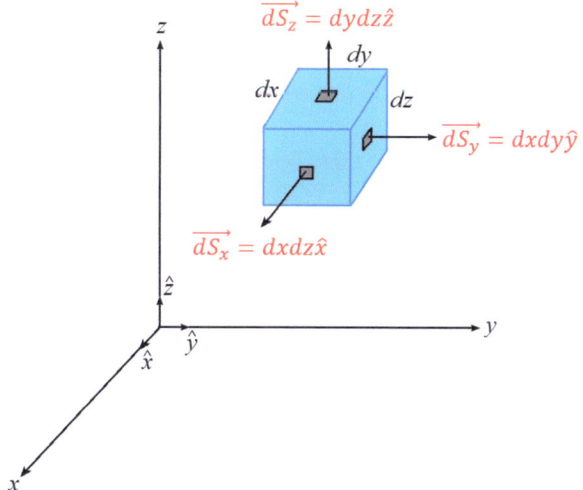

Fig. 2.4 The differential area vectors in the Cartesian coordinate system

2.17. As is illustrated in Fig. 2.5, the differential volume in the Cartesian coordinate system is as follows.

$$dV = (dx)(dy)(dz)$$

$$\Rightarrow dV = dxdydz$$

Choice (3) is the answer.

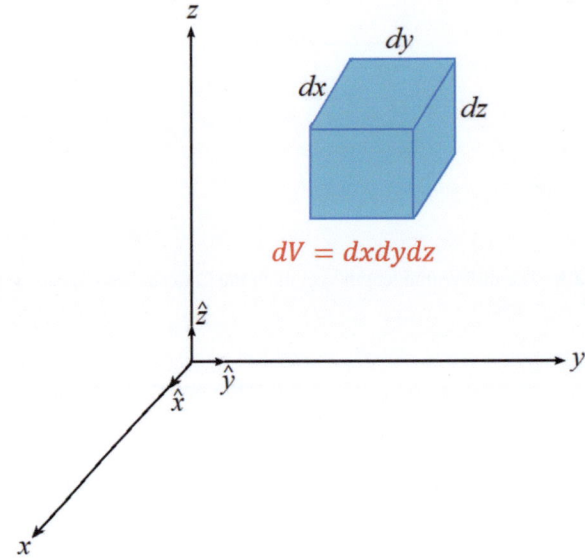

Fig. 2.5 The differential volume in the Cartesian coordinate system

2.5 Cylindrical Coordinate System

2.19. As is shown in Fig. 2.6, the range of variables in the cylindrical coordinate system is as follows.

$$0 \leq r < \infty$$

$$0 \leq \varphi < 2\pi$$

$$-\infty < z < \infty$$

Choice (4) is the answer.

2.5 Cylindrical Coordinate System

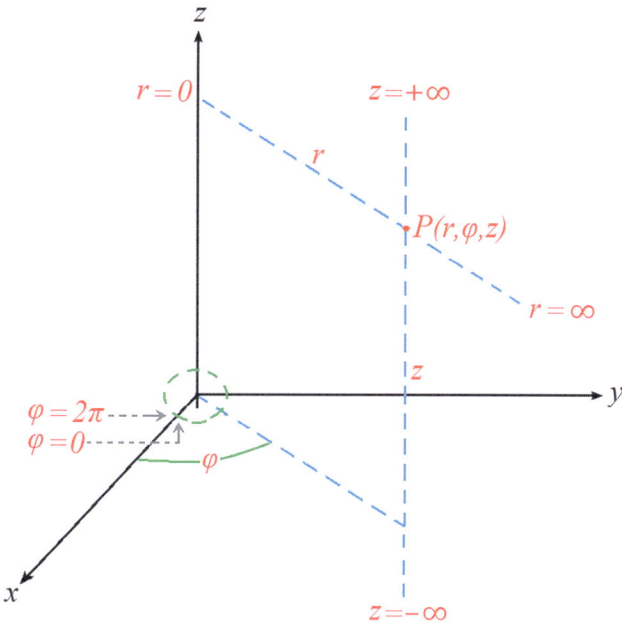

Fig. 2.6 The range of variables in the cylindrical coordinate system

2.20. By using Fig. 2.7 and right-hand rule, the cross product of unit vectors in the cylindrical coordinate system can be calculated as follows.

$$\hat{r} \times \hat{\varphi} = \hat{z} \quad \text{or} \quad \hat{\varphi} \times \hat{r} = -\hat{z}$$

$$\hat{\varphi} \times \hat{z} = \hat{r} \quad \text{or} \quad \hat{z} \times \hat{\varphi} = -\hat{r}$$

$$\hat{z} \times \hat{r} = \hat{\varphi} \quad \text{or} \quad \hat{r} \times \hat{z} = -\hat{\varphi}$$

Choice (2) is the answer.

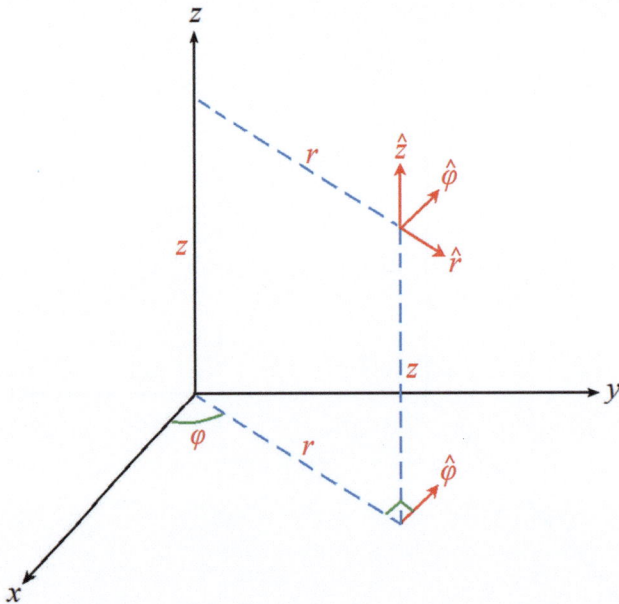

Fig. 2.7 The unit vectors in the cylindrical coordinate system

> **Notes**
> In this problem, the relation below has been used.
> $$\vec{a} \times \vec{b} = -\vec{b} \times \vec{a}$$

2.21. As is illustrated in Fig. 2.8, the overall differential length vector in the cylindrical coordinate system is as follows.

$$\vec{dl} = dr\hat{r} + rd\varphi\hat{\varphi} + dz\hat{z}$$

Choice (1) is the answer.

2.5 Cylindrical Coordinate System

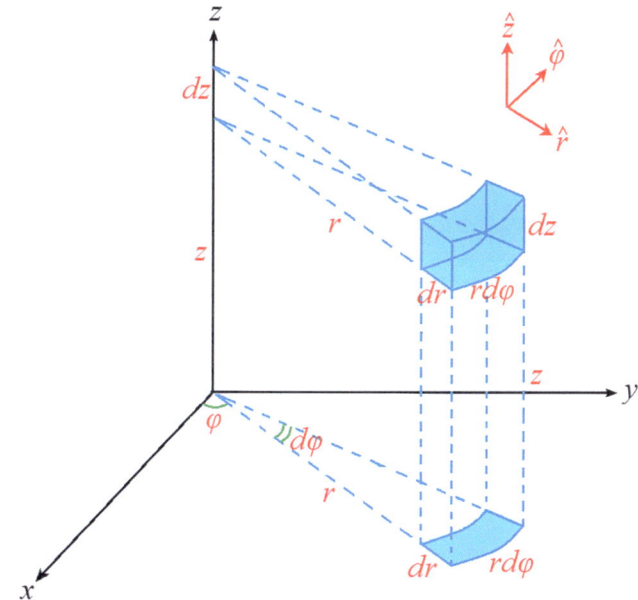

Fig. 2.8 The differential length vectors in the cylindrical coordinate system

2.22. As is illustrated in Fig. 2.9, the three differential area vectors in the cylindrical coordinate system are as follows.

$$\overrightarrow{dS_r} = (rd\varphi)(dz)\hat{r} \Rightarrow \overrightarrow{dS_r} = rd\varphi dz\hat{r}$$

$$\overrightarrow{dS_\varphi} = (dr)(dz)\hat{\varphi} \Rightarrow \overrightarrow{dS_\varphi} = drdz\hat{\varphi}$$

$$\overrightarrow{dS_z} = (dr)(rd\varphi)\hat{z} \Rightarrow \overrightarrow{dS_z} = rdrd\varphi\hat{z}$$

Choice (4) is the answer.

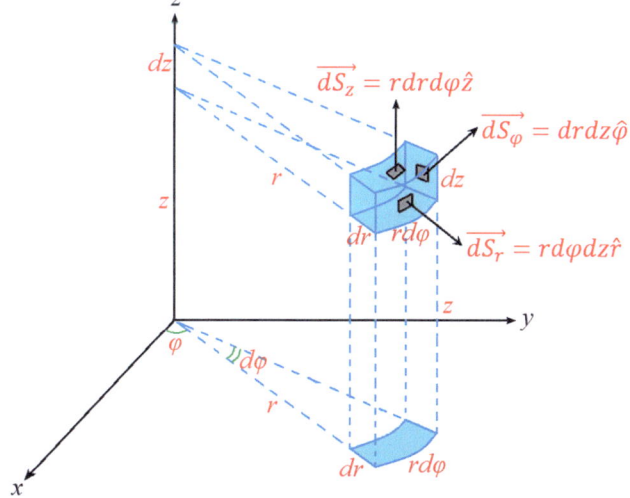

Fig. 2.9 The differential area vectors in the cylindrical coordinate system

2.23. Based on the information given in the problem, we have:

$$r = 1$$

$$30° \leq \varphi \leq 60°$$

$$0 \leq z \leq 1$$

The area is graphically shown in Fig. 2.10. Since r is constant for the surface, we need to use the relation below.

$$S_r = \iint dS_r$$

Thus:

$$S_r = \int_{z=0}^{z=1} \int_{\varphi=\frac{\pi}{6}}^{\varphi=\frac{\pi}{3}} r d\varphi dz = 1 \times \int_{z=0}^{z=1} \int_{\varphi=\frac{\pi}{6}}^{\varphi=\frac{\pi}{3}} d\varphi dz$$

$$\Rightarrow S_r = z \Big|_0^1 \varphi \Big|_{\frac{\pi}{6}}^{\frac{\pi}{3}} = \left(\frac{\pi}{3} - \frac{\pi}{6}\right)(1-0) = \frac{\pi}{6} \times 1$$

$$\Rightarrow S_r = \frac{\pi}{6}$$

Choice (1) is the answer.

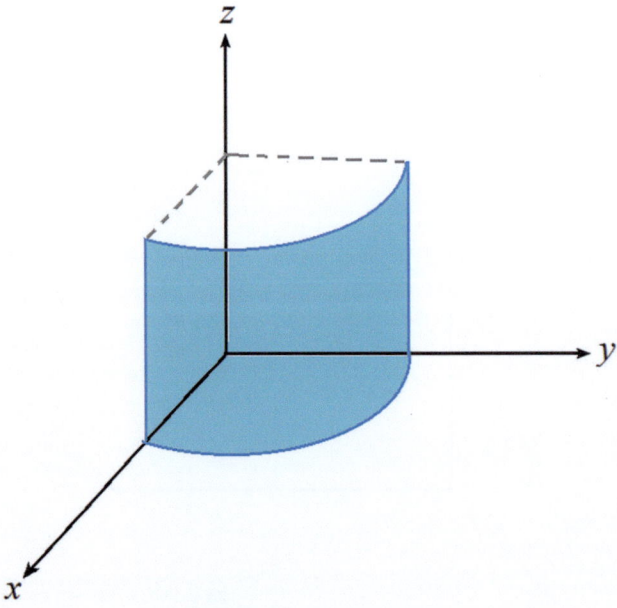

Fig. 2.10 A cylindrical surface

2.25. Based on the information given in the problem, we have:

$$0.1 \leq r \leq 0.5$$

$$\varphi = 90°$$

$$0 \leq z \leq 2$$

The area is graphically shown in Fig. 2.11. Since φ is constant for the surface, we need to use the relation below.

$$S_\varphi = \iint dS_\varphi$$

Therefore:

$$S_\varphi = \int_{z=0}^{z=2} \int_{r=0.1}^{r=0.5} dr\,dz$$

$$\Rightarrow S_\varphi = z\Big|_0^2 \, r\Big|_{0.1}^{0.5} = (2-0)(0.5-0.1) = 2 \times 0.4$$

$$\Rightarrow S_\varphi = 0.8$$

Choice (3) is the answer.

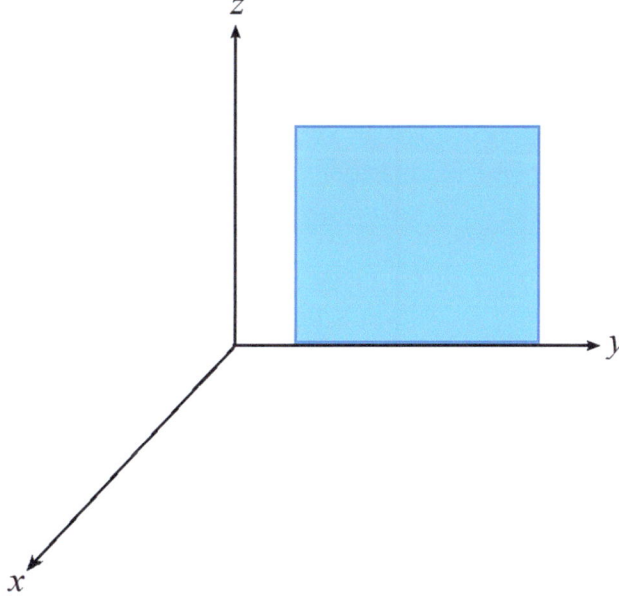

Fig. 2.11 A cylindrical surface

2.28. Based on the information given in the problem, we have:

$$1 \leq r \leq 2$$

$$30° \leq \varphi \leq 60°$$

$$z = 1$$

The area is graphically shown in Fig. 2.12. Since z is constant for the surface, we need to use the relation below.

$$S_z = \iint dS_z$$

Hence:

$$S_z = \int_{\varphi=\frac{\pi}{6}}^{\varphi=\frac{\pi}{3}} \int_{r=1}^{r=2} r\, dr\, d\varphi$$

$$\Rightarrow S_z = \varphi \Big|_{\frac{\pi}{6}}^{\frac{\pi}{3}} \frac{r^2}{2}\Big|_1^2 = \left(\frac{\pi}{3} - \frac{\pi}{6}\right)\left(\frac{2^2}{2} - \frac{1^2}{2}\right) = \frac{\pi}{6} \times \frac{3}{2}$$

$$\Rightarrow S_z = \frac{\pi}{4}$$

Choice (1) is the answer.

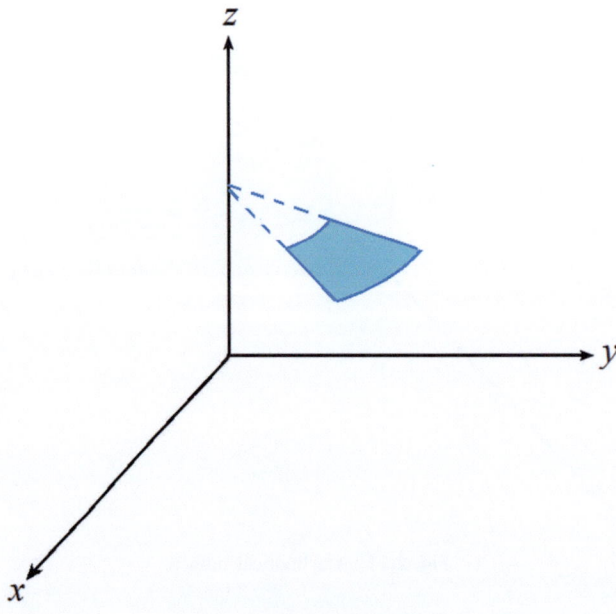

Fig. 2.12 A cylindrical surface

2.5 Cylindrical Coordinate System

> **Notes**
>
> In this problem, the relation below has been used.
>
> $$\int x^n dx = \frac{x^{n+1}}{n+1}$$

2.30. As is illustrated in Fig. 2.13, the differential volume in the cylindrical coordinate system is as follows.

$$dV = (dr)(rd\varphi)(dz)$$

$$\Rightarrow dV = rdrd\varphi dz$$

Choice (2) is the answer.

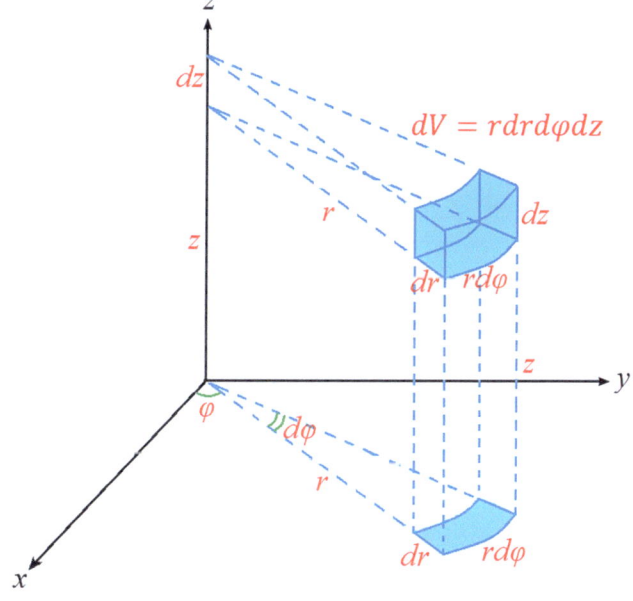

Fig. 2.13 The differential volume in the cylindrical coordinate system

2.31. Based on the information given in the problem, we have:

$$2 \leq r \leq 5$$

$$90° \leq \varphi \leq 180°$$

$$0 \leq z \leq 2$$

The area is graphically shown in Fig. 2.14. The problem can be solved as follows.

$$V = \iiint dV$$

$$V = \int_{z=0}^{z=2} \int_{\varphi=\frac{\pi}{2}}^{\varphi=\pi} \int_{r=2}^{r=5} r\, dr\, d\varphi\, dz$$

$$\Rightarrow V = z\Big|_0^2 \; \varphi\Big|_{\frac{\pi}{2}}^{\pi} \; \frac{r^2}{2}\Big|_2^5$$

$$\Rightarrow V = (2-0)\left(\pi - \frac{\pi}{2}\right)\left(\frac{5^2}{2} - \frac{2^2}{2}\right) = 2 \times \frac{\pi}{2} \times \frac{21}{2}$$

$$\Rightarrow V = \frac{21}{2}\pi$$

Choice (4) is the answer.

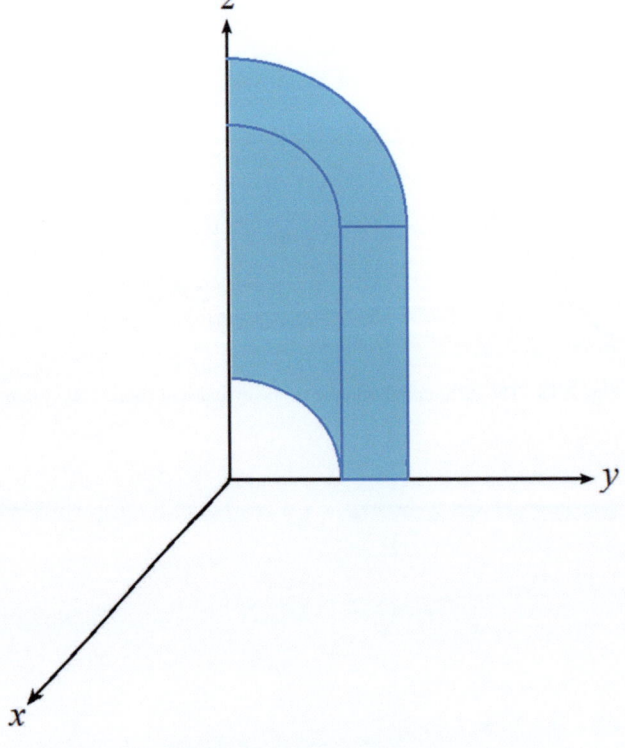

Fig. 2.14 A cylindrical volume

2.6 Spherical Coordinate System

> **Notes**
>
> In this problem, the relation below has been used.
>
> $$\int x^n dx = \frac{x^{n+1}}{n+1}$$

2.6 Spherical Coordinate System

2.34. As is illustrated in Fig. 2.15, the range of variables in the spherical coordinate system is as follows.

$$0 \leq R < \infty$$

$$0 \leq \theta < \pi$$

$$0 \leq \varphi < 2\pi$$

Choice (4) is the answer.

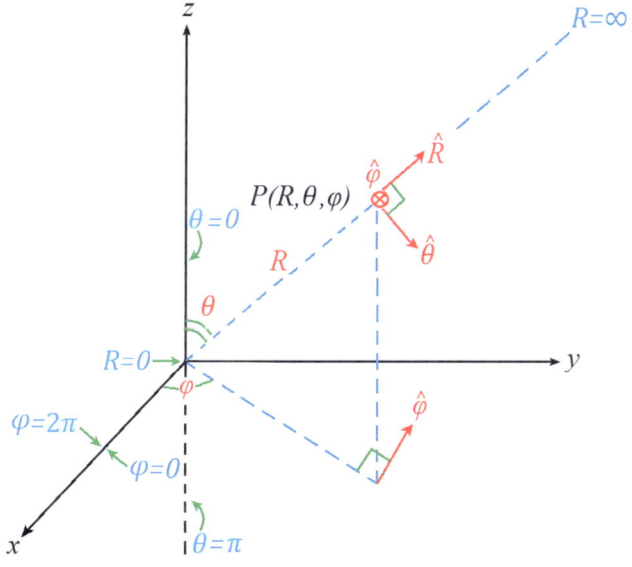

Fig. 2.15 The range of variables in the spherical coordinate system

2.35. By using Fig. 2.16 and right-hand rule, the cross product of unit vectors in the spherical coordinate system can be calculated as follows.

$$\widehat{R} \times \widehat{\theta} = \widehat{\varphi} \quad \text{or} \quad \widehat{\theta} \times \widehat{R} = -\widehat{\varphi}$$

$$\widehat{\theta} \times \widehat{\varphi} = \widehat{R} \quad \text{or} \quad \widehat{\varphi} \times \widehat{\theta} = -\widehat{R}$$

$$\widehat{\varphi} \times \widehat{R} = \widehat{\theta} \quad \text{or} \quad \widehat{R} \times \widehat{\varphi} = -\widehat{\theta}$$

Choice (1) is the answer.

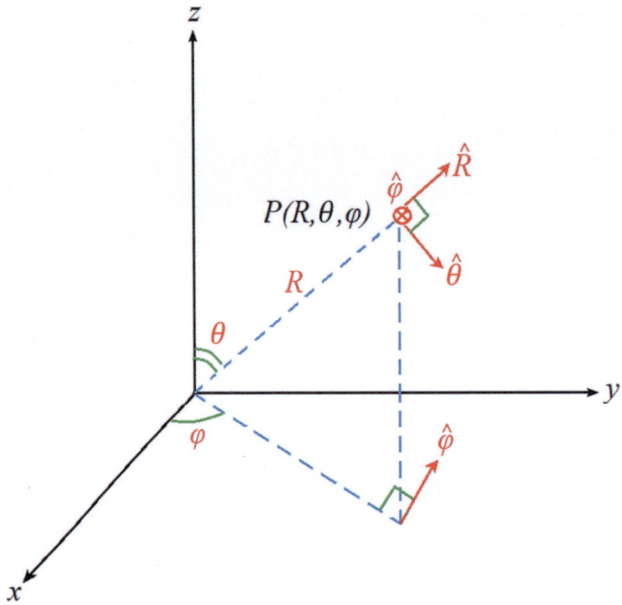

Fig. 2.16 The unit vectors in the spherical coordinate system

> **Notes**
> In this problem, the relation below has been used.
>
> $$\vec{a} \times \vec{b} = -\vec{b} \times \vec{a}$$

2.36. As is illustrated in Fig. 2.17, the overall differential length vector in the spherical coordinate system is as follows.

$$\vec{dl} = dR\widehat{R} + Rd\theta\widehat{\theta} + R\sin\theta\, d\varphi\widehat{\varphi}$$

Choice (1) is the answer.

2.6 Spherical Coordinate System

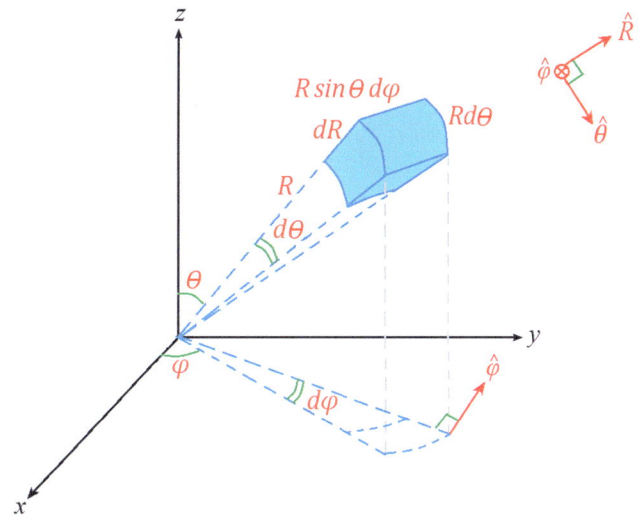

Fig. 2.17 The differential length vectors in the spherical coordinate system

2.37. As is illustrated in Fig. 2.18, the three differential area vectors in the spherical coordinate system are as follows.

$$\overrightarrow{dS_R} = (R\sin\theta\, d\varphi)(Rd\theta)\widehat{R} \Rightarrow \overrightarrow{dS_R} = R^2\sin\theta\, d\theta d\varphi \widehat{R}$$

$$\overrightarrow{dS_\theta} = (R\sin\theta\, d\varphi)(dR)\widehat{\theta} \Rightarrow \overrightarrow{dS_\theta} = R\sin\theta\, dRd\varphi\widehat{\theta}$$

$$\overrightarrow{dS_\varphi} = (dR)(Rd\theta)\widehat{\varphi} \Rightarrow \overrightarrow{dS_\varphi} = RdRd\theta\widehat{\varphi}$$

Choice (4) is the answer.

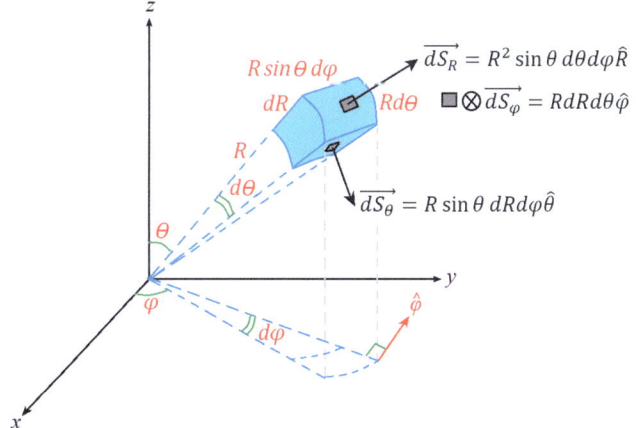

Fig. 2.18 The differential area vectors in the spherical coordinate system

2.38. Based on the information given in the problem, we have:

$$R = 3$$

$$0 \leq \theta \leq \frac{\pi}{4}$$

$$0 \leq \varphi \leq 2\pi$$

The area is graphically shown in Fig. 2.19. Since R is constant for the surface, we need to use the relation below.

$$S_R = \iint dS_R$$

Therefore:

$$S_R = \int_{\theta=0}^{\theta=\frac{\pi}{4}} \int_{\varphi=0}^{\varphi=2\pi} R^2 \sin\theta \, d\theta d\varphi = 3^2 \int_{\theta=0}^{\theta=\frac{\pi}{4}} \int_{\varphi=0}^{\varphi=2\pi} \sin\theta \, d\theta d\varphi$$

$$\Rightarrow S_R = -9\cos\theta \Big|_0^{\frac{\pi}{4}} \varphi \Big|_0^{2\pi} = -9\left(\cos\frac{\pi}{4} - \cos 0\right)(2\pi - 0) = -9 \times \left(\frac{\sqrt{2}}{2} - 1\right) \times 2\pi$$

$$\Rightarrow S_R = 9\pi\left(2 - \sqrt{2}\right)$$

Choice (2) is the answer.

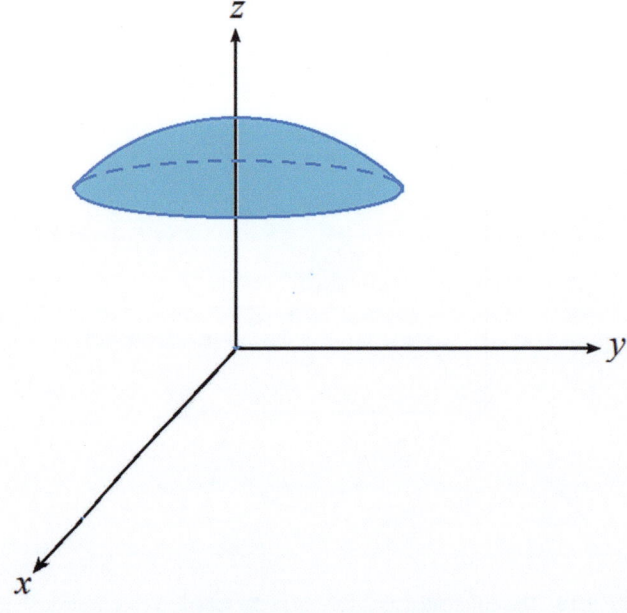

Fig. 2.19 A spherical surface

2.6 Spherical Coordinate System

> **Notes**
>
> In this problem, the relations below have been used.
>
> $$\int \sin\theta \, d\theta = -\cos\theta$$
>
> $$\cos\frac{\pi}{4} = \frac{\sqrt{2}}{2}$$
>
> $$\cos 0 = 1$$

2.40. Based on the information given in the problem, we have:

$$1 \leq R \leq 5$$

$$\theta = 60°$$

$$0° \leq \varphi \leq 360°$$

The area is graphically shown in Fig. 2.20. Since θ is constant for the surface, we need to use the relation below.

$$S_\theta = \iint dS_\theta$$

Therefore:

$$S_\theta = \int_{R=1}^{R=5} \int_{\varphi=0}^{\varphi=2\pi} R\sin\theta \, dRd\varphi = \sin\frac{\pi}{3} \int_{R=1}^{R=5} \int_{\varphi=0}^{\varphi=2\pi} R \, dRd\varphi$$

$$\Rightarrow S_\theta = \frac{\sqrt{3}}{2} \times \frac{R^2}{2}\bigg|_1^5 \varphi \bigg|_0^{2\pi} = \frac{\sqrt{3}}{2} \times \left(\frac{5^2}{2} - \frac{1^2}{2}\right)(2\pi - 0) = \frac{\sqrt{3}}{2} \times (12) \times 2\pi$$

$$\Rightarrow S_\theta = 12\sqrt{3}\pi$$

Choice (1) is the answer.

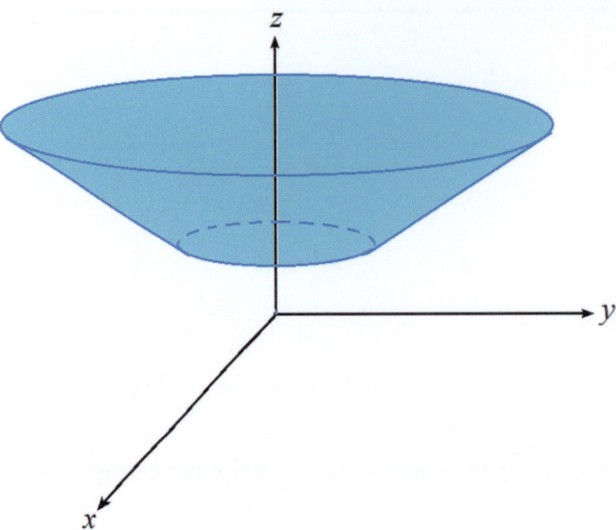

Fig. 2.20 A spherical surface

> **Notes**
> In this problem, the relations below have been used.
>
> $$\int x^n dx = \frac{x^{n+1}}{n+1}$$
>
> $$\sin\frac{\pi}{3} = \frac{\sqrt{3}}{2}$$

2.42. Based on the information given in the problem, we have:

$$1 \leq R \leq 2$$

$$0° \leq \theta \leq 60°$$

$$\varphi = 90°$$

Since φ is constant for the surface, we need to use the relation below.

$$S_\varphi = \iint dS_\varphi$$

2.6 Spherical Coordinate System

Therefore:

$$S_\varphi = \int_{R=1}^{R=2} \int_{\theta=0}^{\theta=\frac{\pi}{3}} R\, dR\, d\theta$$

$$\Rightarrow S_\varphi = \left.\frac{R^2}{2}\right|_1^2 \left.\theta\right|_0 = \left(\frac{2^2}{2} - \frac{1^2}{2}\right)\left(\frac{\pi}{3} - 0\right) = \frac{3}{2} \times \frac{\pi}{3}$$

$$\Rightarrow S_\varphi = \frac{\pi}{2}$$

Choice (3) is the answer (Fig. 2.21).

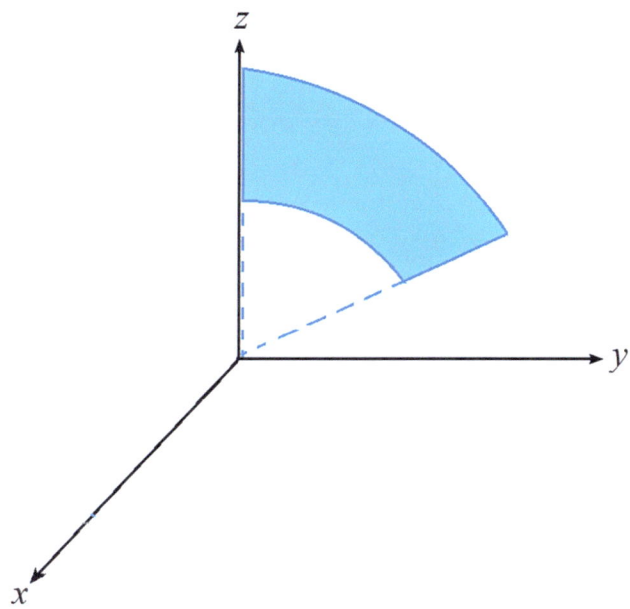

Fig. 2.21 A spherical surface

> **Notes**
> In this problem, the relation below has been used.
> $$\int x^n dx = \frac{x^{n+1}}{n+1}$$

2.44. As is illustrated in Fig. 2.22, the differential volume in the spherical coordinate system is as follows.

$$dV = (dR)(R \sin\theta\, d\varphi)(Rd\theta)$$

$$\Rightarrow dV = R^2 \sin\theta\, dRd\theta d\varphi$$

Choice (2) is the answer.

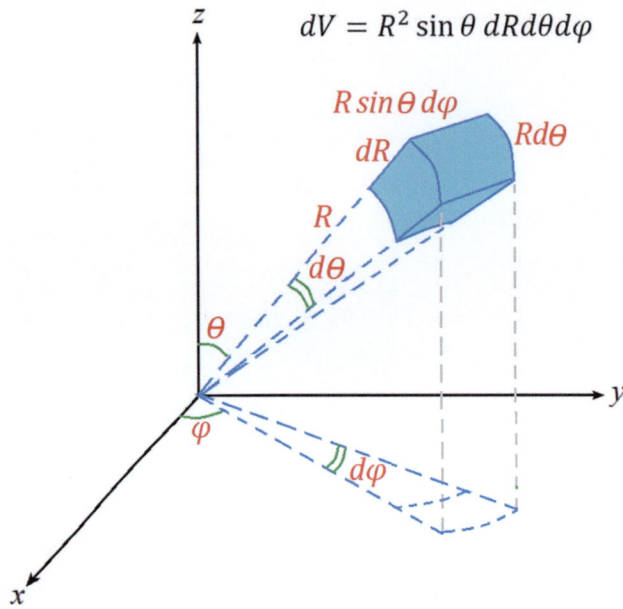

Fig. 2.22 The differential volume in the spherical coordinate system

2.45. Based on the information given in the problem, we have:

$$0 \le R \le 5$$

$$30° \le \theta \le 60°$$

$$0° \le \varphi \le 360°$$

The volume is graphically shown in Fig. 2.23. The problem can be solved as follows.

$$V = \iiint dV$$

2.6 Spherical Coordinate System

$$V = \int_{\varphi=0}^{\varphi=2\pi} \int_{\theta=\frac{\pi}{6}}^{\theta=\frac{\pi}{3}} \int_{R=0}^{R=5} R^2 \sin\theta \, dR \, d\theta \, d\varphi$$

$$\Rightarrow V = \varphi \Big|_0^{2\pi} (-\cos\theta) \Big|_{\frac{\pi}{6}}^{\frac{\pi}{3}} \frac{R^3}{3} \Big|_0^5$$

$$\Rightarrow V = (2\pi - 0)\left(-\cos\frac{\pi}{3} + \cos\frac{\pi}{6}\right)\left(\frac{5^3}{3} - \frac{0^3}{3}\right) = 2\pi \times \left(-\frac{1}{2} + \frac{\sqrt{3}}{2}\right) \times \frac{125}{3}$$

$$\Rightarrow V = \frac{125(\sqrt{3}-1)}{3}\pi$$

Choice (1) is the answer.

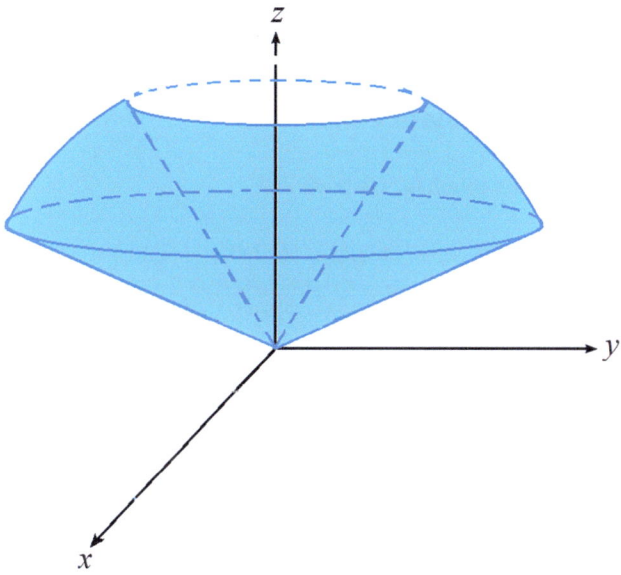

Fig. 2.23 A spherical volume

> **Notes**
>
> In this problem, the relations below have been used.
>
> $$\int \sin\theta\, d\theta = -\cos\theta$$
>
> $$\int x^n\, dx = \frac{x^{n+1}}{n+1}$$
>
> $$\cos\frac{\pi}{3} = \frac{1}{2}$$
>
> $$\cos\frac{\pi}{6} = \frac{\sqrt{3}}{2}$$

2.7 Cartesian to Cylindrical Coordinate Conversion

2.47. Based on the information given in the problem, we have:

$$(x, y, z) = (1, 2, 0)$$

The relations below are used to convert the coordinates of a point from Cartesian (x, y, z) to cylindrical (r, φ, z).

$$r = \sqrt{x^2 + y^2}$$

$$\varphi = \tan^{-1}\left(\frac{y}{x}\right)$$

$$z = z$$

Therefore:

$$r = \sqrt{1^2 + 2^2} = \sqrt{5}$$

$$\varphi = \tan^{-1}\left(\frac{2}{1}\right) = 63.4°$$

$$z = 0$$

$$\Rightarrow (r, \varphi, z) = \left(\sqrt{5}, 63.4°, 0\right)$$

Choice (1) is the answer.

2.8 Cartesian to Spherical Coordinate Conversion

2.50. Based on the information given in the problem, we have:

$$(x, y, z) = (1, 2, 0)$$

The relations below are used to convert the coordinates of a point from Cartesian (x, y, z) to spherical (R, θ, φ).

$$R = \sqrt{x^2 + y^2 + z^2}$$

$$\theta = \tan^{-1}\left(\frac{\sqrt{x^2 + y^2}}{z}\right)$$

$$\varphi = \tan^{-1}\left(\frac{y}{x}\right)$$

Therefore:

$$R = \sqrt{1^2 + 2^2 + 0^2} = \sqrt{5}$$

$$\theta = \tan^{-1}\left(\frac{\sqrt{1^2 + 2^2}}{0}\right) = \tan^{-1}\infty = 90°$$

$$\varphi = \tan^{-1}\left(\frac{2}{1}\right) = 63.4°$$

$$\Rightarrow (R, \theta, \varphi) = \left(\sqrt{5}, 90°, 63.4°\right)$$

Choice (2) is the answer.

2.9 Cylindrical to Cartesian Coordinates Conversion

2.53. Based on the information given in the problem, we have:

$$(r, \varphi, z) = (3, 0°, -2)$$

The relations below are used to convert the coordinates of a point from cylindrical (r, φ, z) to Cartesian (x, y, z).

$$x = r\cos\varphi$$

$$y = r\sin\varphi$$

$$z = z$$

Thus:

$$x = 3\cos 0 = 3$$

$$y = 3\sin 0 = 0$$

$$z = -2$$

$$\Rightarrow (x, y, z) = (3, 0, -2)$$

Choice (1) is the answer.

2.10 Spherical to Cartesian Coordinates Conversion

2.56. Based on the information given in the problem, we have:

$$(R, \theta, \varphi) = \left(\sqrt{3}, 90°, 90°\right)$$

The relations below are used to convert the coordinates of a point from spherical (R, θ, φ) to Cartesian (x, y, z).

$$x = R\sin\theta\cos\varphi$$

$$y = R\sin\theta\sin\varphi$$

$$z = R\cos\theta$$

Thus:

$$x = \sqrt{3}\sin 90° \cos 90° = 0$$

$$y = \sqrt{3}\sin 90° \sin 90° = \sqrt{3}$$

$$z = \sqrt{3}\cos 90° = 0$$

$$\Rightarrow (x, y, z) = \left(0, \sqrt{3}, 0\right)$$

Choice (4) is the answer.

2.11 Spherical to Cylindrical Coordinates Conversion

2.59. Based on the information given in the problem, we have:

$$(R, \theta, \varphi) = \left(\sqrt{3}, 90°, 90°\right)$$

The relations below are used to convert the coordinates of a point from spherical (R, θ, φ) to cylindrical (r, φ, z).

$$r = R \sin \theta$$

$$\varphi = \varphi$$

$$z = R \cos \theta$$

Therefore:

$$r = \sqrt{3} \sin 90° = \sqrt{3}$$

$$\varphi = 90°$$

$$z = \sqrt{3} \cos 90° = 0$$

$$\Rightarrow (r, \varphi, z) = \left(\sqrt{3}, 90°, 0\right)$$

Choice (3) is the answer.

2.12 Cylindrical to Spherical Coordinates Conversion

2.62. Based on the information given in the problem, we have:

$$(r, \varphi, z) = (10, 90°, 10)$$

The relations below are used to convert the coordinates of a point from cylindrical (r, φ, z) to spherical (R, θ, φ).

$$R = \sqrt{r^2 + z^2}$$

$$\theta = \tan^{-1}\left(\frac{r}{z}\right)$$

$$\varphi = \varphi$$

Therefore:

$$R = \sqrt{10^2 + 10^2} = 10\sqrt{2}$$

$$\theta = \tan^{-1}\left(\frac{10}{10}\right) = \tan^{-1} 1 = 45°$$

$$\varphi = 90°$$

$$\Rightarrow (R, \theta, \varphi) = \left(10\sqrt{2}, 45°, 90°\right)$$

Choice (1) is the answer.

2.13 Position Vector

2.65. The position vector of the point $P(x_0, y_0, z_0)$ in Cartesian coordinate system can be calculated as follows.

$$\overrightarrow{OP} = x_0 \hat{x} + y_0 \hat{y} + z_0 \hat{z}$$

Choice (4) is the answer.

2.66. The position vector of the point $P(r_0, \varphi_0, z_0)$ in cylindrical coordinate system can be calculated as follows.

$$\overrightarrow{OP} = r_0 \hat{r} + z_0 \hat{z}$$

Choice (3) is the answer.

2.67. The position vector of the point $P(R_0, \theta_0, \varphi_0)$ in spherical coordinate system can be calculated as follows.

$$\overrightarrow{OP} = R_0 \hat{R}$$

Choice (1) is the answer.

2.68. Based on the information given in the problem, we have:

$$P = (1, 2, 3)$$

The position vector the point $P(x_0, y_0, z_0)$ in the Cartesian coordinate system can be calculated as follows.

$$\overrightarrow{OP} = x_0 \hat{x} + y_0 \hat{y} + z_0 \hat{z}$$

Hence, for this problem, we have:

$$\overrightarrow{OP} = \widehat{x} + 2\widehat{y} + 3\widehat{z}$$

Choice (2) is the answer.

2.14 Distance Vector

2.72. Based on the information given in the problem, we have:

$$P_1 = (-2, 2, -5)$$

$$P_2 = (1, 1, 4)$$

The distance vector from the point $P_1(x_1, y_1, z_1)$ to the point $P_2(x_2, y_2, z_2)$ in the Cartesian coordinate system can be calculated as follows.

$$\overrightarrow{P_1P_2} = (x_2 - x_1)\widehat{x} + (y_2 - y_1)\widehat{y} + (z_2 - z_1)\widehat{z}$$

Therefore:

$$\overrightarrow{P_1P_2} = (1 - (-2))\widehat{x} + (1 - 2)\widehat{y} + (4 - (-5))\widehat{z}$$

$$\Rightarrow \overrightarrow{P_1P_2} = 3\widehat{x} - \widehat{y} + 9\widehat{z}$$

Choice (4) is the answer.

References

1. General Physics II – Practice Problems, Methods, and Solutions, Springer Nature, 2025.
2. General Physics I – Practice Problems, Methods, and Solutions, Springer Nature, 2025.
3. Mathematics of Engineering and Science – Practice Problems, Methods, and Solutions, Springer Nature, 2024.
4. Rahmani-Andebili, M. (2022). Differential Equations – Practice Problems, Methods, and Solutions, Springer Nature.
5. Rahmani-Andebili, M. (2023). Calculus III – Practice Problems, Methods, and Solutions, Springer Nature.
6. Rahmani-Andebili, M. (2023). Calculus II – Practice Problems, Methods, and Solutions, Springer Nature.
7. Rahmani-Andebili, M. (2023). Calculus I (2nd Ed.) – Practice Problems, Methods, and Solutions, Springer Nature.
8. Rahmani-Andebili, M. (2024). Precalculus (2nd Ed.) – Practice Problems, Methods, and Solutions, Springer Nature.

3. Gradient, Divergence, Curl, and Laplacian: Part A

Abstract

In this chapter, the basic and advanced problems of gradient, divergence, curl, and Laplacian operators are studied. The subjects include the gradient of a scalar, divergence of a vector, curl of a vector, and Laplacian of a scalar in the cartesian, cylindrical, and spherical coordinate systems. Herein, different types of problems and exercises are presented that are categorized as follows.

- **Problems with detailed solution**: They have been designed to teach students the subjects in detail. Moreover, they have been categorized in different levels based on their difficulty levels (easy, normal, and hard) and calculation amounts (small, normal, and large).
- **Partially solved exercises**: They have been designed to encourage students to practice problems while guiding them through the problem-solving procedure and hinting the required formulas.
- **Exercises with final answer**: They have been designed to encourage students to practice more by themselves while hinting them by the final answer as well as to help instructors to give tests or quizzes.

3.1 Gradient

Problem

3.1. Which one of the following choices is wrong about the properties of gradient of scalar functions [1–8].
Difficulty level   ● Easy   ○ Normal   ○ Hard
Calculation amount   ● Small   ○ Normal   ○ Large
1) $\nabla(F_1 + F_2) = \nabla F_1 + \nabla F_2$
2) $\nabla(F_1 - F_2) = \nabla F_1 - \nabla F_2$
3) $\nabla(F_1 F_2) = F_1 \nabla F_2 - F_2 \nabla F_1$
4) $\nabla(F^n) = nF^{n-1} \nabla F$

3.2 Gradient of a Scalar in the Cartesian Coordinate System

Problem

3.2. Find the gradient of the scalar function $F(x, y, z) = x^2 + y^2 z$.
Difficulty level   ● Easy   ○ Normal   ○ Hard
Calculation amount   ● Small   ○ Normal   ○ Large

1) $\nabla F = 2x\hat{x} + 2yz\hat{y} + 2y\hat{z}$
2) $\nabla F = 2x\hat{x} + 2z\hat{y} + 2y\hat{z}$
3) $\nabla F = 2\hat{x} + 2yz\hat{y} + y^2\hat{z}$
4) $\nabla F = 2x\hat{x} + 2yz\hat{y} + y^2\hat{z}$

Problem

3.3. In Problem 3.2, calculate the gradient of the function at $(x, y, z) = (1, 1, 1)$.
Difficulty level ● Easy ○ Normal ○ Hard
Calculation amount ● Small ○ Normal ○ Large

1) $\nabla F \big|_{(1,1,1)} = 2\hat{x} + 2\hat{y} + 2\hat{z}$

2) $\nabla F \big|_{(1,1,1)} = 2\hat{x} + 2\hat{y} + \hat{z}$

3) $\nabla F \big|_{(1,1,1)} = \hat{x} + \hat{y} + \hat{z}$

4) $\nabla F \big|_{(1,1,1)} = 0$

Partially Solved Exercise

3.4. (a) Find the gradient of the scalar function below. (b) Moreover, calculate the value of the gradient at $(1, -1, 9)$.

$$F(x, y, z) = \frac{3}{x^2 + y^2}$$

Solution

(a) The gradient of a scalar function in the Cartesian coordinate system can be calculated as follows.

$$\nabla F = \left(\frac{\partial}{\partial x}\hat{x} + \frac{\partial}{\partial y}\hat{y} + \frac{\partial}{\partial z}\hat{z} \right) F$$

Therefore:

$$\nabla F = \left(\frac{\partial}{\partial x}\hat{x} + \frac{\partial}{\partial y}\hat{y} + \frac{\partial}{\partial z}\hat{z} \right) \left(\frac{3}{x^2 + y^2} \right)$$

$$\Rightarrow \nabla F = \frac{\partial}{\partial x}(\qquad)\hat{x} + \frac{\partial}{\partial y}(\qquad)\hat{y} + \frac{\partial}{\partial z}(\qquad)\hat{z}$$

$$\Rightarrow \nabla F = (\qquad)\hat{x} + (\qquad)\hat{y} + (\qquad)\hat{z}$$

$$\Rightarrow \nabla F = -\frac{6x}{(x^2 + y^2)^2}\hat{x} - \frac{6y}{(x^2 + y^2)^2}\hat{y}$$

3.3 Gradient of a Scalar in Cylindrical Coordinate System

(b) Moreover:

$$\nabla F \Big|_{(1,-1,9)} = (\qquad)\hat{x} + (\qquad)\hat{y}$$

$$\Rightarrow \nabla F \Big|_{(1,-1,9)} = -\frac{3}{2}\hat{x} + \frac{3}{2}\hat{y}$$

Exercise

3.5. Calculate the value of ∇F if $F(x, y, z) = e^{-z} \sin 2x \cosh y$.

Final Answer

$\nabla F = 2e^{-z} \cos 2x \cosh y \hat{x} + e^{-z} \sin 2x \sinh y \hat{y} - e^{-z} \sin 2x \cosh y \hat{z}$

3.3 Gradient of a Scalar in Cylindrical Coordinate System

Problem

3.6. Calculate the value of ∇F if $F(r, \varphi, z) = V_0 e^{-2r} \sin 3\varphi$.
Difficulty level   ○ Easy   ● Normal   ○ Hard
Calculation amount   ○ Small   ● Normal   ○ Large
1) $\nabla F = -2V_0 e^{-2r} \sin 3\varphi \hat{r} + 3V_0 e^{-2r} \cos 3\varphi \hat{\varphi} + V_0 e^{-2r} \sin 3\varphi \hat{z}$
2) $\nabla F = -2V_0 e^{-2r} \sin 3\varphi \hat{r} + \frac{3}{r} V_0 e^{-2r} \cos 3\varphi \hat{\varphi} + V_0 e^{-2r} \hat{z}$
3) $\nabla F = -2V_0 e^{-2r} \sin 3\varphi \hat{r} + \frac{3}{r} V_0 e^{-2r} \cos 3\varphi \hat{\varphi}$
4) $\nabla F = -2V_0 e^{-2r} \sin 3\varphi \hat{r} + 3V_0 e^{-2r} \cos 3\varphi \hat{\varphi}$

Problem

3.7. In Problem 3.6, calculate the gradient of the function at $(r, \varphi, z) = \left(1, \frac{\pi}{2}, 3\right)$.
Difficulty level   ● Easy   ○ Normal   ○ Hard
Calculation amount   ● Small   ○ Normal   ○ Large
1) $\nabla F \Big|_{(1,\frac{\pi}{2},3)} = 2V_0 e^{-2} \hat{r} + 3V_0 e^{-2} \hat{\varphi}$
2) $\nabla F \Big|_{(1,\frac{\pi}{2},3)} = -2V_0 e^{-2} \hat{r} + 3V_0 e^{-2} \hat{\varphi}$
3) $\nabla F \Big|_{(1,\frac{\pi}{2},3)} = -2V_0 e^{-2} \hat{r}$
4) $\nabla F \Big|_{(1,\frac{\pi}{2},3)} = 2V_0 e^{-2} \hat{r}$

Partially Solved Exercise

3.8. (a) Find the gradient of the scalar function below. (b) Moreover, calculate the value of the gradient at (1, 0, 0).

$$F(r, \varphi, z) = \frac{z \cos \varphi}{1 + r^2}$$

Solution

(a) The gradient of a scalar function in cylindrical coordinate system can be calculated as follows.

$$\nabla F = \left(\frac{\partial}{\partial r} \hat{r} + \frac{1}{r} \frac{\partial}{\partial \varphi} \hat{\varphi} + \frac{\partial}{\partial z} \hat{z} \right) F$$

Hence:

$$\nabla F = \left(\frac{\partial}{\partial r} \hat{r} + \frac{1}{r} \frac{\partial}{\partial \varphi} \hat{\varphi} + \frac{\partial}{\partial z} \hat{z} \right) (\quad)$$

$$\Rightarrow \nabla F = \frac{\partial}{\partial r} (\quad) \hat{r} + \frac{1}{r} \frac{\partial}{\partial \varphi} (\quad) \hat{\varphi} + \frac{\partial}{\partial z} (\quad) \hat{z}$$

$$\Rightarrow \nabla F = -\frac{2rz \cos \varphi}{(1 + r^2)^2} \hat{r} - \frac{z \sin \varphi}{r(1 + r^2)} \hat{\varphi} + \frac{\cos \varphi}{1 + r^2} \hat{z}$$

(b) Moreover:

$$\nabla F \bigg|_{(1,0,0)} = -\frac{2(\;)(\;)\cos(\;)}{\left(1 + (\;)^2\right)^2} \hat{r} - \frac{(\;) \sin(\;)}{(\;)\left(1 + (\;)^2\right)} \hat{\varphi} + \frac{\cos(\;)}{1 + (\;)^2} \hat{z}$$

$$\Rightarrow \nabla F \bigg|_{(1,0,0)} = \frac{1}{2} \hat{z}$$

Exercise

3.9. Calculate the value of ∇F if $F(r, \varphi, z) = r^2 z \cos 2\varphi$.

Final Answer

$\nabla F = 2rz \cos 2\varphi \hat{r} - 2rz \sin 2\varphi \hat{\varphi} + r^2 \cos 2\varphi \hat{z}$

3.4 Gradient of a Scalar in Spherical Coordinate System

Problem

3.10. Find the gradient of the scalar function below.

$$F(R, \theta, \varphi) = \frac{V_0 a \cos 2\theta}{R}$$

3.4 Gradient of a Scalar in Spherical Coordinate System

Difficulty level ○ Easy ● Normal ○ Hard
Calculation amount ○ Small ● Normal ○ Large

1) $\nabla F = -\dfrac{V_0 a \cos 2\theta}{R^2}\widehat{R} - \dfrac{2V_0 a \sin 2\theta}{R^2}\widehat{\theta} + \dfrac{2V_0 a \cos 2\theta}{R}\widehat{\varphi}$

2) $\nabla F = -\dfrac{V_0 a \cos 2\theta}{R^2}\widehat{R} - \dfrac{2V_0 a \sin 2\theta}{R^2}\widehat{\theta}$

3) $\nabla F = \dfrac{V_0 a \cos 2\theta}{R^2}\widehat{R} + \dfrac{2V_0 a \sin 2\theta}{R^2}\widehat{\theta}$

4) $\nabla F = -\dfrac{V_0 a \cos 2\theta}{R^2}\widehat{R} - \dfrac{2V_0 a \cos 2\theta}{R}\widehat{\varphi}$

Problem

3.11. In Problem 3.10, calculate the gradient of the function at $(R, \theta, \varphi) = (2a, 0, \pi)$.

Difficulty level ● Easy ○ Normal ○ Hard
Calculation amount ● Small ○ Normal ○ Large

1) $\nabla F \big|_{(2a,0,\pi)} = \dfrac{V_0}{4a}\widehat{R} - \dfrac{V_0}{2a}\widehat{\theta}$

2) $\nabla F \big|_{(2a,0,\pi)} = \dfrac{V_0}{4a}\widehat{R}$

3) $\nabla F \big|_{(2a,0,\pi)} = -\dfrac{V_0}{4a}\widehat{R}$

4) $\nabla F \big|_{(2a,0,\pi)} = -\dfrac{V_0}{4a}\widehat{R} - \dfrac{V_0}{2a}\widehat{\theta}$

Partially Solved Exercise

3.12. (a) Find the gradient of the scalar function below. (b) Moreover, calculate the value of the gradient at $\left(1, \dfrac{\pi}{2}, \dfrac{\pi}{2}\right)$.

$$F(R, \theta, \varphi) = R \cos \theta \sin \varphi$$

Solution

(a) The gradient of a scalar function in cylindrical coordinate system can be calculated as follows.

$$\nabla F = \left(\dfrac{\partial}{\partial R}\widehat{R} + \dfrac{1}{R}\dfrac{\partial}{\partial \theta}\widehat{\theta} + \dfrac{1}{R \sin \theta}\dfrac{\partial}{\partial \varphi}\widehat{\varphi}\right) F$$

Hence:

$$\nabla F = \left(\dfrac{\partial}{\partial R}\widehat{R} + \dfrac{1}{R}\dfrac{\partial}{\partial \theta}\widehat{\theta} + \dfrac{1}{R \sin \theta}\dfrac{\partial}{\partial \varphi}\widehat{\varphi}\right)(\qquad)$$

$$\Rightarrow \nabla F = \dfrac{\partial}{\partial R}(\qquad)\widehat{R} + \dfrac{1}{R}\dfrac{\partial}{\partial \theta}(\qquad)\widehat{\theta} + \dfrac{1}{R \sin \theta}\dfrac{\partial}{\partial \varphi}(\qquad)\widehat{\varphi}$$

$$\Rightarrow \nabla F = \cos\theta \sin\varphi \widehat{R} - \sin\theta \sin\varphi \widehat{\theta} + \cot\theta \cos\varphi \widehat{\varphi}$$

b) Moreover:

$$\nabla F \bigg|_{(1,\frac{\pi}{2},\frac{\pi}{2})} = \cos(\ )\sin(\ )\widehat{R} - \sin(\ )\sin(\ )\widehat{\theta} + \cot(\ )\cos(\ )\widehat{\varphi}$$

$$\Rightarrow \nabla F \bigg|_{(1,\frac{\pi}{2},\frac{\pi}{2})} = -\widehat{\theta}$$

Exercise

3.13. Calculate the value of ∇F if $F(R, \theta, \varphi) = 10R\sin^2\theta \cos\varphi$.

Final Answer

$\nabla F = 10\sin^2\theta \cos\varphi \widehat{R} + 10\sin 2\theta \cos\varphi \widehat{\theta} - 10\sin\theta \sin\varphi \widehat{\varphi}$

3.5 Divergence

Problem

3.14. Which one of the following choices is correct about the properties of divergence of the vector fields presented below.
(a) $\nabla \cdot \left(\overrightarrow{F_1} \pm \overrightarrow{F_2}\right) = \nabla \cdot \overrightarrow{F_1} \pm \nabla \cdot \overrightarrow{F_2}$
(b) $\nabla \cdot \left(\overrightarrow{F_1} \cdot \overrightarrow{F_2}\right) = \nabla \cdot \overrightarrow{F_1} \cdot \nabla \cdot \overrightarrow{F_2}$
Difficulty level   ● Easy   ○ Normal   ○ Hard
Calculation amount ● Small   ○ Normal   ○ Large
1) Only Property (a) is correct.
2) Only Property (b) is correct.
3) Both Properties correct.
4) None of them is correct.

Problem

3.15. Which one of the following choices correctly presents the divergence theorem.
Difficulty level   ● Easy   ○ Normal   ○ Hard
Calculation amount ● Small   ○ Normal   ○ Large
1) $\int_V \left(\nabla \cdot \overrightarrow{F}\right) dV = \oint_S \overrightarrow{F} \cdot \overrightarrow{dS}$
2) $\int_S \left(\nabla \times \overrightarrow{F}\right) \cdot \overrightarrow{dS} = \oint_C \overrightarrow{F} \cdot \overrightarrow{dl}$

3.6 Divergence of a Vector in the Cartesian Coordinate System

3) $\int_S \left(\nabla \cdot \vec{F} \right) dS = \oint_l \vec{F} \cdot \vec{dl}$

4) $\int_V \left(\nabla \times \vec{F} \right) dV = \oint_S \vec{F} \cdot \vec{dS}$

Problem

3.16. Which one of the following choices is correct about the fields shown in Fig. 3.1.
Difficulty level   ● Easy   ○ Normal   ○ Hard
Calculation amount ● Small   ○ Normal   ○ Large

1) $\nabla \cdot \vec{F_1} < 0, \nabla \cdot \vec{F_2} > 0, \nabla \cdot \vec{F_3} = 0, \nabla \cdot \vec{F_4} > 0, \nabla \cdot \vec{F_5} < 0$
2) $\nabla \cdot \vec{F_1} < 0, \nabla \cdot \vec{F_2} > 0, \nabla \cdot \vec{F_3} > 0, \nabla \cdot \vec{F_4} > 0, \nabla \cdot \vec{F_5} < 0$
3) $\nabla \cdot \vec{F_1} > 0, \nabla \cdot \vec{F_2} < 0, \nabla \cdot \vec{F_3} > 0, \nabla \cdot \vec{F_4} = 0, \nabla \cdot \vec{F_5} > 0$
4) $\nabla \cdot \vec{F_1} > 0, \nabla \cdot \vec{F_2} < 0, \nabla \cdot \vec{F_3} = 0, \nabla \cdot \vec{F_4} = 0, \nabla \cdot \vec{F_5} > 0$

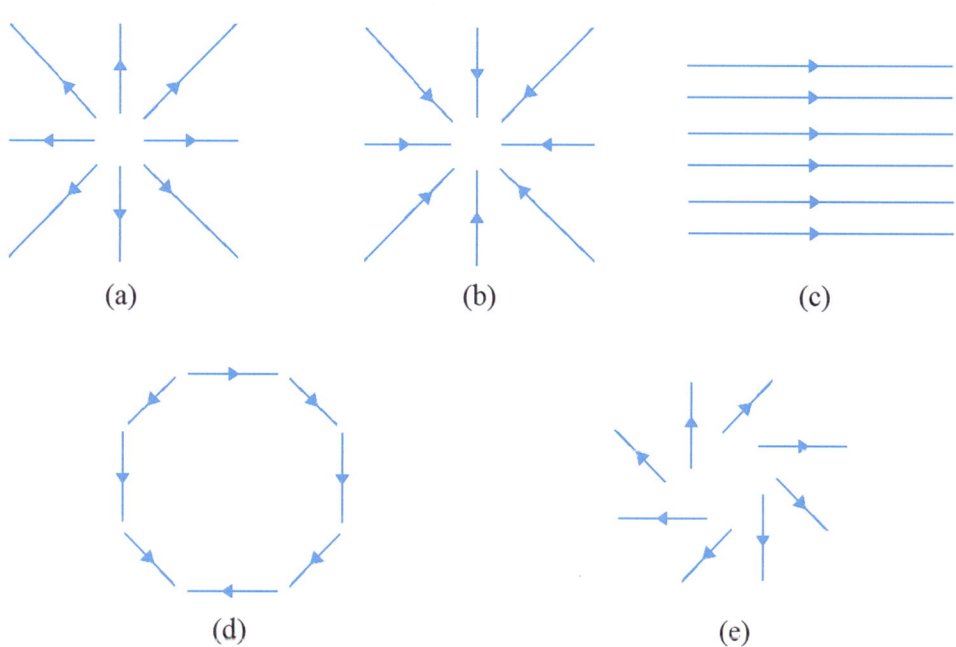

Fig. 3.1 The fields with different divergences

3.6 Divergence of a Vector in the Cartesian Coordinate System

Problem

3.17. Find the divergence of the vector field $\vec{F}(x, y, z) = 3x^2 \hat{x} + 2z \hat{y} + x^2 z \hat{z}$.
Difficulty level   ● Easy   ○ Normal   ○ Hard
Calculation amount ● Small   ○ Normal   ○ Large

1) $\nabla \cdot \vec{F} = 6 + x^2$
2) $\nabla \cdot \vec{F} = x^2 + 2z$
3) $\nabla \cdot \vec{F} = 6x + x^2 z$
4) $\nabla \cdot \vec{F} = 6x + x^2$

Problem

3.18. In Problem 3.17, calculate the divergence of the vector field at $(x, y, z) = (2, -2, 0)$.

Difficulty level ● Easy ○ Normal ○ Hard
Calculation amount ● Small ○ Normal ○ Large

1) $\nabla \cdot \vec{F}\big|_{(2,-2,0)} = 10$

2) $\nabla \cdot \vec{F}\big|_{(2,-2,0)} = 4$

3) $\nabla \cdot \vec{F}\big|_{(2,-2,0)} = 16$

4) $\nabla \cdot \vec{F}\big|_{(2,-2,0)} = 12$

Partially Solved Exercise

3.19. (a) Find the divergence of the vector field $\vec{F}(x,y,z) = -\cos x \sin y \hat{x} + \sin x \cos y \hat{y}$. (b) In addition, calculate the divergence of the vector field at $(x, y, z) = (1, 1, 1)$.

Solution

(a) The divergence of a vector field in the Cartesian coordinate system can be calculated as follows.

$$\nabla \cdot \vec{F} = \frac{\partial}{\partial x} F_x + \frac{\partial}{\partial y} F_y + \frac{\partial}{\partial z} F_z$$

Therefore:

$$\nabla \cdot \vec{F} = \frac{\partial}{\partial x}(\quad) + \frac{\partial}{\partial y}(\quad) + \frac{\partial}{\partial z}(\quad)$$

$$\Rightarrow \nabla \cdot \vec{F} = (\quad) + (\quad) + (\quad)$$

$$\Rightarrow \nabla \cdot \vec{F} = 0$$

(b) Since the vector field is divergenceless everywhere, we have:

$$\nabla \cdot \vec{F}\big|_{(1,1,1)} = 0$$

Exercise

3.20. Find the divergence of the vector field $\vec{F}(x, y, z) = -\sin 2y\hat{x} + \cos 2x\hat{y}$.

Final Answer

$\nabla \cdot \vec{F} = 0$

3.7 Divergence of a Vector in Cylindrical Coordinate System

Problem

3.21. Find the divergence of the vector field $\vec{F}(r, \varphi, z) = r \sin \varphi \hat{r} + r^2 z \hat{\varphi} + z \cos \varphi \hat{z}$.
Difficulty level ○ Easy ● Normal ○ Hard
Calculation amount ○ Small ● Normal ○ Large
1) $\nabla \cdot \vec{F} = 2 \sin \varphi + \cos \varphi$
2) $\nabla \cdot \vec{F} = \sin \varphi$
3) $\nabla \cdot \vec{F} = 2 \sin \varphi + 2 \cos \varphi$
4) $\nabla \cdot \vec{F} = \sin \varphi + 3 \cos \varphi$

Problem

3.22. In Problem 3.21, calculate the divergence of the vector field at $(r, \varphi, z) = (0, 0, 0)$.
Difficulty level ● Easy ○ Normal ○ Hard
Calculation amount ● Small ○ Normal ○ Large
1) $\nabla \cdot \vec{F} \Big|_{(0,0,0)} = 0$
2) $\nabla \cdot \vec{F} \Big|_{(0,0,0)} = 1$
3) $\nabla \cdot \vec{F} \Big|_{(0,0,0)} = 2$
4) $\nabla \cdot \vec{F} \Big|_{(0,0,0)} = 3$

Partially Solved Exercise

3.23. (a) Find the divergence of the vector field $\vec{F}(r, \varphi, z) = r\hat{r} + r \cos \varphi \hat{\varphi}$. (b) In addition, calculate the divergence of the vector field at $(r, \varphi, z) = \left(1, \frac{3\pi}{2}, 1\right)$.

Solution

(a) The divergence of a vector field in cylindrical coordinate system can be calculated as follows.

$$\nabla \cdot \vec{F} = \frac{1}{r}\frac{\partial}{\partial r}(rF_r) + \frac{1}{r}\frac{\partial}{\partial \varphi}(F_\varphi) + \frac{\partial}{\partial z}(F_z)$$

Therefore:

$$\nabla \cdot \vec{F} = \frac{1}{r}\frac{\partial}{\partial r}(\quad) + \frac{1}{r}\frac{\partial}{\partial \varphi}(\quad) + \frac{\partial}{\partial z}(\quad)$$

$$\Rightarrow \nabla \cdot \vec{F} = 2 - \sin\varphi$$

(b)

$$\nabla \cdot \vec{F}\bigg|_{\left(1,\frac{3\pi}{2},1\right)} = 2 - \sin(\quad) =$$

$$\Rightarrow \nabla \cdot \vec{F}\bigg|_{\left(1,\frac{3\pi}{2},1\right)} = 3$$

Exercise

3.24. Find the divergence of the vector field $\vec{F}(r,\varphi,z) = r^2\hat{r} + r^2\sin\varphi\hat{\varphi}$.

Final Answer

$\nabla \cdot \vec{F} = 3r + r\cos\varphi$

3.8 Divergence of a Vector in Spherical Coordinate System

Problem

3.25. Find the divergence of the following vector field.

$$\vec{F}(R,\theta,\varphi) = \frac{a^3\cos\theta}{R^2}\hat{R} - \frac{a^3\sin\theta}{R^2}\hat{\theta}$$

Difficulty level ○ Easy ● Normal ○ Hard 😐
Calculation amount ○ Small ● Normal ○ Large

1) $\nabla \cdot \vec{F} = -\dfrac{2a^3\sin\theta}{R^3}$
2) $\nabla \cdot \vec{F} = -\dfrac{2a^3\cos\theta}{R^3}$
3) $\nabla \cdot \vec{F} = -\dfrac{2a^3\sin\theta}{R^2}$
4) $\nabla \cdot \vec{F} = \dfrac{2a^3\cos\theta}{R^2}$

3.8 Divergence of a Vector in Spherical Coordinate System

Problem

3.26. In Problem 3.25, calculate the divergence of the vector field at $(R, \theta, \varphi) = \left(\frac{a}{2}, 0, \pi\right)$.

Difficulty level ● Easy ○ Normal ○ Hard
Calculation amount ● Small ○ Normal ○ Large

1) $\nabla \cdot \vec{F} \Big|_{\left(\frac{a}{2}, 0, \pi\right)} = -16$

2) $\nabla \cdot \vec{F} \Big|_{\left(\frac{a}{2}, 0, \pi\right)} = 0$

3) $\nabla \cdot \vec{F} \Big|_{\left(\frac{a}{2}, 0, \pi\right)} = -8a$

4) $\nabla \cdot \vec{F} \Big|_{\left(\frac{a}{2}, 0, \pi\right)} = 8a$

Partially Solved Exercise

3.27. (a) Find the divergence of the following vector field. (b) In addition, calculate the divergence of the vector field at $(R, \theta, \varphi) = (0.1, 0, \pi)$.

$$\vec{F}(R, \theta, \varphi) = \frac{1}{R}\widehat{R}$$

Solution

(a) The divergence of a vector field in spherical coordinate system can be calculated as follows.

$$\nabla \cdot \vec{F} = \frac{1}{R^2}\frac{\partial}{\partial R}(R^2 F_R) + \frac{1}{R \sin\theta}\frac{\partial}{\partial \theta}(\sin\theta F_\theta) + \frac{1}{R \sin\theta}\frac{\partial}{\partial \varphi}F_\varphi$$

Therefore:

$$\nabla \cdot \vec{F} = \frac{1}{R^2}\frac{\partial}{\partial R}(\quad) + \frac{1}{R \sin\theta}\frac{\partial}{\partial \theta}(\quad) + \frac{1}{R \sin\theta}\frac{\partial}{\partial \varphi}(\quad)$$

$$\Rightarrow \nabla \cdot \vec{F} = \frac{1}{R^2}\frac{\partial}{\partial R}(\quad) = \frac{1}{R^2} \times (\quad)$$

$$\Rightarrow \nabla \cdot \vec{F} = \frac{1}{R^2}$$

(b)

$$\nabla \cdot \vec{F} \Big|_{(0.1, 0, \pi)} = \frac{1}{(\quad)^2}$$

$$\Rightarrow \nabla \cdot \vec{F}\bigg|_{(0.1, 0, \pi)} = 100$$

Exercise

3.28. Calculate the divergence of the vector field $\vec{F}(R, \theta, \varphi) = Re^R \hat{R}$.

Final Answer

$\nabla \cdot \vec{F} = e^{-R}(3 - R)$

3.9 Curl

Problem

3.29. Which one of the following choices is correct about the fields shown in Fig. 3.2.
Difficulty level ● Easy ○ Normal ○ Hard
Calculation amount ● Small ○ Normal ○ Large
1) $\nabla \times \vec{F_1} \neq 0, \nabla \times \vec{F_2} = 0, \nabla \times \vec{F_3} = 0, \nabla \times \vec{F_4} \neq 0, \nabla \times \vec{F_5} = 0$
2) $\nabla \times \vec{F_1} = 0, \nabla \times \vec{F_2} = 0, \nabla \times \vec{F_3} = 0, \nabla \times \vec{F_4} = 0, \nabla \times \vec{F_5} \neq 0$
3) $\nabla \times \vec{F_1} = 0, \nabla \times \vec{F_2} = 0, \nabla \times \vec{F_3} = 0, \nabla \times \vec{F_4} \neq 0, \nabla \times \vec{F_5} \neq 0$
4) $\nabla \times \vec{F_1} \neq 0, \nabla \times \vec{F_2} \neq 0, \nabla \times \vec{F_3} = 0, \nabla \times \vec{F_4} \neq 0, \nabla \times \vec{F_5} \neq 0$

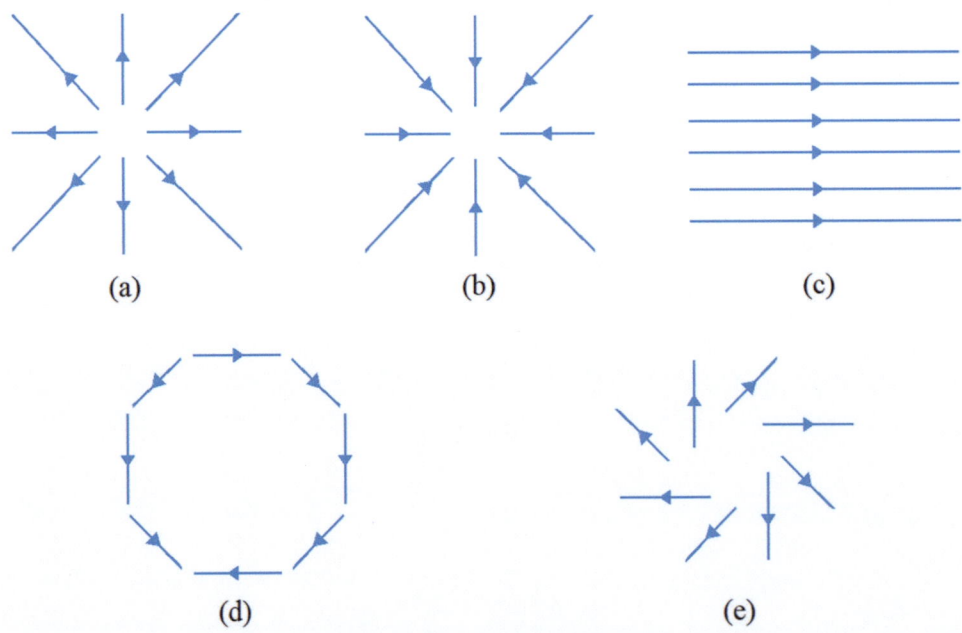

Fig. 3.2 The fields with different curls

3.10 Curl of a Vector in the Cartesian Coordinate System

Problem

3.30. Which one of the following choices is wrong about the properties of curl of vector fields.
Difficulty level ● Easy ○ Normal ○ Hard
Calculation amount ● Small ○ Normal ○ Large

1) $\nabla \times (\overrightarrow{F_1} + \overrightarrow{F_2}) = \nabla \times \overrightarrow{F_1} + \nabla \times \overrightarrow{F_2}$
2) $\nabla \times (\overrightarrow{F_1} - \overrightarrow{F_2}) = \nabla \times \overrightarrow{F_1} - \nabla \times \overrightarrow{F_2}$
3) $\nabla \cdot (\nabla \times \overrightarrow{F}) \neq 0$
4) $\nabla \times (\nabla F) = 0$

Problem

3.31. Which one of the following choices correctly presents the Stokes's theorem.
Difficulty level ● Easy ○ Normal ○ Hard
Calculation amount ● Small ○ Normal ○ Large

1) $\int_V (\nabla \cdot \overrightarrow{F}) dV = \oint_S \overrightarrow{F} \cdot \overrightarrow{dS}$
2) $\int_S (\nabla \times \overrightarrow{F}) \cdot \overrightarrow{dS} = \oint_C \overrightarrow{F} \cdot \overrightarrow{dl}$
3) $\int_S (\nabla \cdot \overrightarrow{F}) dS = \oint_l \overrightarrow{F} \cdot \overrightarrow{dl}$
4) $\int_V (\nabla \times \overrightarrow{F}) dV = \oint_S \overrightarrow{F} \cdot \overrightarrow{dS}$

3.10 Curl of a Vector in the Cartesian Coordinate System

Problem

3.32. Find the curl of the vector field $\overrightarrow{F}(x, y, z) = x^2 yz\widehat{x} + xz\widehat{z}$.
Difficulty level ○ Easy ● Normal ○ Hard
Calculation amount ○ Small ● Normal ○ Large

1) $\nabla \times \overrightarrow{F} = x^2 y\widehat{y} - x^2 \widehat{z}$
2) $\nabla \times \overrightarrow{F} = x^2 y\widehat{y} - x^2 z\widehat{z}$
3) $\nabla \times \overrightarrow{F} = (x^2 y - z)\widehat{y} - x^2 \widehat{z}$
4) $\nabla \times \overrightarrow{F} = (x^2 y - z)\widehat{y} - x^2 z\widehat{z}$

Problem

3.33. In Problem 3.32, calculate the curl of the vector field at $(x, y, z) = (1, 1, 1)$.
Difficulty level ● Easy ○ Normal ○ Hard
Calculation amount ● Small ○ Normal ○ Large

1) $\nabla \times \vec{F}\Big|_{(1,1,1)} = \hat{y} - \hat{z}$

2) $\nabla \times \vec{F}\Big|_{(1,1,1)} = -\hat{y} - \hat{z}$

3) $\nabla \times \vec{F}\Big|_{(1,1,1)} = -\hat{z}$

4) $\nabla \times \vec{F}\Big|_{(1,1,1)} = \hat{y} + \hat{z}$

Partially Solved Exercise

3.34. (a) Find the curl of the vector field $\vec{F}(x, y, z) = x^2\hat{x} + z\hat{y} + x^2z\hat{z}$. (b) In addition, calculate the curl of the vector field at $(x, y, z) = (1, 0, 1)$.

Solution

(a) The curl of a vector field in the Cartesian coordinate system can be calculated as follows.

$$\nabla \times \vec{F} = \begin{vmatrix} \hat{x} & \hat{y} & \hat{z} \\ \frac{\partial}{\partial x} & \frac{\partial}{\partial y} & \frac{\partial}{\partial z} \\ F_x & F_y & F_z \end{vmatrix} = \left(\frac{\partial F_z}{\partial y} - \frac{\partial F_y}{\partial z}\right)\hat{x} + \left(\frac{\partial F_x}{\partial z} - \frac{\partial F_z}{\partial x}\right)\hat{y} + \left(\frac{\partial F_y}{\partial x} - \frac{\partial F_x}{\partial y}\right)\hat{z}$$

Hence:

$$\nabla \times \vec{F} = \left(\frac{\partial}{\partial y}(\quad) - \frac{\partial}{\partial z}(\quad)\right)\hat{x} + \left(\frac{\partial}{\partial z}(\quad) - \frac{\partial}{\partial x}(\quad)\right)\hat{y} + \left(\frac{\partial}{\partial x}(\quad) - \frac{\partial}{\partial y}(\quad)\right)\hat{z}$$

$$\Rightarrow \nabla \times \vec{F} = (\quad)\hat{x} + (\quad)\hat{y} + (\quad)\hat{z}$$

$$\Rightarrow \nabla \times \vec{F} = -\hat{x} - 2xz\hat{y}$$

(b)

$$\nabla \times \vec{F}\Big|_{(1,0,1)} =$$

$$\Rightarrow \nabla \times \vec{F}\Big|_{(1,0,1)} = -\hat{x} - 2\hat{y}$$

3.11 Curl of a Vector in Cylindrical Coordinate System

Problem

3.35. Find the curl of the vector field $\vec{F}(r,\varphi,z) = r\sin\varphi\,\hat{r} + r^2 z\,\hat{\varphi} + z\cos\varphi\,\hat{z}$.
Difficulty level ○ Easy ● Normal ○ Hard
Calculation amount ○ Small ● Normal ○ Large

1) $\nabla \times \vec{F} = -\left(\dfrac{z\sin\varphi}{r} + r^2\right)\hat{r} + (3rz - \cos\varphi)\hat{z}$

2) $\nabla \times \vec{F} = \left(\dfrac{z\sin\varphi}{r} + r^3\right)\hat{r} + (3rz - \cos\varphi)\hat{z}$

3) $\nabla \times \vec{F} = \left(\dfrac{z\sin\varphi}{r} + r\right)\hat{r} + (3rz - \cos\varphi)\hat{z}$

4) $\nabla \times \vec{F} = \left(\dfrac{z\sin\varphi}{r} + r^2\right)\hat{r} + (3rz - \sin\varphi)\hat{z}$

Problem

3.36. In Problem 3.25, calculate the curl of the vector field at $(r,\varphi,z) = (1, 0°, 0)$.
Difficulty level ● Easy ○ Normal ○ Hard
Calculation amount ● Small ○ Normal ○ Large

1) $\nabla \times \vec{F}\Big|_{(1,0°,0)} = \hat{r} - \hat{z}$

2) $\nabla \times \vec{F}\Big|_{(1,0°,0)} = -\hat{r} - \hat{z}$

3) $\nabla \times \vec{F}\Big|_{(1,0°,0)} = -\hat{r} + \hat{z}$

4) $\nabla \times \vec{F}\Big|_{(1,0°,0)} = \hat{r} + \hat{z}$

Partially Solved Exercise

3.37. (a) Find the curl of the vector field below. (b) In addition, calculate the curl of the vector field at $(r,\varphi,z) = (1, 0°, 5)$.

$$\vec{F}(r,\varphi,z) = \frac{\cos\varphi}{r}\hat{z}$$

Solution

(a) The curl of a vector field in cylindrical coordinate system can be calculated as follows.

$$\nabla \times \vec{F} = \frac{1}{r}\begin{vmatrix} \hat{r} & r\hat{\varphi} & \hat{z} \\ \dfrac{\partial}{\partial r} & \dfrac{\partial}{\partial \varphi} & \dfrac{\partial}{\partial z} \\ F_r & F_\varphi & F_z \end{vmatrix} = \left(\frac{1}{r}\frac{\partial F_z}{\partial \varphi} - \frac{\partial F_\varphi}{\partial z}\right)\hat{r} + \left(\frac{\partial F_r}{\partial z} - \frac{\partial F_z}{\partial r}\right)\hat{\varphi} + \frac{1}{r}\left(\frac{\partial (rF_\varphi)}{\partial r} - \frac{\partial F_r}{\partial \varphi}\right)\hat{z}$$

Hence:

$$\nabla \times \vec{F} = \left(\frac{1}{r}\frac{\partial}{\partial \varphi}(\quad) - \frac{\partial}{\partial z}(\quad)\right)\hat{r} + \left(\frac{\partial}{\partial z}(\quad) - \frac{\partial}{\partial r}(\quad)\right)\hat{\varphi} + \frac{1}{r}\left(\frac{\partial}{\partial r}(\quad) - \frac{\partial}{\partial \varphi}(\quad)\right)\hat{z}$$

$$\Rightarrow \nabla \times \vec{F} = (\quad)\hat{r} + (\quad)\hat{\varphi} + \frac{1}{r}(\quad)\hat{z}$$

$$\Rightarrow \nabla \times \vec{F} = \frac{-\sin\varphi}{r^2}\hat{r} + \frac{\cos\varphi}{r^2}\hat{\varphi}$$

(b)

$$\nabla \times \vec{F}\bigg|_{(1,0°,5)} =$$

$$\Rightarrow \nabla \times \vec{F}\bigg|_{(1,0°,5)} = \hat{\varphi}$$

Exercise

3.38. Find the curl of the vector field $\vec{F}(r,\varphi,z) = 10e^{-2r}\cos\varphi\hat{r} + 10\sin\varphi\hat{z}$.

Final Answer

$\nabla \times \vec{F} = 5\hat{r}$

3.12 Curl of a Vector in Spherical Coordinate System

Problem

3.39. Find the curl of the vector field below.

$$\vec{F}(R,\theta,\varphi) = \frac{1}{R^2}\cos\theta\hat{R} + R\sin\theta\cos\varphi\hat{\theta} + \cos\theta\hat{\varphi}$$

Difficulty level ○ Easy ● Normal ○ Hard
Calculation amount ○ Small ● Normal ○ Large

1) $\nabla \times \vec{F} = \left(\frac{\cos 2\theta}{R\sin\theta} + \sin\varphi\right)\hat{R} + \frac{\cos\theta}{R}\hat{\theta} + \left(2\cos\varphi + \frac{1}{R^3}\right)\sin\theta\hat{\varphi}$

2) $\nabla \times \vec{F} = \left(\frac{\cos 2\theta}{R\sin\theta} + \cos\varphi\right)\hat{R} - \frac{\cos\theta}{R}\hat{\theta} + \left(2\cos\varphi + \frac{1}{R^3}\right)\sin\theta\hat{\varphi}$

3) $\nabla \times \vec{F} = \left(\frac{\cos 2\theta}{R\sin\theta} + \sin\varphi\right)\hat{R} - \frac{\cos\theta}{R}\hat{\theta} + \left(2\cos\varphi + \frac{1}{R^3}\right)\sin\theta\hat{\varphi}$

4) $\nabla \times \vec{F} = \left(\frac{\cos 2\theta}{R\sin\theta} + \cos\varphi\right)\hat{R} - \frac{\cos\theta}{R}\hat{\theta} + \left(2\sin\varphi + \frac{1}{R^3}\right)\sin\theta\hat{\varphi}$

3.12 Curl of a Vector in Spherical Coordinate System

Problem

3.40. In Problem 3.39, calculate the curl of the vector field at $(R, \theta, \varphi) = (1, 90°, 0°)$.

Difficulty level ● Easy ○ Normal ○ Hard
Calculation amount ● Small ○ Normal ○ Large

1) $\nabla \times \vec{F}\Big|_{(1,90°,0°)} = \hat{R} - 3\hat{\varphi}$

2) $\nabla \times \vec{F}\Big|_{(1,90°,0°)} = \hat{R} + 3\hat{\varphi}$

3) $\nabla \times \vec{F}\Big|_{(1,90°,0°)} = -\hat{R} - 3\hat{\varphi}$

4) $\nabla \times \vec{F}\Big|_{(1,90°,0°)} = -\hat{R} + 3\hat{\varphi}$

Partially Solved Exercise

3.41. (a) Find the curl of the vector field $\vec{F}(R,\theta,\varphi) = 12\sin\theta\,\hat{\theta}$. (b) In addition, calculate the curl of the vector field at $(R, \theta, \varphi) = (12, 90°, 90°)$.

Solution

(a) The curl of a vector field in cylindrical coordinate system can be calculated as follows.

$$\nabla \times \vec{F} = \frac{1}{R^2 \sin\theta} \begin{vmatrix} \hat{R} & R\hat{\theta} & R\sin\theta\,\hat{\varphi} \\ \dfrac{\partial}{\partial R} & \dfrac{\partial}{\partial \theta} & \dfrac{\partial}{\partial \varphi} \\ F_R & RF_\varphi & R\sin\theta F_\varphi \end{vmatrix}$$

$$\Rightarrow \nabla \times \vec{F} = \frac{1}{R\sin\theta}\left(\frac{\partial(\sin\theta F_\varphi)}{\partial\theta} - \frac{\partial F_\theta}{\partial\varphi}\right)\hat{R} + \frac{1}{R}\left(\frac{1}{\sin\theta}\frac{\partial F_R}{\partial\varphi} - \frac{\partial(RF_\varphi)}{\partial R}\right)\hat{\theta} + \frac{1}{R}\left(\frac{\partial(RF_\theta)}{\partial R} - \frac{\partial F_R}{\partial\theta}\right)\hat{\varphi}$$

Hence:

$$\nabla \times \vec{F} = \frac{1}{R\sin\theta}\left(\frac{\partial}{\partial\theta}(\quad) - \frac{\partial}{\partial\varphi}(\quad)\right)\hat{R} + \frac{1}{R}\left(\frac{1}{\sin\theta}\frac{\partial}{\partial\varphi}(\quad) - \frac{\partial}{\partial R}(\quad)\right)\hat{\theta} + \frac{1}{R}\times\left(\frac{\partial}{\partial R}(\quad) - \frac{\partial}{\partial\theta}(\quad)\right)\hat{\varphi}$$

$$\Rightarrow \nabla \times \vec{F} = \frac{1}{R\sin\theta}(\quad)\hat{R} + \frac{1}{R}(\quad)\hat{\theta} + \frac{1}{R}(\quad)\hat{\varphi}$$

$$\Rightarrow \nabla \times \vec{F} = \frac{12\sin\theta}{R}\hat{\varphi}$$

(b)

$$\nabla \times \vec{F}\Big|_{(12,90°,90°)} =$$

$$\left.\nabla \times \vec{F}\,\right|_{(12,90°,90°)} = \widehat{\varphi}$$

3.13 Laplacian

Problem

3.42. Which one of the following choices is correct about the Laplacian operator.
Difficulty level   ● Easy   ○ Normal   ○ Hard
Calculation amount ● Small   ○ Normal   ○ Large
1) $\nabla \cdot (\nabla F) = \nabla^2 F$
2) $\nabla \times (\nabla F) = \nabla^2 F$
3) $\nabla \times \nabla \times F = \nabla^2 F$
4) $\nabla \cdot \nabla \times F = \nabla^2 F$

3.14 Laplacian of a Scalar in the Cartesian Coordinate System

Problem

3.43. Find the Laplacian of the scalar function $F(x, y, z) = e^{-z} \sin 2x \cos y$.
Difficulty level   ○ Easy   ● Normal   ○ Hard
Calculation amount ○ Small   ● Normal   ○ Large
1) $\nabla^2 F = 4\, e^{-z} \sin 2x \cos y$
2) $\nabla^2 F = 2\, e^{-z} \cos 2x \cos y$
3) $\nabla^2 F = -2\, e^{-z} \cos 2x \cos y$
4) $\nabla^2 F = -4\, e^{-z} \sin 2x \cos y$

Problem

3.44. In Problem 3.43, calculate the Laplacian of the scalar function at $(x, y, z) = \left(\frac{\pi}{4}, 0, 0\right)$.
Difficulty level   ● Easy   ○ Normal   ○ Hard
Calculation amount ● Small   ○ Normal   ○ Large
1) $\left.\nabla^2 F\right|_{\left(\frac{\pi}{4},0,0\right)} = 4$
2) $\left.\nabla^2 F\right|_{\left(\frac{\pi}{4},0,0\right)} = 0$
3) $\left.\nabla^2 F\right|_{\left(\frac{\pi}{4},0,0\right)} = 1$
4) $\left.\nabla^2 F\right|_{\left(\frac{\pi}{4},0,0\right)} = -4$

3.15 Laplacian of a Scalar in Cylindrical Coordinate System

Partially Solved Exercise

3.45. (a) Find the Laplacian of the scalar function $F(x, y, z) = xy^2z^3$. (b) Also, calculate the curl of the vector field at $(x, y, z) = (1, 2, 3)$.

Solution

(a) The curl of a vector field in the Cartesian coordinate system can be calculated as follows.

$$\nabla^2 F = \frac{\partial^2 F}{\partial x^2} + \frac{\partial^2 F}{\partial y^2} + \frac{\partial^2 F}{\partial z^2}$$

Hence:

$$\nabla^2 F = \frac{\partial^2}{\partial x^2}(\quad) + \frac{\partial^2}{\partial y^2}(\quad) + \frac{\partial^2}{\partial z^2}(\quad)$$

$$\Rightarrow \nabla^2 F = \frac{\partial}{\partial x}(\quad) + \frac{\partial}{\partial y}(\quad) + \frac{\partial}{\partial z}(\quad)$$

$$\Rightarrow \nabla^2 F = 2xz^3 + 6xy^2 z$$

(b)

$$\Rightarrow \nabla^2 F \bigg|_{(1,2,3)} =$$

$$\Rightarrow \nabla^2 F \bigg|_{(1,2,3)} = 126$$

Exercise

3.46. Find the Laplacian of the scalar function $F(x, y, z) = xy + yz + zx$.

Final Answer

$\nabla^2 F = 0$

3.15 Laplacian of a Scalar in Cylindrical Coordinate System

Problem

3.47. Find the Laplacian of the scalar function $F(r, \varphi, z) = r^2 z \cos 2\varphi$.
Difficulty level   ○ Easy   ● Normal   ○ Hard
Calculation amount   ○ Small   ● Normal   ○ Large
1) $\nabla^2 F = 0$
2) $\nabla^2 F = z \cos 2\varphi$
3) $\nabla^2 F = -z \cos 2\varphi$
4) $\nabla^2 F = 2r \cos 2\varphi$

Partially Solved Exercise

3.48. (a) Find the Laplacian of the scalar function $F(r, \varphi, z) = r^3 \sin 2\varphi$. (b) Moreover, calculate the curl of the vector field at $(r, \varphi, z) = (1, 45°, 2)$.

Solution

(a) The curl of a vector field in the Cartesian coordinate system can be calculated as follows.

$$\nabla^2 F = \frac{1}{r}\frac{\partial}{\partial r}\left(r\frac{\partial F}{\partial r}\right) + \frac{1}{r^2}\frac{\partial^2 F}{\partial \varphi^2} + \frac{\partial^2 F}{\partial z^2}$$

Therefore:

$$\nabla^2 F = \frac{1}{r}\frac{\partial}{\partial r}\left(r\frac{\partial}{\partial r}(\quad)\right) + \frac{1}{r^2}\frac{\partial^2}{\partial \varphi^2}(\quad) + \frac{\partial^2}{\partial z^2}(\quad)$$

$$\Rightarrow \nabla^2 F = \frac{1}{r}\frac{\partial}{\partial r}(\quad) + \frac{1}{r^2}\frac{\partial}{\partial \varphi}(\quad)$$

$$\Rightarrow \nabla^2 F =$$

$$\Rightarrow \nabla^2 F = 5r \sin 2\varphi$$

(b)

$$\nabla^2 F \Big|_{(1, 45°, 2)} =$$

$$\Rightarrow \nabla^2 F \Big|_{(1, 45°, 2)} = 5$$

Exercise

3.49. Find the Laplacian of the scalar function $F(r, \varphi, z) = 5e^{-r} \cos \varphi$.

Final Answer

$\nabla^2 F = 5e^{-r}\left(1 - \frac{1}{r} - \frac{1}{r^2}\right)\cos \varphi$

3.16 Laplacian of a Scalar in Spherical Coordinate System

Problem

3.50. Find the Laplacian of the scalar function $F(R, \theta, \varphi) = \cos^2\theta R \sin^2\theta \cos \varphi$.
Difficulty level ○ Easy ● Normal ○ Hard
Calculation amount ○ Small ○ Normal ● Large

3.16 Laplacian of a Scalar in Spherical Coordinate System

1) $\nabla^2 F = \dfrac{2\cos\varphi}{R\sin\theta}$
2) $\nabla^2 F = -\dfrac{\cos\varphi}{R\sin\theta}$
3) $\nabla^2 F = 0$
4) $\nabla^2 F = \dfrac{\cos\varphi(\cos^2\theta - \sin^2\theta)}{R\sin\theta}$

Partially Solved Exercise

3.51. Find the Laplacian of the scalar function below.

$$F(R, \theta, \varphi) = \frac{1}{R^2}$$

Solution

The Laplacian of a vector field in spherical coordinate system can be calculated as follows.

$$\nabla^2 F = \frac{1}{R^2}\frac{\partial}{\partial R}\left(R^2 \frac{\partial F}{\partial R}\right) + \frac{1}{R^2 \sin\theta}\frac{\partial}{\partial \theta}\left(\sin\theta \frac{\partial F}{\partial \theta}\right) + \frac{1}{R^2 \sin^2\theta}\frac{\partial^2 F}{\partial \varphi^2}$$

Thus:

$$\nabla^2 F = \frac{1}{R^2}\frac{\partial}{\partial R}\left(R^2 \frac{\partial}{\partial R}(\quad)\right) + \frac{1}{R^2 \sin\theta}\frac{\partial}{\partial \theta}\left(\sin\theta \frac{\partial}{\partial \theta}(\quad)\right) + \frac{1}{R^2 \sin^2\theta}\frac{\partial^2}{\partial \varphi^2}(\quad)$$

$$\Rightarrow \nabla^2 F = \frac{1}{R^2}\frac{\partial}{\partial R}(\quad) + \frac{1}{R^2 \sin\theta}\frac{\partial}{\partial \theta}(\quad) + \frac{1}{R^2 \sin^2\theta}\frac{\partial}{\partial \varphi}(\quad)$$

$$\Rightarrow \nabla^2 F = \frac{1}{R^2}\frac{\partial}{\partial R}(\quad) + (\quad) + (\quad)$$

$$\Rightarrow \nabla^2 F = \frac{1}{R^2}(\quad)$$

$$\Rightarrow \nabla^2 F = \frac{2}{R^4}$$

Exercise

3.52. Find the Laplacian of the scalar function $F(R, \theta, \varphi) = 10e^{-R}\sin\theta$.

Final Answer

$$\nabla^2 F = 10e^{-R}\left[\sin\theta\left(1 - \frac{2}{R}\right) + \frac{\cos^2\theta - \sin^2\theta}{R^2 \sin\theta}\right]$$

References

1. Rahmani-Andebili, M., General Physics II - Practice Problems, Methods, and Solutions, Springer Nature, 2025.
2. Rahmani-Andebili, M., General Physics I - Practice Problems, Methods, and Solutions, Springer Nature, 2025.
3. Rahmani-Andebili, M., Mathematics of Engineering and Science – Practice Problems, Methods, and Solutions, Springer Nature, 2024.
4. Rahmani-Andebili, M., Differential Equations – Practice Problems, Methods, and Solutions, Springer Nature, 2022.
5. Rahmani-Andebili, M., Calculus III – Practice Problems, Methods, and Solutions, Springer Nature, 2023.
6. Rahmani-Andebili, M., Calculus II – Practice Problems, Methods, and Solutions, Springer Nature, 2023.
7. Rahmani-Andebili, M., Calculus I (2nd Ed.) – Practice Problems, Methods, and Solutions, Springer Nature, 2023.
8. Rahmani-Andebili, M., Precalculus (2nd Ed.) – Practice Problems, Methods, and Solutions, Springer Nature, 2024.

4 Gradient, Divergence, Curl, and Laplacian: Part B

Abstract

In this chapter, the problems of the third chapter are fully solved, in detail, step-by-step, and with different methods.

4.1 Gradient

4.1. The following properties are valid for gradient [1–8].

$$\nabla(F_1 + F_2) = \nabla F_1 + \nabla F_2$$

$$\nabla(F_1 - F_2) = \nabla F_1 - \nabla F_2$$

$$\nabla(F_1 F_2) = F_1 \nabla F_2 + F_2 \nabla F_1$$

$$\nabla(F^n) = n F^{n-1} \nabla F$$

Choice (3) is the answer.

4.2 Gradient of a Scalar in the Cartesian Coordinate System

4.2. Based on the information given in the problem, we have:

$$F(x, y, z) = x^2 + y^2 z$$

The gradient of a scalar function in the Cartesian coordinate system is calculated as follows.

$$\nabla F = \left(\frac{\partial}{\partial x}\hat{x} + \frac{\partial}{\partial y}\hat{y} + \frac{\partial}{\partial z}\hat{z}\right) F$$

Therefore:

$$\nabla F = \left(\frac{\partial}{\partial x}\hat{x} + \frac{\partial}{\partial y}\hat{y} + \frac{\partial}{\partial z}\hat{z}\right)(x^2 + y^2 z)$$

$$\Rightarrow \nabla F = \frac{\partial}{\partial x}\left(x^2 + y^2 z\right)\hat{x} + \frac{\partial}{\partial y}\left(x^2 + y^2 z\right)\hat{y} + \frac{\partial}{\partial z}\left(x^2 + y^2 z\right)\hat{z}$$

$$\Rightarrow \nabla F = 2x\hat{x} + 2yz\hat{y} + y^2\hat{z}$$

Choice (4) is the answer.

> **Notes**
> In this problem, the relation below has been used.
>
> $$\frac{d}{dx}x^n = nx^{n-1}$$

4.3. Based on the information given in the problem, we have:

$$(x, y, z) = (1, 1, 1)$$

$$\nabla F = 2x\hat{x} + 2yz\hat{y} + y^2\hat{z}$$

Thus:

$$\nabla F \bigg|_{(1,1,1)} = 2(1)\hat{x} + 2(1)(1)\hat{y} + \left(1^2\right)\hat{z}$$

$$\Rightarrow \nabla F \bigg|_{(1,1,1)} = 2\hat{x} + 2\hat{y} + \hat{z}$$

Choice (2) is the answer.

4.3 Gradient of a Scalar in Cylindrical Coordinate System

4.6. Based on the information given in the problem, we have:

$$F(r, \varphi, z) = V_0 e^{-2r} \sin 3\varphi$$

The gradient of a scalar function in cylindrical coordinate system can be calculated as follows.

4.3 Gradient of a Scalar in Cylindrical Coordinate System

$$\nabla F = \left(\frac{\partial}{\partial r}\hat{r} + \frac{1}{r}\frac{\partial}{\partial \varphi}\hat{\varphi} + \frac{\partial}{\partial z}\hat{z}\right) F$$

Hence:

$$\nabla F = \left(\frac{\partial}{\partial r}\hat{r} + \frac{1}{r}\frac{\partial}{\partial \varphi}\hat{\varphi} + \frac{\partial}{\partial z}\hat{z}\right) V_0 e^{-2r} \sin 3\varphi$$

$$\Rightarrow \nabla F = \frac{\partial}{\partial r}\left(V_0 e^{-2r}\sin 3\varphi\right)\hat{r} + \frac{1}{r}\frac{\partial}{\partial \varphi}\left(V_0 e^{-2r}\sin 3\varphi\right)\hat{\varphi} + \frac{\partial}{\partial z}\left(V_0 e^{-2r}\sin 3\varphi\right)\hat{z}$$

$$\Rightarrow \nabla F = -2V_0 e^{-2r}\sin 3\varphi \hat{r} + \frac{3}{r}V_0 e^{-2r}\cos 3\varphi \hat{\varphi}$$

Choice (3) is the answer.

Notes
In this problem, the relations below have been used.

$$\frac{d}{dx}x^n = nx^{n-1}$$

$$\frac{d}{dx}e^{u(x)} = u'(x)e^{u(x)}$$

$$\frac{d}{dx}\sin(u(x)) = u'(x)\cos(u(x))$$

4.7. Based on the information given in the problem, we have:

$$(r, \varphi, z) = \left(1, \frac{\pi}{2}, 3\right)$$

$$\nabla F = -2V_0 e^{-2r}\sin 3\varphi \hat{r} + \frac{3}{r}V_0 e^{-2r}\cos 3\varphi \hat{\varphi}$$

Therefore:

$$\nabla F \bigg|_{\left(1, \frac{\pi}{2}, 3\right)} = -2V_0 e^{-2}\sin\left(\frac{3\pi}{2}\right)\hat{r} + \frac{3}{1}V_0 e^{-2r}\cos\left(\frac{3\pi}{2}\right)\hat{\varphi}$$

$$\Rightarrow \nabla F \bigg|_{\left(1, \frac{\pi}{2}, 3\right)} = 2V_0 e^{-2}\hat{r}$$

Choice (4) is the answer.

4.4 Gradient of a Scalar in Spherical Coordinate System

4.10. Based on the information given in the problem, we have:

$$F(R, \theta, \varphi) = \frac{V_0 a \cos 2\theta}{R}$$

The gradient of a scalar function in cylindrical coordinate system can be calculated as follows.

$$\nabla F = \left(\frac{\partial}{\partial R} \widehat{R} + \frac{1}{R} \frac{\partial}{\partial \theta} \widehat{\theta} + \frac{1}{R \sin \theta} \frac{\partial}{\partial \varphi} \widehat{\varphi} \right) F$$

Hence:

$$\nabla F = \left(\frac{\partial}{\partial R} \widehat{R} + \frac{1}{R} \frac{\partial}{\partial \theta} \widehat{\theta} + \frac{1}{R \sin \theta} \frac{\partial}{\partial \varphi} \widehat{\varphi} \right) \frac{V_0 a \cos 2\theta}{R}$$

$$\nabla F = \frac{\partial}{\partial R} \left(\frac{V_0 a \cos 2\theta}{R} \right) \widehat{R} + \frac{1}{R} \frac{\partial}{\partial \theta} \left(\frac{V_0 a \cos 2\theta}{R} \right) \widehat{\theta} + \frac{1}{R \sin \theta} \frac{\partial}{\partial \varphi} \left(\frac{V_0 a \cos 2\theta}{R} \right) \widehat{\varphi}$$

$$\Rightarrow \nabla F = -\frac{V_0 a \cos 2\theta}{R^2} \widehat{R} - \frac{2 V_0 a \sin 2\theta}{R^2} \widehat{\theta}$$

Choice (2) is the answer.

> **Notes**
> In this problem, the relations below have been used.
>
> $$\frac{d}{dx} \left(\frac{u(x)}{v(x)} \right) = \frac{u'(x) v(x) - v'(x) u(x)}{v^2(x)}$$
>
> $$\frac{d}{dx} \cos(u(x)) = -u'(x) \sin(u(x))$$

4.11. Based on the information given in the problem, we have:

$$(R, \theta, \varphi) = (2a, 0, \pi)$$

$$\nabla F = -\frac{V_0 a \cos 2\theta}{R^2} \widehat{R} - \frac{2 V_0 a \sin 2\theta}{R^2} \widehat{\theta}$$

Therefore:

4.5 Divergence

$$\nabla F \bigg|_{(2a,0,\pi)} = -\frac{V_0 a \cos 0}{(2a)^2}\hat{R} - \frac{2V_0 a \sin 0}{(2a)^2}\hat{\theta}$$

$$\Rightarrow \nabla F \bigg|_{(2a,0,\pi)} = -\frac{V_0}{4a}\hat{R}$$

Choice (3) is the answer.

4.5 Divergence

4.14. The following properties are valid for divergence.

$$\nabla \cdot \left(\overrightarrow{F_1} + \overrightarrow{F_2}\right) = \nabla \cdot \overrightarrow{F_1} + \nabla \cdot \overrightarrow{F_2}$$

$$\nabla \cdot \left(\overrightarrow{F_1} - \overrightarrow{F_2}\right) = \nabla \cdot \overrightarrow{F_1} - \nabla \cdot \overrightarrow{F_2}$$

Choice (1) is the answer.

4.15. The divergence theorem is as follows.

$$\int_V \left(\nabla \cdot \overrightarrow{F}\right) dV = \oint_S \overrightarrow{F} \cdot \overrightarrow{dS}$$

It states that the surface integral of a vector field over a closed surface, which is called the flux through the surface, is equal to the volume integral of the divergence over the region enclosed by the surface. Choice (1) is the answer.

4.16. Figs. 4.1 (a) and (e) illustrate positive divergences because the arrows have a general tendency to leave the area, that is, diverge or go away from it. Thus:

$$\nabla \cdot \overrightarrow{F_1} > 0$$

$$\nabla \cdot \overrightarrow{F_5} > 0$$

Fig. 4.1 (b) shows a negative divergence because the arrows have a general tendency to cluster and converge around the area. Hence:

$$\nabla \cdot \overrightarrow{F_2} < 0$$

Figs. 4.1 (c) and (d) illustrate zero divergences everywhere because although the arrows show some movements, the volume rate flowing into any closed surface is equal to the volume rate flowing out. Therefore:

$$\nabla \cdot \overrightarrow{F_3} = 0$$

$$\nabla \cdot \overrightarrow{F_4} = 0$$

Choice (4) is the answer.

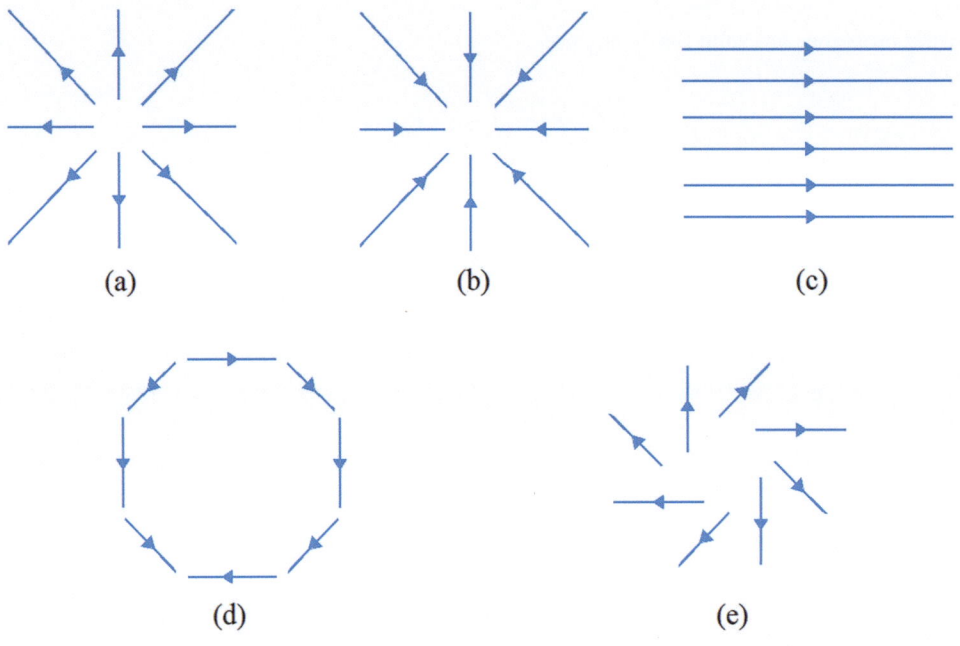

Fig. 4.1 The fields with different divergences

4.6 Divergence of a Vector in the Cartesian Coordinate System

4.17. Based on the information given in the problem, we have:

$$\overrightarrow{F}(x, y, z) = 3x^2 \widehat{x} + 2z\widehat{y} + x^2 z\widehat{z}$$

The divergence of a vector field in the Cartesian coordinate system can be calculated as follows.

4.6 Divergence of a Vector in the Cartesian Coordinate System

$$\nabla \cdot \vec{F} = \frac{\partial}{\partial x} F_x + \frac{\partial}{\partial y} F_y + \frac{\partial}{\partial z} F_z$$

Therefore:

$$\nabla \cdot \vec{F} = \frac{\partial}{\partial x}(3x^2) + \frac{\partial}{\partial y}(2z) + \frac{\partial}{\partial z}(x^2 z)$$

$$\Rightarrow \nabla \cdot \vec{F} = 6x + 0 + x^2$$

$$\Rightarrow \nabla \cdot \vec{F} = 6x + x^2$$

Choice (4) is the answer.

> **Notes**
> In this problem, the relation below has been used.
>
> $$\frac{d}{dx} x^n = n x^{n-1}$$

4.18. Based on the information given in the problem, we have:

$$(x, y, z) = (2, -2, 0)$$

$$\nabla \cdot \vec{F} = 6x + x^2$$

Hence:

$$\nabla \cdot \vec{F} \Big|_{(2, -2, 0)} = 6(2) + (2)^2$$

$$\Rightarrow \nabla \cdot \vec{F} \Big|_{(2, -2, 0)} = 16$$

Choice (3) is the answer.

4.7 Divergence of a Vector in Cylindrical Coordinate System

4.21. Based on the information given in the problem, we have:

$$\vec{F}(r,\varphi,z) = r\sin\varphi\,\hat{r} + r^2 z\,\hat{\varphi} + z\cos\varphi\,\hat{z}$$

The divergence of a vector field in cylindrical coordinate system can be calculated as follows.

$$\nabla\cdot\vec{F} = \frac{1}{r}\frac{\partial}{\partial r}(rF_r) + \frac{1}{r}\frac{\partial}{\partial \varphi}(F_\varphi) + \frac{\partial}{\partial z}(F_z)$$

Therefore:

$$\nabla\cdot\vec{F} = \frac{1}{r}\frac{\partial}{\partial r}(r^2\sin\varphi) + \frac{1}{r}\frac{\partial}{\partial \varphi}(r^2 z) + \frac{\partial}{\partial z}(z\cos\varphi)$$

$$\Rightarrow \nabla\cdot\vec{F} = 2\sin\varphi + 0 + \cos\varphi$$

$$\Rightarrow \nabla\cdot\vec{F} = 2\sin\varphi + \cos\varphi$$

Choice (1) is the answer.

> **Notes**
> In this problem, the relation below has been used.
>
> $$\frac{d}{dx}x^n = nx^{n-1}$$

4.22. Based on the information given in the problem, we have:

$$(r,\varphi,z) = (0,0,0)$$

$$\nabla\cdot\vec{F} = 2\sin\varphi + \cos\varphi$$

Therefore:

$$\nabla\cdot\vec{F}\bigg|_{(0,0,0)} = 2\sin 0 + \cos 0 = 0 + 1$$

$$\Rightarrow \nabla\cdot\vec{F}\bigg|_{(0,0,0)} = 1$$

Choice (2) is the answer.

4.8 Divergence of a Vector in Spherical Coordinate System

4.25. Based on the information given in the problem, we have:

$$\vec{F}(R, \theta, \varphi) = \frac{a^3 \cos\theta}{R^2}\hat{R} - \frac{a^3 \sin\theta}{R^2}\hat{\theta}$$

The divergence of a vector field in spherical coordinate system can be calculated as follows.

$$\nabla \cdot \vec{F} = \frac{1}{R^2}\frac{\partial}{\partial R}(R^2 F_R) + \frac{1}{R\sin\theta}\frac{\partial}{\partial \theta}(\sin\theta F_\theta) + \frac{1}{R\sin\theta}\frac{\partial}{\partial \varphi}F_\varphi$$

Therefore:

$$\nabla \cdot \vec{F} = \frac{1}{R^2}\frac{\partial}{\partial R}\left(R^2 \frac{a^3 \cos\theta}{R^2}\right) + \frac{1}{R\sin\theta}\frac{\partial}{\partial \theta}\left(-\sin\theta \frac{a^3 \sin\theta}{R^2}\right) + \frac{1}{R\sin\theta}\frac{\partial}{\partial \varphi}0$$

$$\Rightarrow \nabla \cdot \vec{F} = \frac{1}{R^2}\frac{\partial}{\partial R}(a^3 \cos\theta) + \frac{1}{R\sin\theta}\frac{\partial}{\partial \theta}\left(-\frac{a^3 \sin^2\theta}{R^2}\right) + \frac{1}{R\sin\theta}\frac{\partial}{\partial \varphi}0$$

$$\Rightarrow \nabla \cdot \vec{F} = 0 + \frac{1}{R\sin\theta}\left(-\frac{2a^3 \cos\theta \sin\theta}{R^2}\right) + 0$$

$$\Rightarrow \nabla \cdot \vec{F} = -\frac{2a^3 \cos\theta}{R^3}$$

Choice (2) is the answer.

Notes

In this problem, the relations below have been used.

$$\frac{d}{dx}c = 0$$

$$\frac{d}{dx}\sin^n(u(x)) = nu'(x)\cos(u(x))\sin^{n-1}(u(x))$$

4.26. Based on the information given in the problem, we have:

$$(R, \theta, \varphi) = \left(\frac{a}{2}, 0, \pi\right)$$

$$\nabla \cdot \vec{F} = -\frac{2a^3 \cos \theta}{R^3}$$

Therefore:

$$\nabla \cdot \vec{F} \bigg|_{\left(\frac{a}{2}, 0, \pi\right)} = -\frac{2a^3 \cos 0}{\left(\frac{a}{2}\right)^3}$$

$$\Rightarrow \nabla \cdot \vec{F} \bigg|_{\left(\frac{a}{2}, 0, \pi\right)} = -16$$

Choice (1) is the answer.

4.9 Curl

4.29. The curl of a vector field at a given point measures the tendency of particles to rotate about the axis. Herein, the curl is represented by a vector whose length and direction denote the magnitude and axis of the maximum circulation.

Hence, Figs. 4.2 (a)–(c) illustrate zero curl because the arrows do not have any circulation. In other words:

$$\nabla \times \vec{F_1} = 0$$

$$\nabla \times \vec{F_2} = 0$$

$$\nabla \times \vec{F_3} = 0$$

However, Figs. 4.2 (d) and (e) have non-zero curl since the arrows rotate. In other words:

$$\nabla \times \vec{F_4} \neq 0$$

$$\nabla \times \vec{F_5} \neq 0$$

4.9 Curl

Choice (3) is the answer.

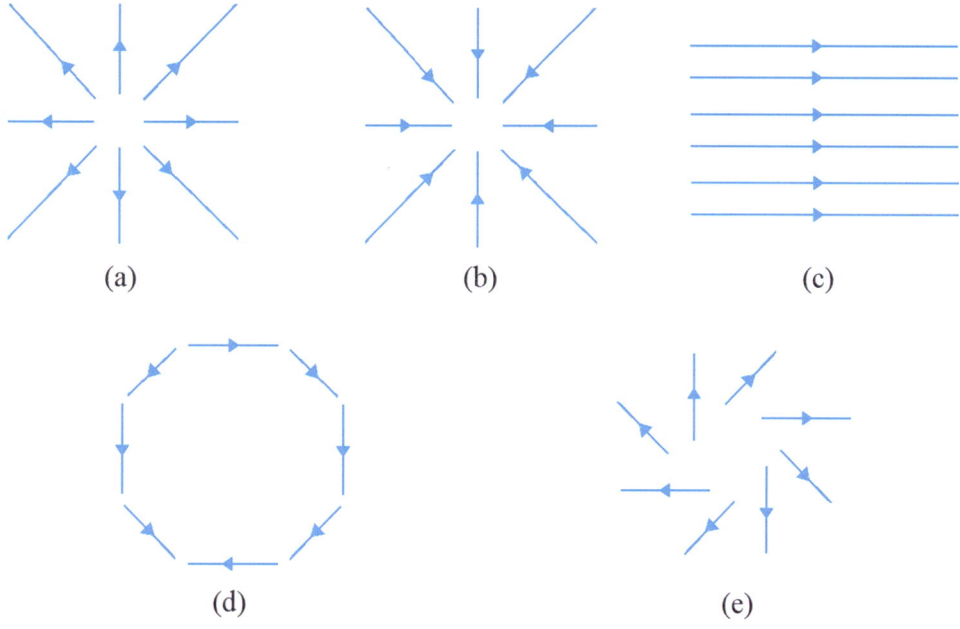

Fig. 4.2 The fields with different curls

4.30. The following identities are available for the curl operator.

$$\nabla \times \left(\vec{F_1} + \vec{F_2} \right) = \nabla \times \vec{F_1} + \nabla \times \vec{F_2}$$

$$\nabla \times \left(\vec{F_1} - \vec{F_2} \right) = \nabla \times \vec{F_1} - \nabla \times \vec{F_2}$$

$$\nabla \cdot \left(\nabla \times \vec{F} \right) = 0$$

$$\nabla \times \left(\nabla F \right) = 0$$

Choice (3) is the answer.

4.31. The Stokes's theorem is as follows.

$$\int_S (\nabla \times \vec{F}) \cdot \vec{dS} = \oint_C \vec{F} \cdot \vec{dl}$$

In fact, Stokes's theorem equates a surface integral of the curl of a vector field to a three-dimensional line integral of the vector field around the boundary of the surface. It basically states that the surface integral of curl of the vector field over a surface is the circulation of the field around the boundary of the surface.

Choice (2) is the answer.

4.10 Curl of a Vector in the Cartesian Coordinate System

4.32. Based on the information given in the problem, we have:

$$\vec{F}(x, y, z) = x^2 yz \hat{x} + xz \hat{z}$$

The curl of a vector field in the Cartesian coordinate system can be calculated as follows.

$$\nabla \times \vec{F} = \begin{vmatrix} \hat{x} & \hat{y} & \hat{z} \\ \frac{\partial}{\partial x} & \frac{\partial}{\partial y} & \frac{\partial}{\partial z} \\ F_x & F_y & F_z \end{vmatrix} = \left(\frac{\partial F_z}{\partial y} - \frac{\partial F_y}{\partial z}\right)\hat{x} + \left(\frac{\partial F_x}{\partial z} - \frac{\partial F_z}{\partial x}\right)\hat{y} + \left(\frac{\partial F_y}{\partial x} - \frac{\partial F_x}{\partial y}\right)\hat{z}$$

Hence:

$$\nabla \times \vec{F} = \left(\frac{\partial}{\partial y}(xz) - \frac{\partial}{\partial z}(0)\right)\hat{x} + \left(\frac{\partial}{\partial z}(x^2 yz) - \frac{\partial}{\partial x}(xz)\right)\hat{y} + \left(\frac{\partial}{\partial x}(0) - \frac{\partial}{\partial y}(x^2 yz)\right)\hat{z}$$

$$\Rightarrow \nabla \times \vec{F} = (0-0)\hat{x} + (x^2 y - z)\hat{y} + (0 - x^2 z)\hat{z}$$

$$\Rightarrow \nabla \times \vec{F} = (x^2 y - z)\hat{y} - x^2 z \hat{z}$$

Choice (4) is the answer.

4.33. Based on the information given in the problem, we have:

$$(x, y, z) = (1, 1, 1)$$

$$\nabla \times \vec{F} = (x^2 y - z)\hat{y} - x^2 z \hat{z}$$

Therefore:

$$\nabla \times \vec{F} \Big|_{(1,1,1)} = \left((1)^2(1) - (1)\right)\hat{y} - (1)^2(1)\hat{z}$$

$$\Rightarrow \nabla \times \vec{F} \Big|_{(1,1,1)} = -\hat{z}$$

Choice (3) is the answer.

4.11 Curl of a Vector in Cylindrical Coordinate System

4.35. Based on the information given in the problem, we have:

$$\vec{F}(r, \varphi, z) = r \sin \varphi \hat{r} + r^2 z \hat{\varphi} + z \cos \varphi \hat{z}$$

The curl of a vector field in cylindrical coordinate system can be calculated as follows.

$$\nabla \times \vec{F} = \frac{1}{r} \begin{vmatrix} \hat{r} & r\hat{\varphi} & \hat{z} \\ \frac{\partial}{\partial r} & \frac{\partial}{\partial \varphi} & \frac{\partial}{\partial z} \\ F_r & F_\varphi & F_z \end{vmatrix} = \left(\frac{1}{r}\frac{\partial F_z}{\partial \varphi} - \frac{\partial F_\varphi}{\partial z}\right)\hat{r} + \left(\frac{\partial F_r}{\partial z} - \frac{\partial F_z}{\partial r}\right)\hat{\varphi} + \frac{1}{r}\left(\frac{\partial (rF_\varphi)}{\partial r} - \frac{\partial F_r}{\partial \varphi}\right)\hat{z}$$

Hence:

$$\nabla \times \vec{F} = \left(\frac{1}{r}\frac{\partial}{\partial \varphi}(z\cos \varphi) - \frac{\partial}{\partial z}(r^2 z)\right)\hat{r} + \left(\frac{\partial}{\partial z}(r \sin \varphi) - \frac{\partial}{\partial r}(z\cos \varphi)\right)\hat{\varphi} + \frac{1}{r}\left(\frac{\partial}{\partial r}(r^3 z) - \frac{\partial}{\partial \varphi}(r \sin \varphi)\right)\hat{z}$$

$$\Rightarrow \nabla \times \vec{F} = \left(\frac{-z \sin \varphi}{r} - r^2\right)\hat{r} + (0 - 0)\hat{\varphi} + \frac{1}{r}\left(3r^2 z - r \cos \varphi\right)\hat{z}$$

$$\Rightarrow \nabla \times \vec{F} = -\left(\frac{z \sin \varphi}{r} + r^2\right)\hat{r} + (3rz - \cos \varphi)\hat{z}$$

Choice (1) is the answer.

Notes

In this problem, the relations below have been used.

$$\frac{d}{dx}\cos(u(x)) = -u'(x)\sin(u(x))$$

$$\frac{d}{dx}x^n = nx^{n-1}$$

$$\frac{d}{dx}\sin(u(x)) = u'(x)\cos(u(x))$$

4.36. Based on the information given in the problem, we have:

$$(r, \varphi, z) = (1, 0°, 0)$$

$$\nabla \times \vec{F} = -\left(\frac{z\sin\varphi}{r} + r^2\right)\hat{r} + (3rz - \cos\varphi)\hat{z}$$

Therefore:

$$\nabla \times \vec{F}\bigg|_{(1,0°,0)} = -\left(\frac{(0)\sin 0}{1} + 1^2\right)\hat{r} + (3(1)(0) - \cos 0)\hat{z}$$

$$\Rightarrow \nabla \times \vec{F}\bigg|_{(1,0°,0)} = -\hat{r} - \hat{z}$$

Choice (2) is the answer.

4.12 Curl of a Vector in Spherical Coordinate System

4.39. Based on the information given in the problem, we have:

$$\vec{F}(r, \varphi, z) = \frac{1}{R^2}\cos\theta\hat{R} + R\sin\theta\cos\varphi\hat{\theta} + \cos\theta\hat{\varphi}$$

The curl of a vector field in cylindrical coordinate system can be calculated as follows.

4.12 Curl of a Vector in Spherical Coordinate System

$$\nabla \times \vec{F} = \frac{1}{R^2 \sin\theta} \begin{vmatrix} \widehat{R} & R\widehat{\theta} & R\sin\theta\widehat{\varphi} \\ \frac{\partial}{\partial R} & \frac{\partial}{\partial \theta} & \frac{\partial}{\partial \varphi} \\ F_R & RF_\varphi & R\sin\theta F_\varphi \end{vmatrix}$$

$$\Rightarrow \nabla \times \vec{F} = \frac{1}{R\sin\theta}\left(\frac{\partial(\sin\theta F_\varphi)}{\partial \theta} - \frac{\partial F_\theta}{\partial \varphi}\right)\widehat{R} + \frac{1}{R}\left(\frac{1}{\sin\theta}\frac{\partial F_R}{\partial \varphi} - \frac{\partial(RF_\varphi)}{\partial R}\right)\widehat{\theta} + \frac{1}{R}\left(\frac{\partial(RF_\theta)}{\partial R} - \frac{\partial F_R}{\partial \theta}\right)\widehat{\varphi}$$

Hence:

$$\nabla \times \vec{F} = \frac{1}{R\sin\theta}\left(\frac{\partial(\sin\theta\cos\theta)}{\partial \theta} - \frac{\partial}{\partial \varphi}(R\sin\theta\cos\varphi)\right)\widehat{R} + \frac{1}{R}\left(\frac{1}{\sin\theta}\frac{\partial}{\partial \varphi}\left(\frac{1}{R^2}\cos\theta\right) - \frac{\partial(R\cos\theta)}{\partial R}\right)\widehat{\theta} + \frac{1}{R}$$
$$\times \left(\frac{\partial}{\partial R}(R^2\sin\theta\cos\varphi) - \frac{\partial}{\partial \theta}\left(\frac{1}{R^2}\cos\theta\right)\right)\widehat{\varphi}$$

$$\Rightarrow \nabla \times \vec{F} = \frac{1}{R\sin\theta}(\cos 2\theta + R\sin\theta\sin\varphi)\widehat{R} + \frac{1}{R}(0 - \cos\theta)\widehat{\theta} + \frac{1}{R}\left(2R\sin\theta\cos\varphi + \frac{1}{R^2}\sin\theta\right)\widehat{\varphi}$$

$$\Rightarrow \nabla \times \vec{F} = \left(\frac{\cos 2\theta}{R\sin\theta} + \sin\varphi\right)\widehat{R} - \frac{\cos\theta}{R}\widehat{\theta} + \left(2\cos\varphi + \frac{1}{R^3}\right)\sin\theta\widehat{\varphi}$$

Choice (3) is the answer.

> **Notes**
> In this problem, the relations below have been used.
>
> $$\frac{d}{dx}\sin(u(x)) = u'(x)\cos(u(x))$$
>
> $$\frac{d}{dx}\cos(u(x)) = -u'(x)\sin(u(x))$$
>
> $$\frac{d}{dx}x^n = nx^{n-1}$$

4.40. Based on the information given in the problem, we have:

$$(R, \theta, \varphi) = (1, 90°, 0°)$$

$$\nabla \times \vec{F} = \left(\frac{\cos 2\theta}{R\sin\theta} + \sin\varphi\right)\widehat{R} - \frac{\cos\theta}{R}\widehat{\theta} + \left(2\cos\varphi + \frac{1}{R^3}\right)\sin\theta\widehat{\varphi}$$

Therefore:

$$\nabla \times \vec{F}\Big|_{(1,90°,0°)} = \left(\frac{\cos 180°}{(1)\sin 90°} + \sin 0°\right)\widehat{R} - \frac{\cos 90°}{1}\widehat{\theta} + \left(2\cos 0° + \frac{1}{1^3}\right)\sin 90°\widehat{\varphi}$$

$$\Rightarrow \nabla \times \vec{F}\Big|_{(1,90°,0°)} = \left(\frac{-1}{(1)(1)} + 0\right)\widehat{R} - \frac{0}{1}\widehat{\theta} + \left(2(1) + \frac{1}{1^3}\right)(1)\widehat{\varphi}$$

$$\Rightarrow \nabla \times \vec{F}\Big|_{(1,90°,0°)} = -\widehat{R} + 3\widehat{\varphi}$$

Choice (4) is the answer.

4.13 Laplacian

4.42. The following identity is available for the Laplacian operator.

$$\nabla \cdot (\nabla F) = \nabla^2 F$$

Choice (1) is the answer.

4.14 Laplacian of a Scalar in the Cartesian Coordinate System

4.43. Based on the information given in the problem, we have:

$$F(x,y,z) = e^{-z}\sin 2x \cos y$$

The curl of a vector field in the Cartesian coordinate system can be calculated as follows.

$$\nabla^2 F = \frac{\partial^2 F}{\partial x^2} + \frac{\partial^2 F}{\partial y^2} + \frac{\partial^2 F}{\partial z^2}$$

Hence:

$$\nabla^2 F = \frac{\partial^2}{\partial x^2}(e^{-z}\sin 2x \cos y) + \frac{\partial^2}{\partial y^2}(e^{-z}\sin 2x \cos y) + \frac{\partial^2}{\partial z^2}(e^{-z}\sin 2x \cos y)$$

4.14 Laplacian of a Scalar in the Cartesian Coordinate System

$$\Rightarrow \nabla^2 F = \frac{\partial}{\partial x}(2e^{-z}\cos 2x \cos y) + \frac{\partial}{\partial y}(-e^{-z}\sin 2x \sin y) + \frac{\partial}{\partial z}(-e^{-z}\sin 2x \cos y)$$

$$\Rightarrow \nabla^2 F = -4e^{-z}\sin 2x \cos y - e^{-z}\sin 2x \cos y + e^{-z}\sin 2x \cos y$$

$$\Rightarrow \nabla^2 F = -4e^{-z}\sin 2x \cos y$$

Choice (4) is the answer.

Notes

In this problem, the relations below have been used.

$$\frac{d}{dx}\sin(u(x)) = u'(x)\cos(u(x))$$

$$\frac{d}{dx}\cos(u(x)) = -u'(x)\sin(u(x))$$

$$\frac{d}{dx}e^{u(x)} = u'(x)e^{u(x)}$$

4.44. Based on the information given in the problem, we have:

$$(x, y, z) = \left(\frac{\pi}{4}, 0, 0\right)$$

$$\nabla^2 F = -4e^{-z}\sin 2x \cos y$$

Therefore:

$$\nabla^2 F \bigg|_{\left(\frac{\pi}{4}, 0, 0\right)} = -4e^0 \sin\left(2 \times \frac{\pi}{4}\right)\cos 0$$

$$\Rightarrow \nabla^2 F \bigg|_{\left(\frac{\pi}{4}, 0, 0\right)} = -4$$

Choice (4) is the answer.

4.15 Laplacian of a Scalar in Cylindrical Coordinate System

4.47. Based on the information given in the problem, we have:

$$F(r, \varphi, z) = r^2 z \cos 2\varphi$$

The Laplacian of a vector field in cylindrical coordinate system can be calculated as follows.

$$\nabla^2 F = \frac{1}{r} \frac{\partial}{\partial r}\left(r \frac{\partial F}{\partial r}\right) + \frac{1}{r^2} \frac{\partial^2 F}{\partial \varphi^2} + \frac{\partial^2 F}{\partial z^2}$$

Therefore:

$$\nabla^2 F = \frac{1}{r} \frac{\partial}{\partial r}\left(r \frac{\partial}{\partial r}(r^2 z \cos 2\varphi)\right) + \frac{1}{r^2} \frac{\partial^2}{\partial \varphi^2}(r^2 z \cos 2\varphi) + \frac{\partial^2}{\partial z^2}(r^2 z \cos 2\varphi)$$

$$\Rightarrow \nabla^2 F = \frac{1}{r} \frac{\partial}{\partial r}(2r^2 z \cos 2\varphi) + \frac{1}{r^2} \frac{\partial}{\partial \varphi}(-2r^2 z \sin 2\varphi) + \frac{\partial}{\partial z}(r^2 \cos 2\varphi)$$

$$\Rightarrow \nabla^2 F = \frac{1}{r}(4rz\cos 2\varphi) + \frac{1}{r^2}(-4r^2 z \cos 2\varphi) + 0$$

$$\Rightarrow \nabla^2 F = 4z \cos 2\varphi - 4z \cos 2\varphi$$

$$\Rightarrow \nabla^2 F = 0$$

Choice (1) is the answer.

> **Notes**
> In this problem, the relations below have been used.
>
> $$\frac{d}{dx} \sin(u(x)) = u'(x) \cos(u(x))$$
>
> $$\frac{d}{dx} \cos(u(x)) = -u'(x) \sin(u(x))$$
>
> $$\frac{d}{dx} x^n = n x^{n-1}$$

4.16 Laplacian of a Scalar in Spherical Coordinate System

4.50. Based on the information given in the problem, we have:

$$F(R, \theta, \varphi) = R \sin \theta \cos \varphi$$

4.16 Laplacian of a Scalar in Spherical Coordinate System

The Laplacian of a vector field in spherical coordinate system can be calculated as follows.

$$\nabla^2 F = \frac{1}{R^2} \frac{\partial}{\partial R}\left(R^2 \frac{\partial F}{\partial R}\right) + \frac{1}{R^2 \sin\theta} \frac{\partial}{\partial \theta}\left(\sin\theta \frac{\partial F}{\partial \theta}\right) + \frac{1}{R^2 \sin^2\theta} \frac{\partial^2 F}{\partial \varphi^2}$$

Thus:

$$\nabla^2 F = \frac{1}{R^2} \frac{\partial}{\partial R}\left(R^2 \frac{\partial}{\partial R}(R\sin\theta\cos\varphi)\right) + \frac{1}{R^2 \sin\theta} \frac{\partial}{\partial \theta}\left(\sin\theta \frac{\partial}{\partial \theta}(R\sin\theta\cos\varphi)\right) + \frac{1}{R^2 \sin^2\theta} \frac{\partial^2}{\partial \varphi^2}(R\sin\theta\cos\varphi)$$

$$\Rightarrow \nabla^2 F = \frac{1}{R^2} \frac{\partial}{\partial R}(R^2 \sin\theta\cos\varphi) + \frac{1}{R^2 \sin\theta} \frac{\partial}{\partial \theta}(R\sin\theta\cos\theta\cos\varphi) + \frac{1}{R^2 \sin^2\theta} \frac{\partial}{\partial \varphi}(-R\sin\theta\sin\varphi)$$

$$\Rightarrow \nabla^2 F = \frac{1}{R^2}(2R\sin\theta\cos\varphi) + \frac{R\cos\varphi(\cos^2\theta - \sin^2\theta)}{R^2 \sin\theta} + \frac{1}{R^2 \sin^2\theta}(-R\sin\theta\cos\varphi)$$

$$\Rightarrow \nabla^2 F = \frac{2\sin\theta\cos\varphi}{R} + \frac{\cos\varphi(\cos^2\theta - \sin^2\theta)}{R\sin\theta} - \frac{\cos\varphi}{R\sin\theta}$$

$$\Rightarrow \nabla^2 F = \frac{2\sin^2\theta\cos\varphi + \cos\varphi(\cos^2\theta - \sin^2\theta) - \cos\varphi}{R\sin\theta}$$

$$\Rightarrow \nabla^2 F = \frac{\cos\varphi(2\sin^2\theta + \cos^2\theta - \sin^2\theta - 1)}{R\sin\theta}$$

$$\Rightarrow \nabla^2 F = \frac{\cos\varphi(\sin^2\theta + \cos^2\theta - 1)}{R\sin\theta}$$

$$\Rightarrow \nabla^2 F = \frac{\cos\varphi(1 - 1)}{R\sin\theta}$$

$$\Rightarrow \nabla^2 F = 0$$

Choice (3) is the answer.

> **Notes**
> In this problem, the relations below have been used.
>
> $$\frac{d}{dx}\sin(u(x)) = u'(x)\cos(u(x))$$
>
> $$\frac{d}{dx}\cos(u(x)) = -u'(x)\sin(u(x))$$
>
> $$\frac{d}{dx}x^n = nx^{n-1}$$
>
> $$\frac{d}{dx}(u(x)v(x)) = u'(x)v(x) + v'(x)u(x)$$

References

1. General Physics II – Practice Problems, Methods, and Solutions, Springer Nature, 2025.
2. General Physics I – Practice Problems, Methods, and Solutions, Springer Nature, 2025.
3. Mathematics of Engineering and Science – Practice Problems, Methods, and Solutions, Springer Nature, 2024.
4. Rahmani-Andebili, M. (2022). Differential Equations – Practice Problems, Methods, and Solutions, Springer Nature.
5. Rahmani-Andebili, M. (2023). Calculus III – Practice Problems, Methods, and Solutions, Springer Nature.
6. Rahmani-Andebili, M. (2023). Calculus II – Practice Problems, Methods, and Solutions, Springer Nature.
7. Rahmani-Andebili, M. (2023). Calculus I (2nd Ed.) – Practice Problems, Methods, and Solutions, Springer Nature.
8. Rahmani-Andebili, M. (2024). Precalculus (2nd Ed.) – Practice Problems, Methods, and Solutions, Springer Nature.

Electric Field and Electric Flux: Part A

Abstract

In this chapter, the basic and advanced problems of electric flux and electric field are studied. The subjects include electric flux and field of uniform and nonuniform charge distributions in the Cartesian, cylindrical, and spherical coordinate systems. Herein, different types of problems and exercises are presented that are categorized as follows.

- **Problems with detailed solution**: They have been designed to teach students the subjects in detail. Moreover, they have been categorized in different levels based on their difficulty levels (easy, normal, and hard) and calculation amounts (small, normal, and large).
- **Partially solved exercises**: They have been designed to encourage students to practice problems while guiding them through the problem-solving procedure and hinting the required formulas.
- **Exercises with final answer**: They have been designed to encourage students to practice more by themselves while hinting them by the final answer as well as to help instructors to give tests or quizzes.

5.1 Electric Flux

Problem

5.1. The charge q is placed at the center of a cube with the side length a. Calculate the electric flux passing from each face of the cube [1–8] (Fig. 5.1).

Difficulty level ● Easy ○ Normal ○ Hard
Calculation amount ● Small ○ Normal ○ Large

1) $\psi = \dfrac{q}{6}$
2) $\psi = \dfrac{q}{6\varepsilon_0}$
3) $\psi = \dfrac{qa^2}{6\varepsilon_0}$
4) $\psi = \dfrac{q}{\varepsilon_0}$

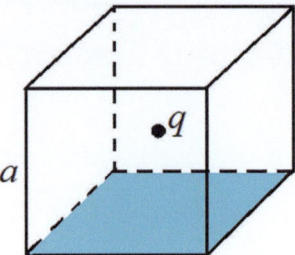

Fig. 5.1 A cube including the charge q at its center

Exercise

5.2. In Problem 5.1, calculate the electric flux passing from all faces of the cube.

Final Answer

$\psi = \dfrac{q}{\varepsilon_0}$

Problem

5.3. Figure 5.2 illustrates a sphere with radius and uniform volume charge density $\dfrac{a}{2}$ and ρ, respectively. The sphere is inside a cube with the side length $4a$. Calculate the electric flux passing through all the faces of the cube.

Difficulty level   ● Easy   ○ Normal   ○ Hard
Calculation amount ● Small ○ Normal   ○ Large

1) $\psi = \rho(4a)^3$
2) $\psi = \dfrac{\rho \pi a^3}{6\varepsilon_0}$
3) $\psi = \dfrac{\rho \pi a^3}{9\varepsilon_0}$
4) $\psi = \dfrac{\rho \pi a^3}{6}$

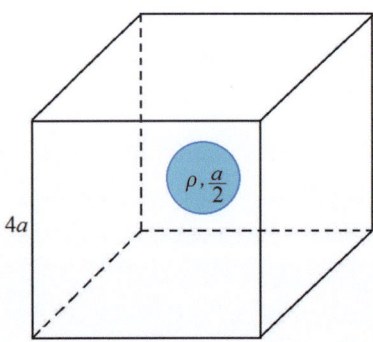

Fig. 5.2 A cube including the charged sphere at its center

Problem

5.4. The charge q is placed at the corner of a cube with the side length a. Calculate the electric flux passing through the highlighted surface of the cube (Fig. 5.3).

Difficulty level ○ Easy ○ Normal ● Hard
Calculation amount ● Small ○ Normal ○ Large

1) $\psi = \dfrac{q}{6\varepsilon_0}$
2) $\psi = \dfrac{q}{8\varepsilon_0}$
3) $\psi = \dfrac{q}{24\varepsilon_0}$
4) $\psi = \dfrac{q}{36\varepsilon_0}$

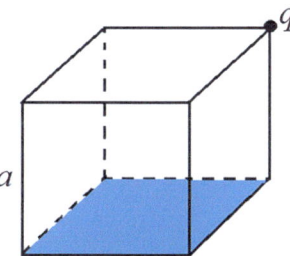

Fig. 5.3 A cube including the charge q at its corner

Problem

5.5. As is shown in Fig. 5.4, two charges of equal magnitude but different types are placed on the normal axis of a ring. Calculate the net electric flux passing from the ring.

Difficulty level ○ Easy ○ Normal ● Hard
Calculation amount ○ Small ● Normal ○ Large

1) $\psi = 0$
2) $\psi = \dfrac{Q}{\varepsilon_0}\left(1 - \dfrac{a}{\sqrt{a^2 + R^2}}\right)$
3) $\psi = \dfrac{Q}{\varepsilon_0}$
4) $\psi = \dfrac{Q}{2\varepsilon_0}\left(1 - \dfrac{a}{\sqrt{a^2 + R^2}}\right)$

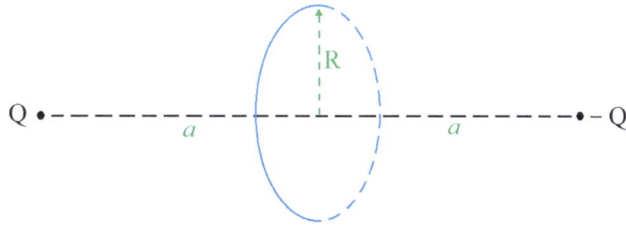

Fig. 5.4 The electric flux passing from the ring due to the charges

Exercise

5.6. Solve Problem 5.5 by assuming that both charges are Q.

Final Answer

$\psi = 0$

Problem

5.7. The point charge q has been placed at the origin. Calculate the magnitude of electric field at $(0, 1, 0)$ if 1 Vm electric flux is passing through the imaginary circular surface (Fig. 5.5).

Difficulty level   ○ Easy   ○ Normal   ● Hard
Calculation amount   ○ Small   ● Normal   ○ Large

1) $E = \dfrac{1}{\pi(2 + \sqrt{2})}$

2) $E = \dfrac{1}{\pi(2 - \sqrt{2})}$

3) $E = \dfrac{1}{2 + \sqrt{2}}$

4) $E = \dfrac{1}{2 - \sqrt{2}}$

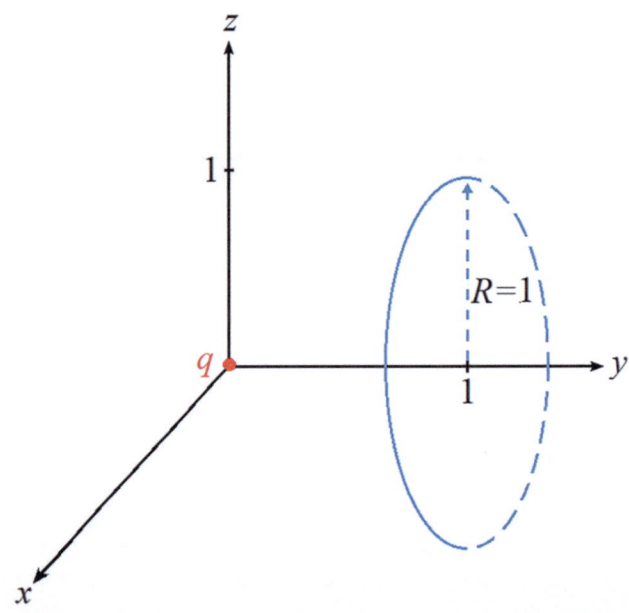

Fig. 5.5 The electric flux passing through the imaginary circular surface

5.2 Electric Field in the Cartesian Coordinate System

Problem

5.8. The electric field in an environment is $\vec{E} = -e^x(\sin y \hat{x} + \cos y \hat{y})$. Calculate the equation of field streamlines at the surface $z = 0$.

5.2 Electric Field in the Cartesian Coordinate System

Difficulty level ○ Easy ● Normal ○ Hard
Calculation amount ○ Small ● Normal ○ Large

1) $\cos y = e^{-(x+c)}$
2) $\cos y = e^{(x+c)}$
3) $\sin y = e^{-(x+c)}$
4) $\sin y = e^{(x+c)}$

Partially Solved Exercise

5.9. Calculate the equation of field streamlines at the surface $z = 0$ if the electric field in the environment is $\vec{E} = -x^2\hat{x} + y\hat{y}$.

Solution

As we know, the equation of streamlines of the field $\vec{E} = E_x\hat{x} + E_y\hat{y}$ can be calculated as follows.

$$\frac{dx}{E_x} = \frac{dy}{E_y}$$

Therefore:

$$\frac{dx}{(\quad)} = \frac{dy}{(\quad)}$$

$$\Rightarrow \int (\quad) + c = (\quad)$$

$$\Rightarrow y = e^{(\quad)}$$

$$\to y = e^c e^{\frac{1}{x}}$$

$$\stackrel{e^c \triangleq k}{\Rightarrow} y = k e^{\frac{1}{x}}$$

Problem

5.10. Calculate the electric field on top and bottom of a very large surface with the surface charge density ρ_s (Fig. 5.6).

Difficulty level ○ Easy ● Normal ○ Hard
Calculation amount ○ Small ● Normal ○ Large

1) $\vec{E} = \begin{cases} +\dfrac{\rho_s}{2\varepsilon_0}\hat{z} & z > 0 \\ -\dfrac{\rho_s}{2\varepsilon_0}\hat{z} & z < 0 \end{cases}$

2) $\vec{E} = \begin{cases} +\dfrac{\rho_s}{\varepsilon_0}\hat{z} & z > 0 \\ -\dfrac{\rho_s}{\varepsilon_0}\hat{z} & z < 0 \end{cases}$

3) $\vec{E} = \begin{cases} -\dfrac{\rho_s}{\varepsilon_0}\hat{z} & z > 0 \\ -\dfrac{\rho_s}{\varepsilon_0}\hat{z} & z < 0 \end{cases}$

4) $\vec{E} = \begin{cases} +\dfrac{\rho_s}{2\varepsilon_0}\hat{z} & z > 0 \\ +\dfrac{\rho_s}{2\varepsilon_0}\hat{z} & z < 0 \end{cases}$

Fig. 5.6 A very large surface with the surface charge density ρ_s

Partially Solved Exercise

5.11. In the Problem 5.10, calculate the magnitude of electric field around the surface if $\rho_s = 10.6~\mu C/m^2$ and $\varepsilon_0 = 8.85 \times 10^{-12} \frac{C^2}{N.m^2}$.

Solution

The magnitude of electric fields on the top and bottom of a very large surface with the surface charge density of ρ_s can be calculated as follows.

$$E = \frac{\rho_s}{2\varepsilon_0}$$

Therefore:

$$E = \frac{(\quad)}{(\quad) \times (\quad)}$$

$$\Rightarrow E = 6 \times 10^5~N/C$$

Problem

5.12. Calculate the electric field between two large parallel surfaces with the surface charge densities of ρ_s and $-\rho_s$, as is illustrated in Fig. 5.7.

Difficulty level ○ Easy ● Normal ○ Hard
Calculation amount ○ Small ● Normal ○ Large

1) $\vec{E} = \frac{\rho_s}{2\varepsilon_0}\hat{y}$
2) $\vec{E} = \frac{\rho_s}{\varepsilon_0}\hat{y}$
3) $\vec{E} = -\frac{\rho_s}{2\varepsilon_0}\hat{y}$
4) $\vec{E} = -\frac{\rho_s}{\varepsilon_0}\hat{y}$

5.2 Electric Field in the Cartesian Coordinate System

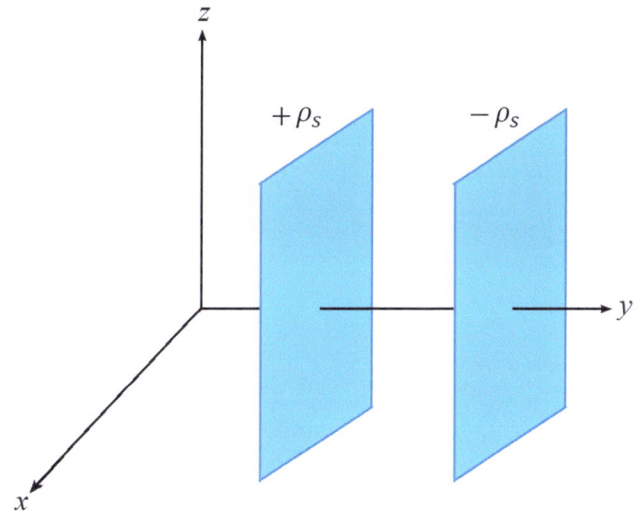

Fig. 5.7 Two large parallel surfaces with the surface charge densities ρ_s and $-\rho_s$

Problem

5.13. As is shown in Fig. 5.8, a thin rod with length l and total charge q, which is uniformly distrusted across it, is placed on x-axis in free space. Calculate the electric field at the origin.

Difficulty level ○ Easy ● Normal ○ Hard
Calculation amount ○ Small ● Normal ○ Large

1) $\vec{E} = -\dfrac{q}{4\pi\varepsilon_0 l^2}\hat{x}$

2) $\vec{E} = 0$

3) $\vec{E} = \dfrac{q}{4\pi\varepsilon_0 l^2}\hat{x}$

4) $\vec{E} = -\dfrac{q}{8\pi\varepsilon_0 l^2}\hat{x}$

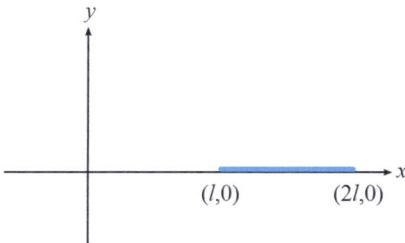

Fig. 5.8 A thin rod with charge q, uniformly distrusted across it, placed in free space

Exercise

5.14. In Problem 5.13, calculate the electric field as a function of linear charge density ρ_l.

Final Answer

$\vec{E} = -\dfrac{\rho_l}{8\pi\varepsilon_0 l}\hat{x}$

Partially Solved Exercise

5.15. In Problem 5.13, assume that the linear charge density is $\rho_l = x$. Calculate the electric field at the origin.

Solution

As is shown in Fig. 5.9, a differential length produces a differential electric field in x-axis which is as follows.

$$\overrightarrow{dE} = -\frac{dq}{4\pi\varepsilon_0 x^2}\hat{x}$$

For $dq = \rho_l dx$, we have:

$$\overrightarrow{dE} = -\frac{\rho_l dx}{4\pi\varepsilon_0 x^2}\hat{x}$$

For $\rho_l = x$, we have:

$$\overrightarrow{dE} = -\frac{dx}{4\pi\varepsilon_0 x}\hat{x}$$

The total electric field can be calculated as follows.

$$\vec{E} = \int_{x=l}^{x=2l} \overrightarrow{dE}$$

$$\Rightarrow \vec{E} = \int_{x=l}^{x=2l} (\qquad)\hat{x}$$

$$\Rightarrow \vec{E} = -\frac{1}{4\pi\varepsilon_0} \int_{x=l}^{x=2l} (\qquad)\hat{x}$$

$$\Rightarrow \vec{E} = -\frac{1}{4\pi\varepsilon_0} \Big[(\qquad) \Big]_{l}^{2l} \hat{x}$$

$$\Rightarrow \vec{E} = -\frac{1}{4\pi\varepsilon_0} ((\qquad) - (\qquad))\hat{x}$$

$$\Rightarrow \vec{E} = -\frac{\ln 2}{4\pi\varepsilon_0}\hat{x}$$

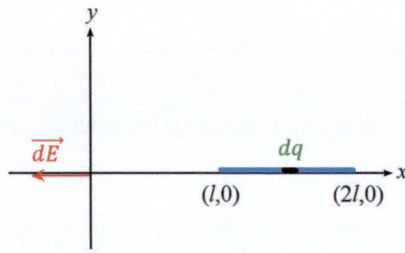

Fig. 5.9 Illustrating a differential length that produces a differential electric field on x-axis

5.3 Electric Field in Cylindrical Coordinate System

Problem

5.16. As is shown in Fig. 5.10, a thin ring with radius a and total charge Q, which is uniformly distrusted across it, is placed at the origin in free space. Calculate the magnitude of electric field at the center of the ring.

Difficulty level ● Easy ○ Normal ○ Hard
Calculation amount ● Small ○ Normal ○ Large

1) $E = 0$
2) $E = \dfrac{Q}{4\pi\varepsilon_0 a^3}$
3) $E = \dfrac{Q}{4\pi\varepsilon_0 a^2}$
4) $E = \dfrac{Q}{4\pi\varepsilon_0 a}$

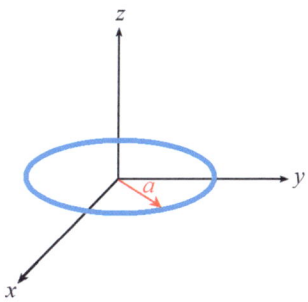

Fig. 5.10 A thin ring with radius a and total charge Q which is uniformly distrusted across it

Problem

5.17. The linear charge density of the semicircle, which is placed at the origin and shown in Fig. 5.11, is $\rho_l(x) = 6x^2$ in the SI system. Calculate the electric field at the origin.

Difficulty level ○ Easy ○ Normal ● Hard
Calculation amount ○ Small ● Normal ○ Large

1) $\vec{E} = -\dfrac{2a}{\pi\varepsilon_0}\hat{y}$
2) $\vec{E} = -\dfrac{a}{\pi\varepsilon_0}\hat{y}$
3) $\vec{E} = -\dfrac{a}{2\pi\varepsilon_0}\hat{y}$
4) $\vec{E} = -\dfrac{\sqrt{2}a}{\pi\varepsilon_0}\hat{y}$

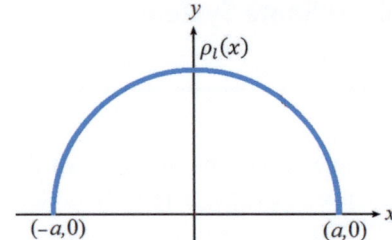

Fig. 5.11 A semicircle with linear charge density $\rho_l(x) = 6x^2$

Partially Solved Exercise

5.18. Solve Problem 5.17 by assuming that $\rho_l = \rho_0$.

Solution

As is shown in Fig. 5.12, a differential length produces a differential electric field that can be decomposed into two parts, one in x-axis and the other in y-axis.

$$\overrightarrow{dE_x} = -\frac{dq}{4\pi\varepsilon_0 a^2}\cos\varphi \hat{x}$$

$$\overrightarrow{dE_y} = -\frac{dq}{4\pi\varepsilon_0 a^2}\sin\varphi \hat{y}$$

Herein, the differential charge can be calculated as follows.

$$dq = \rho_0 dl = \rho_0 a d\varphi$$

Due to the symmetric nature of the problem, the total electric field in x-axis is zero. In other words:

$$\overrightarrow{E_x} = 0$$

However, the total electric field in y-axis can be calculated as follows.

$$\overrightarrow{E_y} = \int \overrightarrow{dE_y}$$

$$\Rightarrow \overrightarrow{E_y} = \int_{\varphi=0}^{\varphi=\pi} -\frac{\rho_0 a d\varphi}{4\pi\varepsilon_0 a^2}\sin\varphi \hat{y}$$

$$\Rightarrow \overrightarrow{E_y} = -\frac{\rho_0}{4\pi\varepsilon_0 a}\int_{\varphi=0}^{\varphi=\pi}\sin\varphi d\varphi \hat{y}$$

$$\Rightarrow \overrightarrow{E_y} = -\frac{\rho_0}{4\pi\varepsilon_0 a}\left[-\cos\varphi\right]_0^{\pi}\hat{y}$$

$$\Rightarrow \overrightarrow{E_y} = -\frac{\rho_0}{4\pi\varepsilon_0 a}(-\cos\pi + \cos 0)\hat{y}$$

5.3 Electric Field in Cylindrical Coordinate System

$$\Rightarrow \vec{E_y} = -\frac{\rho_0}{4\pi\varepsilon_0 a}(-(-1)+1)\hat{y}$$

$$\Rightarrow \vec{E_y} = -\frac{\rho_0}{2\pi\varepsilon_0 a}\hat{y}$$

Finally, the total electric field is calculated as follows.

$$\vec{E} = \vec{E_x} + \vec{E_y}$$

$$\Rightarrow \vec{E} = -\frac{\rho_0}{2\pi\varepsilon_0 a}\hat{y}$$

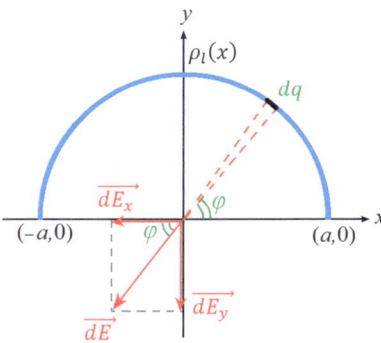

Fig. 5.12 A semicircle with linear charge density $\rho_l = \rho_0$

Problem

5.19. A very long thin rod with linear charge density ρ_l is available. Calculate the electric field around it at the distance r (Fig. 5.13).

Difficulty level   ○ Easy   ● Normal   ○ Hard
Calculation amount   ○ Small   ● Normal   ○ Large

1) $\vec{E} = \frac{\rho_l}{4\pi\varepsilon_0 r}\hat{r}$

2) $\vec{E} = \frac{\rho_l}{2\pi\varepsilon_0 r}\hat{r}$

3) $\vec{E} = \frac{\rho_l r}{2\pi\varepsilon_0}\hat{r}$

4) $\vec{E} = 0$

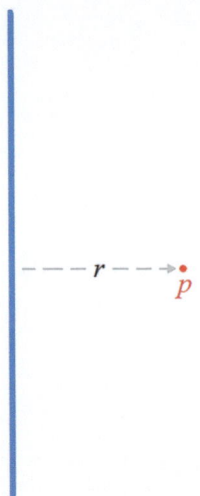

Fig. 5.13 A very long thin rod with linear charge density ρ_l

Problem

5.20. Figure 5.14 shows a single conducting solid cylinder with radius a, very long length l, and total charge Q. Calculate the electric field in the regions $r \leq a$ and $r > a$?

Difficulty level   ○ Easy   ● Normal   ○ Hard
Calculation amount   ○ Small   ● Normal   ○ Large

1) $\vec{E} = \begin{cases} 0 & r \leq a \\ 0 & r > a \end{cases}$

2) $\vec{E} = \begin{cases} \dfrac{Q}{2\pi\varepsilon_0 r l}\hat{r} & r \leq a \\ 0 & r > a \end{cases}$

3) $\vec{E} = \begin{cases} 0 & r \leq a \\ \dfrac{Q}{4\pi\varepsilon_0 r l}\hat{r} & r > a \end{cases}$

4) $\vec{E} = \begin{cases} 0 & r \leq a \\ \dfrac{Q}{2\pi\varepsilon_0 r l}\hat{r} & r > a \end{cases}$

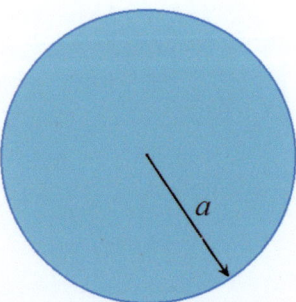

Fig. 5.14 A conducting solid cylinder with radius a, length l, and total charge Q

Exercise

5.21. In Problem 5.20, calculate the electric field based on the surface charge density ρ_s.

Final Answer

$$\vec{E} = \begin{cases} 0 & r \leq a \\ \dfrac{\rho_s a}{\varepsilon_0 r}\hat{r} & r > a \end{cases}$$

Problem

5.22. Figure 5.15 shows a single solid cylinder with radius a, very long length l, and total charge Q which is uniformly distributed across it. Calculate the electric field in the regions $r \leq a$ and $r > a$?

Difficulty level — ○ Easy ○ Normal ● Hard
Calculation amount — ○ Small ○ Normal ● Large

1) $\vec{E} = \begin{cases} 0 & r \leq a \\ \dfrac{Q}{2\pi\varepsilon_0 r l}\hat{r} & r > a \end{cases}$

2) $\vec{E} = \begin{cases} \dfrac{Q}{2\pi\varepsilon_0 a l}\hat{r} & r \leq a \\ \dfrac{Q}{2\pi\varepsilon_0 a l}\hat{r} & r > a \end{cases}$

3) $\vec{E} = \begin{cases} \dfrac{Qr}{2\pi\varepsilon_0 a l}\hat{r} & r \leq a \\ \dfrac{Q}{2\pi\varepsilon_0 r l}\hat{r} & r > a \end{cases}$

4) $\vec{E} = \begin{cases} \dfrac{Qr}{2\pi\varepsilon_0 a^2 l}\hat{r} & r \leq a \\ \dfrac{Q}{2\pi\varepsilon_0 r l}\hat{r} & r > a \end{cases}$

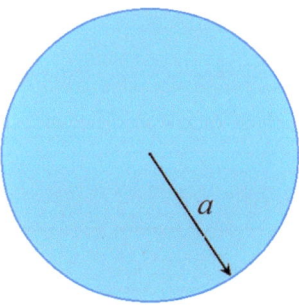

Fig. 5.15 A solid cylinder with radius a, length l, and total charge Q uniformly distributed across it

Exercise

5.23. In Problem 5.22, calculate the electric field as a function of volume charge density ρ_v.

Final Answer

$$\vec{E} = \begin{cases} \dfrac{\rho_v r}{2\varepsilon_0}\hat{r} & r \leq a \\ \dfrac{\rho_v a^2}{2\varepsilon_0 r}\hat{r} & r > a \end{cases}$$

Problem

5.24. As is shown in Fig. 5.16, two cylinders with radius a, length l, uniform volume charge densities ρ_0 and $-\rho_0$, and centers at $(0, 0, -d)$ and $(0, 0, d)$ in the Cartesian coordinate system are available. Calculate the electric field in the common area of the cylinders.

Difficulty level   ○ Easy   ○ Normal   ● Hard
Calculation amount   ○ Small   ○ Normal   ● Large

1) $\vec{E} = 0$
2) $\vec{E} = \dfrac{\rho_0 d}{2\varepsilon_0}(\hat{x} + \hat{y})$
3) $\vec{E} = \dfrac{\rho_0 d}{\varepsilon_0}(\hat{x} + \hat{y} + \hat{z})$
4) $\vec{E} = \dfrac{\rho_0 d}{\varepsilon_0}\hat{z}$

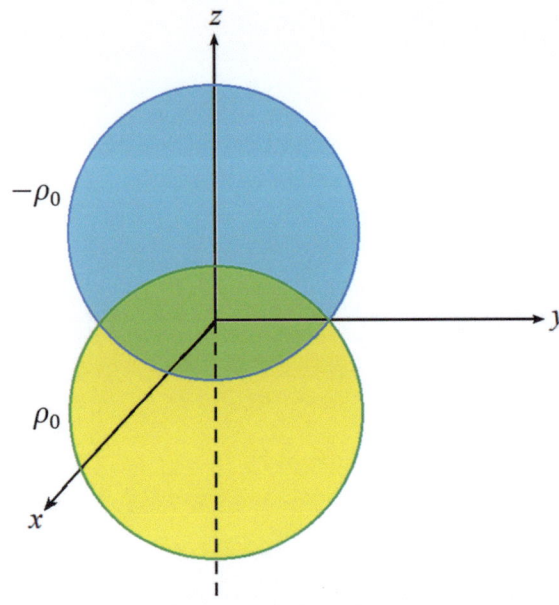

Fig. 5.16 Two cylinders with radius a, length l, uniform volume charge densities ρ_0 and $-\rho_0$, and centers at $(0, 0, -d)$ and $(0, 0, d)$

Problem

5.25. Figure 5.17 shows a conducting solid cylinder with radius a, very long length l, and total charge Q which is inside a conducting hollow cylinder with charge $-Q$ and inner and outer radiuses b and c, respectively. The solid and hollow cylinders are concentric. Which one of the following choices is correct about the electric field in the region $R \leq a$?

Difficulty level   ● Easy   ○ Normal   ○ Hard
Calculation amount   ● Small   ○ Normal   ○ Large

1) $\vec{E} = \dfrac{Q}{4\pi\varepsilon_0 rl}\hat{r}$
2) $\vec{E} = \dfrac{Q}{2\pi\varepsilon_0 rl}\hat{r}$
3) $\vec{E} = \dfrac{Q}{2\pi\varepsilon_0 r}\hat{r}$
4) $\vec{E} = 0$

5.3 Electric Field in Cylindrical Coordinate System

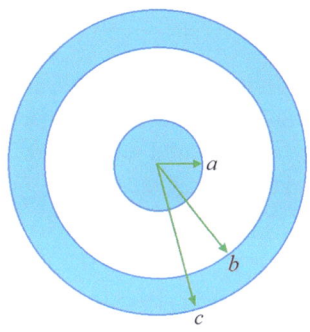

Fig. 5.17 A conducting solid cylinder with radius a, length l, and total charge Q placed inside a conducting hollow cylinder with charge $-Q$

Problem

5.26. In Problem 5.25, which one of the following choices is correct about the electric field in the region $a < r < b$?
Difficulty level   ○ Easy   ● Normal   ○ Hard
Calculation amount ○ Small ● Normal   ○ Large

1) $\vec{E} = \dfrac{Q}{4\pi\varepsilon_0 r l}\hat{r}$
2) $\vec{E} = \dfrac{Q}{2\pi\varepsilon_0 r l}\hat{r}$
3) $\vec{E} = \dfrac{Q}{2\pi\varepsilon_0 r}\hat{r}$
4) $\vec{E} = 0$

Problem

5.27. In Problem 5.25, which one of the following choices is correct about the electric field in the region $b < r < c$?
Difficulty level   ● Easy   ○ Normal   ○ Hard
Calculation amount ● Small ○ Normal   ○ Large

1) $\vec{E} = 0$
2) $\vec{E} = \dfrac{Q}{2\pi\varepsilon_0 r l}\hat{r}$
3) $\vec{E} = \dfrac{Q}{2\pi\varepsilon_0 r}\hat{r}$
4) $\vec{E} = \dfrac{Q}{\pi\varepsilon_0 r l}\hat{r}$

Problem

5.28. In Problem 5.25, which one of the following choices is correct about the electric field in the region $r > c$?
Difficulty level   ● Easy   ○ Normal   ○ Hard
Calculation amount ● Small ○ Normal   ○ Large

1) $\vec{E} = \dfrac{Q}{4\pi\varepsilon_0 r l}\hat{r}$
2) $\vec{E} = \dfrac{Q}{2\pi\varepsilon_0 r l}\hat{r}$
3) $\vec{E} = 0$
4) $\vec{E} = \dfrac{Q}{\pi\varepsilon_0 r}\hat{r}$

Problem

5.29. Figure 5.18 shows a solid cylinder with radius a, very long length l, and total charge Q which is uniformly distributed across it. This solid cylinder is placed inside a hollow cylinder with inner and outer radiuses b and c, respectively, and charge $-Q$ that is uniformly distributed across it. The solid and hollow cylinders are concentric. Which one of the following choices is correct about the electric field in the region $r \leq a$?

Difficulty level ○ Easy ○ Normal ● Hard
Calculation amount ○ Small ○ Normal ● Large

1) $\vec{E} = 0$
2) $\vec{E} = \dfrac{Qr}{2\pi\varepsilon_0 a^2 l}\hat{r}$
3) $\vec{E} = \dfrac{Qr^2}{2\pi\varepsilon_0 a^2}\hat{r}$
4) $\vec{E} = \dfrac{Q}{2\pi\varepsilon_0 a l}\hat{r}$

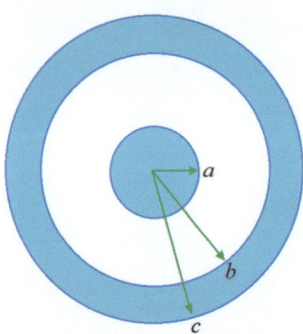

Fig. 5.18 A solid cylinder with radius a, length l, and total charge Q, uniformly distributed across it, placed inside a hollow cylinder with charge $-Q$ uniformly distributed across it

Problem

5.30. In Problem 5.29, which one of the following choices is correct about the electric field in the region $a < r < b$?

Difficulty level ○ Easy ● Normal ○ Hard
Calculation amount ○ Small ● Normal ○ Large

1) $\vec{E} = \dfrac{Q}{2\pi\varepsilon_0 r l}\hat{r}$
2) $\vec{E} = \dfrac{Qr}{4\pi\varepsilon_0 l}\hat{r}$
3) $\vec{E} = \dfrac{Qr}{2\pi\varepsilon_0 l}\hat{r}$
4) $\vec{E} = 0$

Problem

5.31. In Problem 5.29, which one of the following choices is correct about the electric field in the region $b < r < c$?

Difficulty level ○ Easy ○ Normal ● Hard
Calculation amount ○ Small ○ Normal ● Large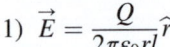

5.3 Electric Field in Cylindrical Coordinate System

1) $\vec{E} = \dfrac{Q}{2\pi\varepsilon_0 l}\left(\dfrac{c^2}{r(c^2-b^2)} - \dfrac{r}{c^2-b^2}\right)\hat{r}$

2) $\vec{E} = \dfrac{Q}{4\pi\varepsilon_0 l}\left(\dfrac{c^2}{r(c^2-b^2)} - \dfrac{r}{c^2-b^2}\right)\hat{r}$

3) $\vec{E} = \dfrac{Q}{4\pi\varepsilon_0 l}\left(\dfrac{c^2}{r(c^2-b^2)} - \dfrac{r^2}{c^2-b^2}\right)\hat{r}$

4) $\vec{E} = \dfrac{Q}{2\pi\varepsilon_0 l}\left(\dfrac{c^2}{r(c^2-b^2)} - \dfrac{r^2}{c^2-b^2}\right)\hat{r}$

Problem

5.32. In Problem 5.29, which one of the following choices is correct about the electric field in the region $r > c$?
Difficulty level ● Easy ○ Normal ○ Hard
Calculation amount ● Small ○ Normal ○ Large

1) $\vec{E} = \dfrac{Q}{2\pi\varepsilon_0 r l}\hat{r}$

2) $\vec{E} = 0$

3) $\vec{E} = \dfrac{Qr}{2\pi\varepsilon_0 l}\hat{r}$

4) $\vec{E} = \dfrac{Q}{2\pi\varepsilon_0 l}\left(\dfrac{c^2}{r(c^2-b^2)} - \dfrac{r^2}{c^2-b^2}\right)\hat{r}$

Problem

5.33. The charge Q has been uniformly distributed in a solid cylinder with radius a and length l as is shown in Fig. 5.19. If a cylindrical hole with radius $\dfrac{a}{2}$ is created inside the solid cylinder, calculate the electric field at point p.
Difficulty level ○ Easy ○ Normal ● Hard
Calculation amount ○ Small ○ Normal ● Large

1) $\vec{E_p} = \dfrac{Q}{8\pi\varepsilon_0 a l}\hat{y}$

2) $\vec{E_p} = \dfrac{Q}{28\pi\varepsilon_0 a l}\hat{y}$

3) $\vec{E_p} = \dfrac{5Q}{56\pi\varepsilon_0 a l}\hat{y}$

4) $\vec{E_p} = \dfrac{5Q}{28\pi\varepsilon_0 a l}\hat{y}$

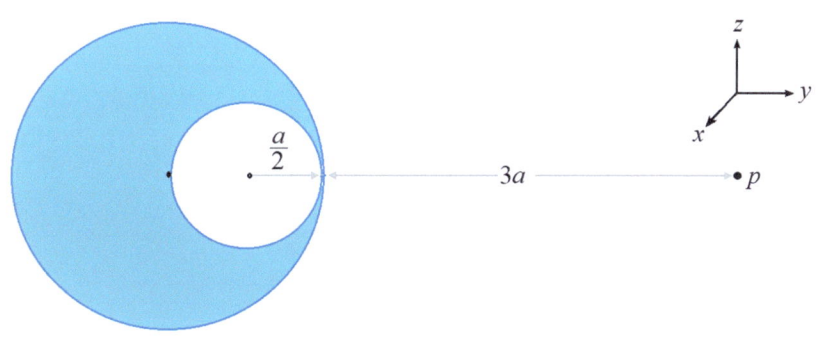

Fig. 5.19 A solid cylinder with length l including a cylindrical hole, where the charge Q has been uniformly distributed across it

Problem

5.34. The volume charge distribution of a solid cylinder with radius a and length l, shown in Fig. 5.20, is ρ. Moreover, the radius of the cylindrical hole is $\frac{a}{2}$. Calculate the electric field at point p which is at the center of the cylindrical hole.

Difficulty level   ○ Easy   ○ Normal   ● Hard
Calculation amount   ○ Small   ● Normal   ○ Large

1) $\vec{E} = \frac{\rho a}{4\varepsilon_0}\hat{y}$
2) $\vec{E} = 0$
3) $\vec{E} = \frac{\rho a}{6\varepsilon_0}\hat{y}$
4) $\vec{E} = \frac{\rho r}{\varepsilon_0}\hat{y}$

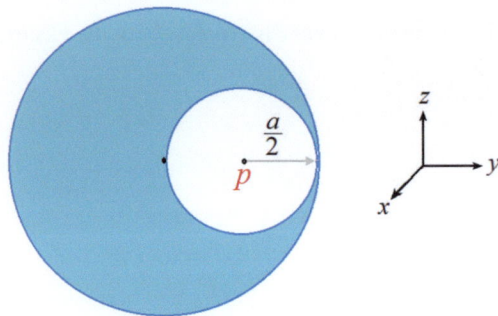

Fig. 5.20 A solid cylinder with length l including a cylindrical hole, where the charge Q has been uniformly distributed across it

Problem

5.35. As is illustrated in Fig. 5.21, a solid cylinder with radius a, length l, and nonuniform volume charge density $\frac{1}{r}$ is available. Calculate the electric field inside the cylinder.

Difficulty level   ○ Easy   ● Normal   ○ Hard
Calculation amount   ○ Small   ● Normal   ○ Large

1) $\vec{E} = \frac{1}{2r\varepsilon_0}\hat{r}$
2) $\vec{E} = \frac{r}{\varepsilon_0}\hat{r}$
3) $\vec{E} = \frac{1}{2\varepsilon_0}\hat{r}$
4) $\vec{E} = \frac{1}{\varepsilon_0}\hat{r}$

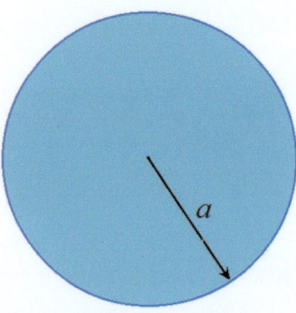

Fig. 5.21 A solid cylinder with nonuniform volume charge density

5.4 Electric Field in Spherical Coordinate System

Problem

5.36. As is shown in Fig. 5.22, the thin and long rod with total charge Q is aligned with z-axis and a long cylindrical shell with radius a, length l, and surface charge density ρ_{s1} is symmetrically aligned with that. In addition, another long cylinder shell with radius b and surface charge density ρ_{s2} is symmetrically aligned with z-axis. Calculate the electric field at $r > b$.

Difficulty level   ○ Easy   ● Normal   ○ Hard
Calculation amount   ○ Small   ● Normal   ○ Large

1) $\vec{E} = \left(\dfrac{Q}{2\pi\varepsilon_0 rl} + \dfrac{\rho_{s1} a}{\varepsilon_0 r} + \dfrac{\rho_{s2} b}{\varepsilon_0 r} \right) \hat{r}$

2) $\vec{E} = \left(\dfrac{Q}{2\pi\varepsilon_0 rl} + \dfrac{\rho_{s1} a^2}{\varepsilon_0 r^2} + \dfrac{\rho_{s2} b^2}{\varepsilon_0 r^2} \right) \hat{r}$

3) $\vec{E} = \left(\dfrac{Q}{2\pi\varepsilon_0 rl} + \dfrac{\rho_{s1}}{\varepsilon_0 r^2} + \dfrac{\rho_{s2}}{\varepsilon_0 r^2} \right) \hat{r}$

4) $\vec{E} = \left(\dfrac{Qr}{2\pi\varepsilon_0 l} + \dfrac{\rho_{s1} r}{\varepsilon_0 a} + \dfrac{\rho_{s2} r}{\varepsilon_0 b} \right) \hat{r}$

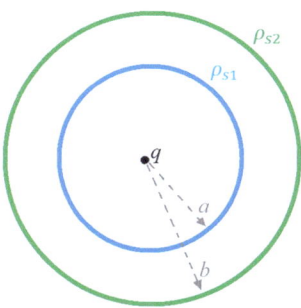

Fig. 5.22 A system including a point charge placed at the origin and two concentric cylindrical shells centered at the origin

5.4 Electric Field in Spherical Coordinate System

Problem

5.37. Figure 5.23 shows a single conducting solid sphere with radius a and total charge Q. Calculate the electric field in the regions $R \leq a$ and $R > a$?

Difficulty level   ○ Easy   ● Normal   ○ Hard
Calculation amount   ○ Small   ● Normal   ○ Large

1) $\vec{E} = \begin{cases} 0 & R \leq a \\ 0 & R > a \end{cases}$

2) $\vec{E} = \begin{cases} 0 & R \leq a \\ \dfrac{Q}{4\pi\varepsilon_0 R^2} \hat{R} & R > a \end{cases}$

3) $\vec{E} = \begin{cases} \dfrac{Q}{4\pi\varepsilon_0 R^2} \hat{R} & R \leq a \\ \dfrac{Q}{4\pi\varepsilon_0 R^2} \hat{R} & R > a \end{cases}$

4) $\vec{E} = \begin{cases} \dfrac{Q}{4\pi\varepsilon_0 R} \hat{R} & R \leq a \\ \dfrac{Q}{4\pi\varepsilon_0 R^2} \hat{R} & R > a \end{cases}$

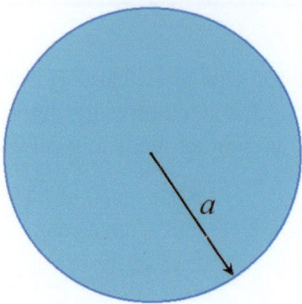

Fig. 5.23 A conducting solid sphere with radius a and total charge Q

Exercise

5.38. In Problem 5.37, calculate the electric field as a function of volume charge density ρ_s.

Final Answer

$$\vec{E} = \begin{cases} 0 & R \leq a \\ \dfrac{\rho_s a^2}{\varepsilon_0 R^2}\hat{R} & R > a \end{cases}$$

Problem

5.39. Figure 5.24 shows a single solid sphere with radius a and total charge Q which is uniformly distributed across it. Calculate the electric field in the regions $R \leq a$ and $R > a$?

Difficulty level ○ Easy ○ Normal ● Hard
Calculation amount ○ Small ○ Normal ● Large

1) $\vec{E} = \begin{cases} \dfrac{Q}{4\pi\varepsilon_0 a^3}\hat{R} & R \leq a \\ \dfrac{Q}{4\pi\varepsilon_0 R^2}\hat{R} & R > a \end{cases}$

2) $\vec{E} = \begin{cases} \dfrac{QR}{4\pi\varepsilon_0 a^2}\hat{R} & R \leq a \\ \dfrac{Q}{4\pi\varepsilon_0 R^2}\hat{R} & R > a \end{cases}$

3) $\vec{E} = \begin{cases} \dfrac{QR}{4\pi\varepsilon_0 a^3}\hat{R} & R \leq a \\ \dfrac{Q}{4\pi\varepsilon_0 R^2}\hat{R} & R > a \end{cases}$

4) $\vec{E} = \begin{cases} \dfrac{Q}{4\pi\varepsilon_0 R^2}\hat{R} & R \leq = a \\ \dfrac{Q}{4\pi\varepsilon_0 R^2}\hat{R} & R > a \end{cases}$

5.4 Electric Field in Spherical Coordinate System

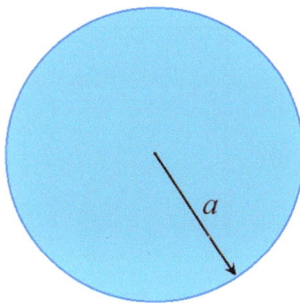

Fig. 5.24 A solid sphere with radius a and total charge Q uniformly distributed across it

Exercise

5.40. In Problem 5.39, calculate the electric field as a function of volume charge density ρ_v.

Final Answer

$$\vec{E} = \begin{cases} \dfrac{\rho_v R}{3\varepsilon_0}\hat{R} & R \leq a \\ \dfrac{\rho_v a^3}{3\varepsilon_0 R^2}\hat{R} & R > a \end{cases}$$

Problem

5.41. As is shown in Fig. 5.25, two spheres with radius a, uniform volume charge densities ρ_0 and $-\rho_0$, and centers at $(0, 0, d)$ and $(0, 0, -d)$ in the Cartesian coordinate system are available. Calculate the electric field in the common area of the spheres.

Difficulty level   ○ Easy   ○ Normal   ● Hard
Calculation amount   ○ Small   ○ Normal   ● Large

1) $\vec{E} = 0$
2) $\vec{E} = \dfrac{2\rho_0 d}{3\varepsilon_0}\hat{z}$
3) $\vec{E} = \dfrac{\rho_0 d}{3\varepsilon_0}(\hat{x} + \hat{y})$
4) $\vec{E} = \dfrac{\rho_0 d}{\varepsilon_0}(\hat{x} + \hat{y} + \hat{z})$

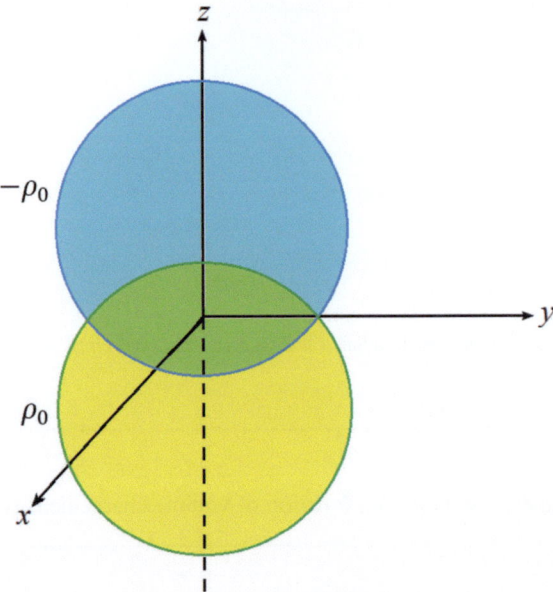

Fig. 5.25 Two spheres with radius a uniform volume charge densities ρ_0 and $-\rho_0$, and centers at $(0, 0, -d)$ and $(0, 0, d)$

Problem

5.42. Figure 5.26 shows a conducting solid sphere with radius a and total charge Q which is inside a conducting hollow sphere with charge $-Q$ and inner and outer radiuses b and c, respectively. The solid and hollow spheres are concentric. Which one of the following choices is correct about the electric field in the region $R \leq a$?

Difficulty level   ● Easy   ○ Normal   ○ Hard
Calculation amount   ● Small   ○ Normal   ○ Large

1) $\vec{E} = \dfrac{Q}{4\pi\varepsilon_0 R^2}\hat{R}$
2) $\vec{E} = \dfrac{Q}{\pi\varepsilon_0 R^2}\hat{R}$
3) $\vec{E} = \dfrac{Q}{2\pi\varepsilon_0 R^2}\hat{R}$
4) $\vec{E} = 0$

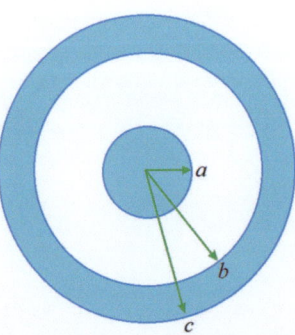

Fig. 5.26 A conducting solid sphere with radius a and total charge Q placed inside a conducting hollow sphere with charge $-Q$

5.4 Electric Field in Spherical Coordinate System

Problem

5.43. In Problem 5.42, which one of the following choices is correct about the electric field in the region $a < R < b$?
Difficulty level ○ Easy ● Normal ○ Hard
Calculation amount ○ Small ● Normal ○ Large

1) $\vec{E} = 0$
2) $\vec{E} = \dfrac{Q}{2\pi\varepsilon_0 R^2}\hat{R}$
3) $\vec{E} = \dfrac{Q}{4\pi\varepsilon_0 R^2}\hat{R}$
4) $\vec{E} = \dfrac{Q}{4\pi R^2}\hat{R}$

Problem

5.44. In Problem 5.42, which one of the following choices is correct about the electric field in the region $b < R < c$?
Difficulty level ● Easy ○ Normal ○ Hard
Calculation amount ● Small ○ Normal ○ Large

1) $\vec{E} = 0$
2) $\vec{E} = \dfrac{Q}{4\pi R^2}\hat{R}$
3) $\vec{E} = \dfrac{Q}{\pi R^2}\hat{R}$
4) $\vec{E} = \dfrac{Q}{\pi\varepsilon_0 R^2}\hat{R}$

Problem

5.45. In Problem 5.42, which one of the following choices is correct about the electric field in the region $R > c$?
Difficulty level ● Easy ○ Normal ○ Hard
Calculation amount ● Small ○ Normal ○ Large

1) $\vec{E} = \dfrac{Q}{4\pi\varepsilon_0 R^2}\hat{R}$
2) $\vec{E} = \dfrac{Q}{\pi\varepsilon_0 R^2}\hat{R}$
3) $\vec{E} = \dfrac{Q}{2\pi\varepsilon_0 R^2}\hat{R}$
4) $\vec{E} = 0$

Problem

5.46. Figure 5.27 shows a solid sphere with radius a and total charge Q which is uniformly distributed across it. This solid sphere is placed inside a hollow sphere with inner and outer radiuses b and c, respectively, and charge $-Q$ that is uniformly distributed across it. The solid and hollow spheres are concentric. Which one of the following choices is correct about the electric field in the region $R \le a$?
Difficulty level ○ Easy ○ Normal ● Hard
Calculation amount ○ Small ○ Normal ● Large

1) $\vec{E} = \dfrac{Q}{4\pi\varepsilon_0 R^2}\hat{R}$

2) $\vec{E} = \dfrac{QR}{4\pi\varepsilon_0 a^3}\hat{R}$

3) $\vec{E} = \dfrac{Q}{4\pi\varepsilon_0 a^3}\hat{R}$

4) $\vec{E} = 0$

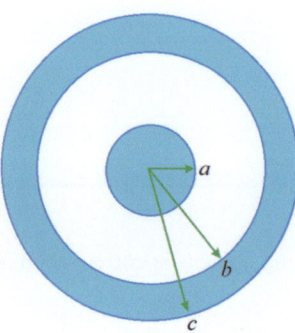

Fig. 5.27 A solid sphere with radius a and total charge Q, uniformly distributed across it, placed inside a hollow sphere with charge $-Q$ uniformly distributed across it

Problem

5.47. In Problem 5.46, which one of the following choices is correct about the electric field in the region $a < R < b$?
Difficulty level   ○ Easy   ● Normal   ○ Hard
Calculation amount   ○ Small   ● Normal   ○ Large

1) $\vec{E} = \dfrac{Q}{4\pi\varepsilon_0 R^2}\hat{R}$

2) $\vec{E} = \dfrac{Qa}{4\pi\varepsilon_0 R^2}\hat{R}$

3) $\vec{E} = \dfrac{QR}{4\pi\varepsilon_0 a^3}\hat{R}$

4) $\vec{E} = \dfrac{Q}{4\pi\varepsilon_0 a^3}\hat{R}$

Problem

5.48. In Problem 5.46, which one of the following choices is correct about the electric field in the region $b < R < c$?
Difficulty level   ○ Easy   ○ Normal   ● Hard
Calculation amount   ○ Small   ○ Normal   ● Large

1) $\vec{E} = 0$

2) $\vec{E} = \dfrac{Q}{4\pi\varepsilon_0}\left(\dfrac{c^3}{R^2(c^3 - b^3)}\right)\hat{R}$

3) $\vec{E} = \dfrac{Q}{4\pi\varepsilon_0}\left(\dfrac{c^3}{R^2(c^3 - b^3)} - \dfrac{R^3}{c^3 - b^3}\right)\hat{R}$

4) $\vec{E} = \dfrac{Q}{4\pi\varepsilon_0}\left(\dfrac{c^3}{R^2(c^3 - b^3)} - \dfrac{R}{c^3 - b^3}\right)\hat{R}$

Problem

5.49. In Problem 5.46, which one of the following choices is correct about the electric field in the region $R > c$?

Difficulty level ● Easy ○ Normal ○ Hard
Calculation amount ● Small ○ Normal ○ Large

1) $\vec{E} = 0$
2) $\vec{E} = \dfrac{Q}{4\pi\varepsilon_0 R^2}\hat{R}$
3) $\vec{E} = \dfrac{Q}{2\pi\varepsilon_0 R^2}\hat{R}$
4) $\vec{E} = -\dfrac{Q}{4\pi\varepsilon_0 R^2}\hat{R}$

Problem

5.50. The charge Q has been uniformly distributed in a solid sphere with radius a as is shown in Fig. 5.28. If a spherical hole with radius $\dfrac{a}{2}$ is created inside the solid sphere, calculate the electric field at point p.

Difficulty level ○ Easy ○ Normal ● Hard
Calculation amount ○ Small ○ Normal ● Large

1) $\vec{E_p} = \dfrac{49Q}{784\pi\varepsilon_0 a^2}\hat{y}$
2) $\vec{E_p} = -\dfrac{47Q}{392\pi\varepsilon_0 a^2}\hat{y}$
3) $\vec{E_p} = \dfrac{41Q}{392\pi\varepsilon_0 a^2}\hat{y}$
4) $\vec{E_p} = \dfrac{47Q}{784\pi\varepsilon_0 a^2}\hat{y}$

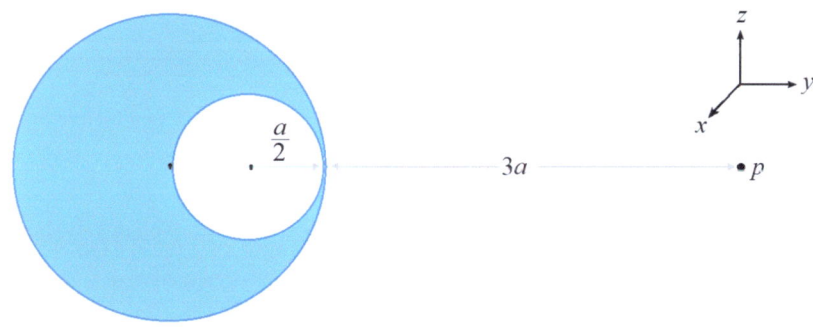

Fig. 5.28 A solid sphere including a spherical hole, where the charge Q has been uniformly distributed across it

Problem

5.51. The volume charge distribution of the solid sphere with radius a, shown in Fig. 5.29, is ρ. Moreover, the radius of the spherical hole is $\dfrac{a}{2}$. Calculate the electric field at point p which is the center of the spherical hole.

Difficulty level ○ Easy ○ Normal ● Hard
Calculation amount ○ Small ○ Normal ● Large

1) $\vec{E} = 0$
2) $\vec{E} = \dfrac{\rho a}{3\varepsilon_0}\hat{y}$
3) $\vec{E} = \dfrac{\rho a}{6\varepsilon_0}\hat{y}$
4) $\vec{E} = \dfrac{2\rho a}{3\varepsilon_0}\hat{y}$

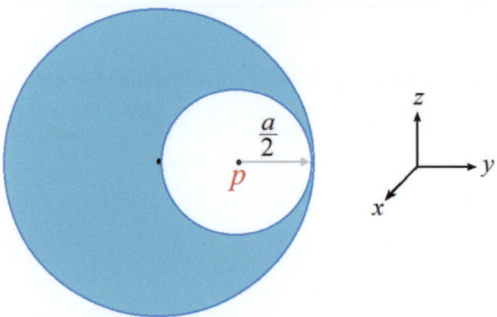

Fig. 5.29 A solid sphere including a spherical hole, where the charge Q has been uniformly distributed across it

Exercise

5.52. In Problem 5.51, if the spherical hole is filled with a material with the volume charge distribution $-\rho$, calculate the electric field at point p which is the center of the small sphere.

Final Answer

$\vec{E_p} = \dfrac{\rho a}{6\varepsilon_0}\hat{y}$

Problem

5.53. A solid sphere with radius a and nonuniform volume charge density $\dfrac{1}{R}$ is available. Calculate the electric field inside the sphere (Fig. 5.30).

Difficulty level ○ Easy ● Normal ○ Hard
Calculation amount ○ Small ● Normal ○ Large

1) $\vec{E} = \dfrac{1}{2\varepsilon_0}\hat{R}$
2) $\vec{E} = \dfrac{R}{2\varepsilon_0 a}\hat{R}$
3) $\vec{E} = \dfrac{a^2}{2\varepsilon_0 R^2}\hat{R}$
4) $\vec{E} = \dfrac{a}{2\varepsilon_0 R}\hat{R}$

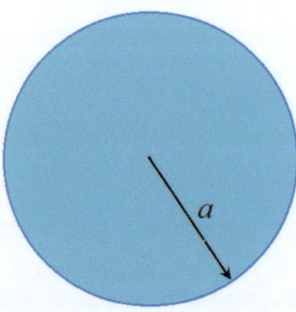

Fig. 5.30 A solid sphere with nonuniform volume charge density

Problem

5.54. As is shown in Fig. 5.31, the point charge Q is placed at the origin and a spherical shell with radius a and surface charge density ρ_{s1} is centered at the origin. In addition, another spherical shell with radius b and surface charge density ρ_{s2} is centered at the origin. Calculate the electric field at $R > b$.

Difficulty level ○ Easy ● Normal ○ Hard
Calculation amount ○ Small ● Normal ○ Large

1) $\vec{E} = \left(\dfrac{q}{4\pi\varepsilon_0 R^2} + \dfrac{\rho_{s1}}{4\pi\varepsilon_0 R^2} + \dfrac{\rho_{s2}}{4\pi\varepsilon_0 R^2} \right) \widehat{R}$

2) $\vec{E} = \left(\dfrac{q}{4\pi\varepsilon_0 R^2} + \dfrac{\rho_{s1} a^2}{\varepsilon_0 R^2} + \dfrac{\rho_{s2} b^2}{\varepsilon_0 R^2} \right) \widehat{R}$

3) $\vec{E} = \left(\dfrac{q}{4\pi\varepsilon_0 R^2} + \dfrac{\rho_{s1}}{\varepsilon_0 R^2} + \dfrac{\rho_{s2}}{\varepsilon_0 R^2} \right) \widehat{R}$

4) $\vec{E} = \left(\dfrac{q}{4\pi\varepsilon_0 R^2} + \dfrac{(\rho_{s1} + \rho_{s2})(a^2 + b^2)}{4\pi\varepsilon_0 R^2} \right) \widehat{R}$

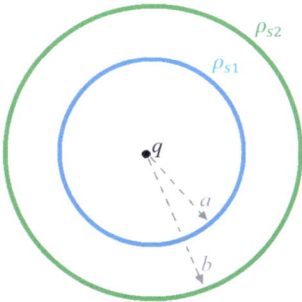

Fig. 5.31 A system including a point charge placed at the origin and two concentric spherical shells centered at the origin

Problem

5.55. Figure 5.32 illustrates a system including three point charges placed inside the two spherical holes of a conducting sphere. Which one of the following choices is correct about the magnitude of electric force (F_{13}) between the charges q_1 and q_3 and the magnitude of electric force (F_{12}) between the charges q_1 and q_2.

Difficulty level ○ Easy ○ Normal ● Hard
Calculation amount ● Small ○ Normal ○ Large

1) $F_{13} = \dfrac{q_1 q_3}{16\pi\varepsilon_0 b^2}, F_{12} = \dfrac{q_1 q_2}{4\pi\varepsilon_0 b^2}$

2) $F_{13} = \dfrac{q_1 q_3}{64\pi\varepsilon_0 b^2}, F_{12} = \dfrac{q_1 q_2}{\pi\varepsilon_0 b^2}$

3) $F_{13} = 0, F_{12} = \dfrac{q_1 q_2}{4\pi\varepsilon_0 b^2}$

4) $F_{13} = 0, F_{12} = \dfrac{q_1 q_2}{\pi\varepsilon_0 b^2}$

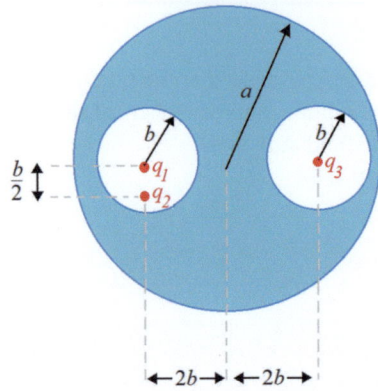

Fig. 5.32 A system including three point charges placed inside the two spherical holes of a conducting sphere

Problem

5.56. The electric field resulted from a volume charge distribution is as follows.

$$\vec{E} = \frac{1}{R^2} e^{-\frac{R}{R_0}} \hat{R}$$

Calculate the total charge of this charge distribution.

Difficulty level ○ Easy ○ Normal ● Hard
Calculation amount ○ Small ○ Normal ● Large
1) $Q = 0$
2) $Q = 4\pi\varepsilon_0$
3) $Q = 8\pi\varepsilon_0$
4) $Q = 4\pi\varepsilon_0 \left(e^{-\frac{R}{R_0}} - 1 \right)$

References

1. Rahmani-Andebili, M., General Physics II – Practice Problems, Methods, and Solutions, Springer Nature, 2025.
2. Rahmani-Andebili, M., General Physics I – Practice Problems, Methods, and Solutions, Springer Nature, 2025.
3. Rahmani-Andebili, M., Mathematics of Engineering and Science – Practice Problems, Methods, and Solutions, Springer Nature, 2024.
4. Rahmani-Andebili, M., Differential Equations – Practice Problems, Methods, and Solutions, Springer Nature, 2022.
5. Rahmani-Andebili, M., Calculus III – Practice Problems, Methods, and Solutions, Springer Nature, 2023.
6. Rahmani-Andebili, M., Calculus II – Practice Problems, Methods, and Solutions, Springer Nature, 2023.
7. Rahmani-Andebili, M., Calculus I (2nd Ed.) – Practice Problems, Methods, and Solutions, Springer Nature, 2023.
8. Rahmani-Andebili, M., Precalculus (2nd Ed.) – Practice Problems, Methods, and Solutions, Springer Nature, 2024.

Electric Field and Electric Flux: Part B

Abstract

In this chapter, the problems of the fifth chapter are fully solved, in detail, step-by-step, and with different methods.

6.1 Electric Flux

6.1. Based on Gauss's law, the net electric flux passing through a closed surface is equal to the net electric charge enclosed by the surface divided by the free space permittivity (ε_0) [1–8]. In other words:

$$\psi = \frac{q}{\varepsilon_0}$$

The electric flux passing through one of the surfaces of the cube is $\frac{1}{6}$ times of the total flux. Therefore:

$$\psi = \frac{q}{6\varepsilon_0}$$

Choice (2) is the answer (Fig. 6.1).

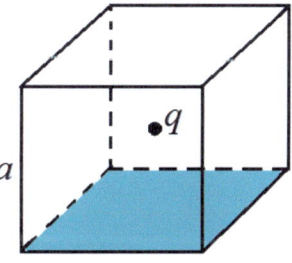

Fig. 6.1 A cube including the charge q at its center

6.3. According to Gauss's law, the net electric flux passing through a closed surface is equal to the net electric charge enclosed by the surface divided by the free space permittivity (ε_0). In other words:

$$\psi = \frac{Q}{\varepsilon_0}$$

Therefore:

$$\psi = \frac{\rho V}{\varepsilon_0} = \frac{\rho \left[\frac{4}{3}\pi \left(\frac{a}{2}\right)^3\right]}{\varepsilon_0}$$

$$\Rightarrow \psi = \frac{\rho \pi a^3}{6\varepsilon_0}$$

Choice (2) is the answer (Fig. 6.2).

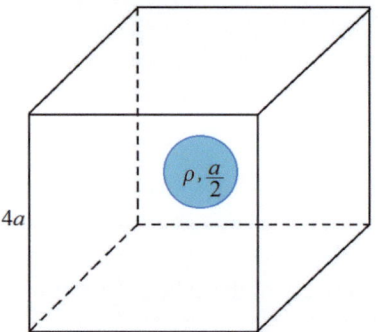

Fig. 6.2 A cube including the charged sphere at its center

6.4. As is illustrated in Fig. 6.3 (b), the cube with the side length a can be considered as one of the eight cubes making a large cube with the side length $2a$. As can be noticed, the charge q is now at the center of the large cube.

Based on Gauss's law, the total electric flux passing through a closed surface is equal to the net electric charge enclosed by the surface divided by the free space permittivity (ε_0). In other words:

$$\psi_{total} = \frac{Q}{\varepsilon_0}$$

The electric flux passing through the highlighted surface of the larger cube is $\frac{1}{24}$ times of the total flux. Hence:

$$\psi = \frac{q}{24\varepsilon_0}$$

Choice (3) is the answer.

6.1 Electric Flux

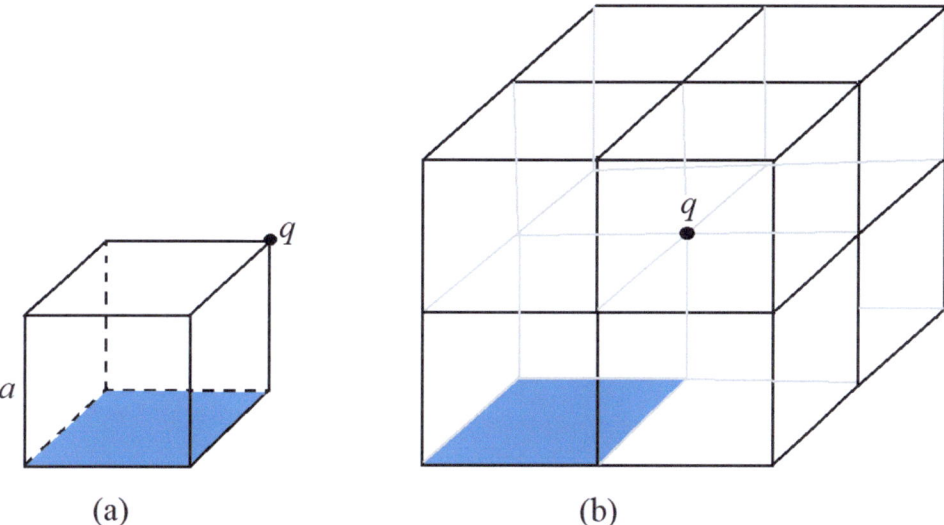

Fig. 6.3 (a) A cube including the charge q at its corner. (b) Considering the small cube as one of the eight cubes of the large cube

6.5. The electric flux, caused by the point charge Q, passing through an area can be calculated as follows.

$$\psi = \frac{Q}{2\varepsilon_0}(1 - \cos\theta)$$

where, θ is the zenith angle in spherical coordinate system.

Therefore, by using Fig. 6.4, for the positive charge, we have:

$$\overrightarrow{\psi_1} = \frac{Q}{2\varepsilon_0}\left(1 - \frac{a}{\sqrt{a^2 + R^2}}\right)\hat{x}$$

Moreover, for the negative charge, we have:

$$\overrightarrow{\psi_2} = \frac{-Q}{2\varepsilon_0}\left(1 - \frac{a}{\sqrt{a^2 + R^2}}\right)(-\hat{x}) = \frac{Q}{2\varepsilon_0}\left(1 - \frac{a}{\sqrt{a^2 + R^2}}\right)\hat{x}$$

Thus, the total electric flux passing from the ring can be calculated as follows.

$$\overrightarrow{\psi} = \overrightarrow{\psi_1} + \overrightarrow{\psi_2} = \frac{Q}{\varepsilon_0}\left(1 - \frac{a}{\sqrt{a^2 + R^2}}\right)\hat{x}$$

$$\Rightarrow \psi = \frac{Q}{\varepsilon_0}\left(1 - \frac{a}{\sqrt{a^2 + R^2}}\right)$$

Choice (2) is the answer.

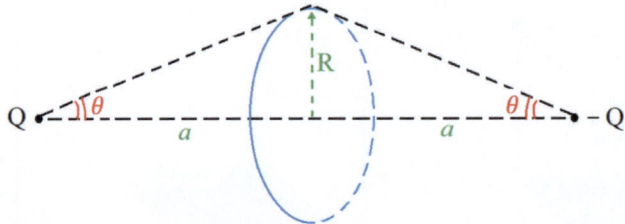

Fig. 6.4 The electric flux passing from the ring due to the charges

---

6.7. Based on the information given in the problem, we have:

$$P = (0, 1, 0) \tag{6.1}$$

$$\psi = 1 \; V.\, m \tag{6.2}$$

The electric flux, caused by the point charge Q, passing through an area can be calculated as follows.

$$\psi = \frac{Q}{2\varepsilon_0}(1 - \cos\theta) \tag{6.3}$$

where, θ is the zenith angle in spherical coordinate system.

Therefore, by using Fig. 6.5, we have:

$$\psi = \frac{Q}{2\varepsilon_0}\left(1 - \frac{1}{\sqrt{1^2 + 1^2}}\right) \tag{6.4}$$

Solving (6.2) and (6.4):

$$1 = \frac{Q}{2\varepsilon_0}\left(1 - \frac{1}{\sqrt{1^2 + 1^2}}\right) \Rightarrow 1 = \frac{Q}{2\varepsilon_0}\left(1 - \frac{\sqrt{2}}{2}\right) \tag{6.5}$$

$$\Rightarrow Q = \frac{2\varepsilon_0}{\frac{2 - \sqrt{2}}{2}} = \frac{4\varepsilon_0}{2 - \sqrt{2}} \tag{6.6}$$

As we know, the magnitude of electric field of a point charge at distance R, can be calculated as follows.

$$E = \frac{Q}{4\pi\varepsilon_0 R^2} \tag{6.7}$$

Solving (6.1), (6.6), and (6.7) and using Fig. 6.6:

$$E = \frac{\frac{4\varepsilon_0}{2-\sqrt{2}}}{4\pi\varepsilon_0(1)^2}$$

$$\Rightarrow E = \frac{1}{\pi(2-\sqrt{2})}$$

Choice (2) is the answer.

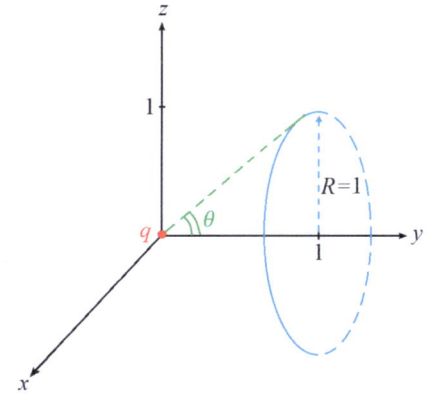

Fig. 6.5 The electric flux passing through the imaginary circular surface

6.2 Electric Field in Cartesian Coordinate System

6.8. Based on the information given in the problem, we have:

$$\vec{E} = -e^x(\sin y\, \hat{x} + \cos y\, \hat{y}) \tag{6.8}$$

$$z = 0 \tag{6.9}$$

The equation of streamlines of the field $\vec{E} = E_x\hat{x} + E_y\hat{y}$ can be calculated as follows.

$$\frac{dx}{E_x} = \frac{dy}{E_y} \tag{6.10}$$

Solving (6.8) and (6.10):

$$\frac{dx}{\sin y} = \frac{dy}{\cos y}$$

$$\Rightarrow dx = \frac{\sin y}{\cos y} dy$$

$$\Rightarrow x + c = -\ln(\cos y)$$

$$\Rightarrow \ln(\cos y) = -(x+c)$$

$$\Rightarrow \cos y = e^{-(x+c)}$$

Choice (1) is the answer.

> **Notes**
> In this problem, the relations below have been used.
>
> $$\int \frac{du}{u(x)} = \ln u(x) + c$$
>
> $$\ln a = b \Rightarrow a = e^b$$

6.10. By applying Gauss's law on the cuboid Gaussian surface with base area of A, shown in Fig. 6.6, we have:

$$Q_{in} = \int \vec{D} \cdot \vec{dS}$$

$$\Rightarrow \rho_s A = \underbrace{\int D\hat{z} \cdot dS\hat{z}}_{\text{top surface}} + \underbrace{\int -D\hat{z} \cdot (-dS\hat{z})}_{\text{bottom surface}} + \underbrace{0}_{\text{the other four surfaces}}$$

$$\Rightarrow \rho_s A = 2D \int dS$$

$$\Rightarrow \rho_s A = 2DA$$

$$\Rightarrow D = \frac{\rho_s}{2}$$

$$\Rightarrow E = \frac{\rho_s}{2\varepsilon_0}$$

$$\Rightarrow \vec{E} = \begin{cases} +\dfrac{\rho_s}{2\varepsilon_0}\hat{z} & z > 0 \\ -\dfrac{\rho_s}{2\varepsilon_0}\hat{z} & z < 0 \end{cases}$$

Choice (1) is the answer.

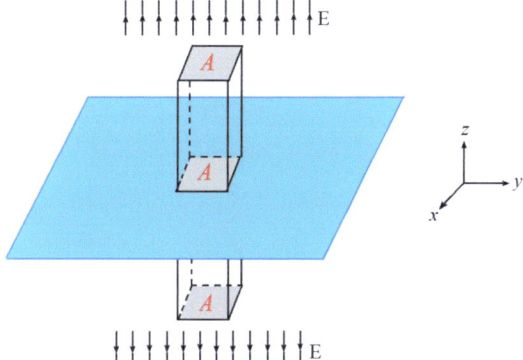

Fig. 6.6 Applying Gauss's law on the cuboid Gaussian surface around the large surface

---

6.12. As we know, the magnitude of electric field around a very large surface with the surface charge density of ρ_s is as follows.

$$E = \frac{\rho_s}{2\varepsilon_0}$$

Moreover, the direction of electric field for positive surface charge density is outward and for negative surface charge density is inward.

Thus, the electric field of the surface (with the charge density of $+\rho_s$) between the surfaces is as follows.

$$\vec{E_+} = \frac{\rho_s}{2\varepsilon_0}\hat{y}$$

In addition, the electric field of the surface (with the charge density of $-\rho_s$) between the surfaces is as follows.

$$\vec{E_-} = \frac{\rho_s}{2\varepsilon_0}\hat{y}$$

Hence, the total electric field between the surfaces can be calculated as follows.

$$\vec{E} = \vec{E_+} + \vec{E_-}$$

$$\Rightarrow \vec{E} = \frac{\rho_s}{2\varepsilon_0}\hat{y} + \frac{\rho_s}{2\varepsilon_0}\hat{y}$$

$$\Rightarrow \vec{E} = \frac{\rho_s}{\varepsilon_0}\hat{y}$$

Choice (2) is the answer (Fig. 6.7).

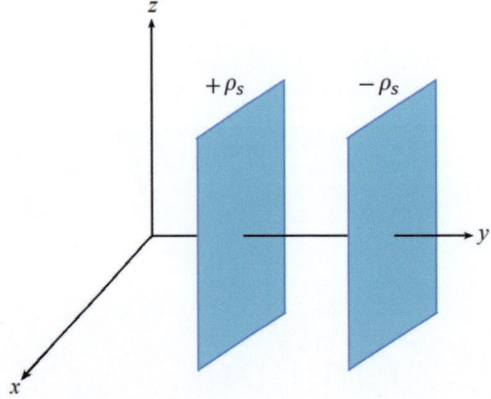

Fig. 6.7 Two large parallel surfaces with the surface charge densities ρ_s and $-\rho_s$

6.13. As is shown in Fig. 6.8, a differential length produces a differential electric field on x-axis which is as follows.

$$\vec{dE} = -\frac{dq}{4\pi\varepsilon_0 x^2}\hat{x}$$

Herein, the linear charge density can be presented as follows.

$$\rho_l = \frac{q}{l}$$

Moreover, the differential charge can be calculated as follows.

$$dq = \rho_l dx = \frac{q}{l}dx$$

Thus:

$$\vec{E} = \int_{x=l}^{x=2l} \vec{dE}$$

$$\Rightarrow \vec{E} = \int_{x=l}^{x=2l} -\frac{\frac{q}{l}dx}{4\pi\varepsilon_0 x^2}\hat{x}$$

6.3 Electric Field in Cylindrical Coordinate System

$$\Rightarrow \vec{E} = -\frac{q}{4\pi\varepsilon_0 l} \int_{x=l}^{x=2l} \frac{dx}{x^2} \hat{x}$$

$$\Rightarrow \vec{E} = -\frac{q}{4\pi\varepsilon_0 l} \left[-\frac{1}{x} \right]_l^{2l} \hat{x}$$

$$\Rightarrow \vec{E} = \frac{q}{4\pi\varepsilon_0 l} \left(\frac{1}{2l} - \frac{1}{l} \right) \hat{x}$$

$$\Rightarrow \vec{E} = -\frac{q}{8\pi\varepsilon_0 l^2} \hat{x}$$

Choice (4) is the answer.

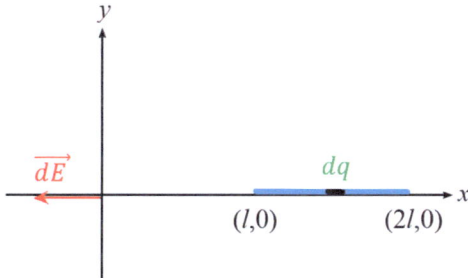

Fig. 6.8 Showing a differential length that produces a differential electric field on x-axis

> **Notes**
> In this problem, the relation below has been used.
> $$\int x^n dx = \frac{x^{n+1}}{n+1}$$

6.3 Electric Field in Cylindrical Coordinate System

6.16. Since the charge is uniformly distributed across the ring and the ring is symmetric with respect to the location (center of ring) that electric field is calculated, the total electric field is zero. In other words:

$$E = 0$$

Choice (1) is the answer (Fig. 6.9).

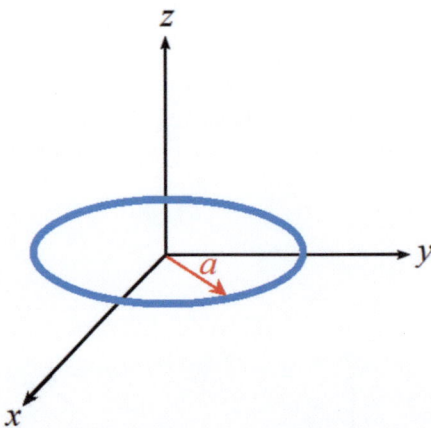

Fig. 6.9 A thin ring with radius a and total charge Q which is uniformly distrusted across it

---

6.17. Based on the information given in the problem, we have:

$$\rho_l(x) = 6x^2$$

As is shown in Fig. 6.10, a differential length produces a differential electric field that can be decomposed into two parts, one in x-axis and the other in y-axis.

$$\overrightarrow{dE_x} = -\frac{dq}{4\pi\varepsilon_0 a^2} \cos\varphi \hat{x}$$

$$\overrightarrow{dE_y} = -\frac{dq}{4\pi\varepsilon_0 a^2} \sin\varphi \hat{y}$$

Herein, the differential charge can be calculated as follows.

$$dq = \rho_l dl = 6x^2 a d\varphi$$

From Fig. 6.10, we have:

$$x = a\cos\varphi$$

Therefore:

$$dq = \rho_l dl = 6a^3 \cos^2\varphi d\varphi$$

6.3 Electric Field in Cylindrical Coordinate System

Due to the symmetric nature of the problem, the total electric field in x-axis is zero. In other words:

$$\vec{E_x} = 0$$

However, the total electric field in y-axis can be calculated as follows.

$$\vec{E_y} = \int \vec{dE_y}$$

$$\Rightarrow \vec{E_y} = \int_{\varphi=0}^{\varphi=\pi} -\frac{6a^3 \cos^2\varphi d\varphi}{4\pi\varepsilon_0 a^2} \sin\varphi \hat{y}$$

$$\Rightarrow \vec{E_y} = \frac{6a}{4\pi\varepsilon_0} \int_{\varphi=0}^{\varphi=\pi} -\cos^2\varphi \sin\varphi d\varphi \hat{y}$$

$$\Rightarrow \vec{E_y} = \frac{6a}{4\pi\varepsilon_0} \left[\frac{\cos^3\varphi}{3}\right]_0^{\pi} \hat{y}$$

$$\Rightarrow \vec{E_y} = \frac{a}{2\pi\varepsilon_0} (\cos^3\pi - \cos^3 0)\hat{y}$$

$$\Rightarrow \vec{E_y} = \frac{a}{2\pi\varepsilon_0} \left((-1)^3 - (1)^3\right)\hat{y}$$

$$\Rightarrow \vec{E_y} = -\frac{a}{\pi\varepsilon_0}\hat{y}$$

Hence, the total electric field is:

$$\vec{E} = \vec{E_x} + \vec{E_y}$$

$$\Rightarrow \vec{E} = -\frac{a}{\pi\varepsilon_0}\hat{y}$$

Choice (2) is the answer.

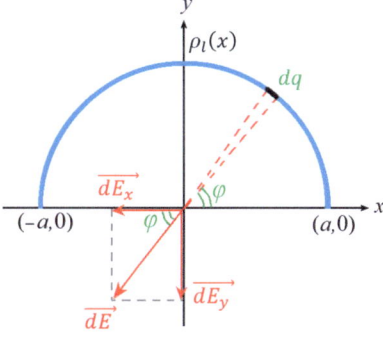

Fig. 6.10 A semicircle with linear charge density $\rho_l(x) = 6x^2$

> **Notes**
>
> In this problem, the relations below have been used.
>
> $$\int u^n(x) du = \frac{u^{n+1}(x)}{n+1}$$
>
> $$\cos \pi = -1$$
>
> $$\cos 0 = 1$$

6.19. As is shown in Fig. 6.11, a cylindrical Gaussian surface is drawn around the rod. Now, by applying Gauss's law we have:

$$Q_{in} = \int \vec{D} \cdot \overrightarrow{dS}$$

$$\Rightarrow \int \rho_l dl = \int \varepsilon_0 \vec{E} \cdot \overrightarrow{dS}$$

$$\Rightarrow \rho_l \int_{z=0}^{z=l} dz = \varepsilon_0 E_r r \int_{z=0}^{z=l} \int_{\varphi=0}^{2\pi} d\varphi dz$$

$$\Rightarrow \rho_l \int_{z=0}^{z=l} dz = \varepsilon_0 E_r r \left[\varphi\right]_0^{2\pi} \int_{z=0}^{z=l} dz$$

$$\Rightarrow \rho_l = \varepsilon_0 E_r r 2\pi$$

$$\Rightarrow E_r = \frac{\rho_l}{2\pi\varepsilon_0 r}$$

$$\Rightarrow \vec{E} = \frac{\rho_l}{2\pi\varepsilon_0 r}\hat{r}$$

Choice (2) is the answer.

6.3 Electric Field in Cylindrical Coordinate System

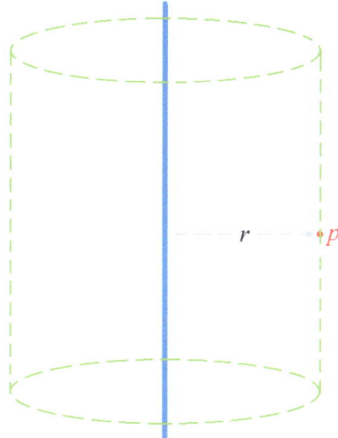

Fig. 6.11 Applying Gauss's law on the cylindrical Gaussian surface around the long thin rod

6.20. The problem can be solved by applying Gauss's law as follows.

$$Q_{in} = \int \vec{D} \cdot \vec{dS}$$

$$\Rightarrow Q_{in} = \int \varepsilon_0 \vec{E} \cdot \vec{dS}$$

Applying Gauss's law for the Gaussian surface ($r \leq a$) shown in Fig. 6.12 will result in zero electric field because electric charge inside a conductor is zero. Therefore:

$$\vec{E_1} = 0$$

By applying Gauss's law for the Gaussian surface ($r > a$) shown in Fig. 6.12, we have:

$$Q = \int_{z=0}^{l} \int_{\varphi=0}^{2\pi} \varepsilon_0 E_r r d\varphi dz$$

$$\Rightarrow Q = \varepsilon_0 E_r r \left[\varphi\right]_0^{2\pi} \left[z\right]_0^{l}$$

$$\Rightarrow Q = \varepsilon_0 E_r r (2\pi - 0)(l - 0)$$

$$\Rightarrow Q = \varepsilon_0 E_r r 2\pi l$$

$$\Rightarrow E_r = \frac{Q}{2\pi\varepsilon_0 rl}$$

$$\Rightarrow \vec{E_2} = \frac{Q}{2\pi\varepsilon_0 rl}\hat{r}$$

Therefore:

$$\vec{E} = \begin{cases} 0 & r \leq a \\ \dfrac{Q}{2\pi\varepsilon_0 rl}\hat{r} & r > a \end{cases}$$

Choice (4) is the answer.

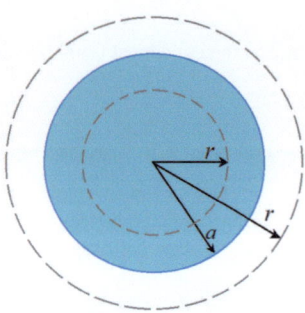

Fig. 6.12 Applying Gauss's law for the cylindrical Gaussian surfaces around the conducting solid cylinder

---

6.22. The volume charge density of the solid cylinder can be calculated as follows.

$$\rho = \frac{Q}{\pi a^2 l}$$

By applying Gauss's law for the spherical surface ($r \leq a$) shown in Fig. 6.13, we have:

$$Q_{in} = \int \vec{D} \cdot \vec{dS}$$

$$\Rightarrow \int \rho dV = \int \varepsilon_0 \vec{E} \cdot \vec{dS}$$

$$\Rightarrow \int_{z=0}^{l} \int_{\varphi=0}^{2\pi} \int_{r=0}^{r} \frac{Q}{\pi a^2 l} r\, dr\, d\varphi\, dz = \int_{z=0}^{l} \int_{\varphi=0}^{2\pi} \varepsilon_0 E_r r\, d\varphi\, dz$$

$$\Rightarrow \frac{Q}{\pi a^2 l} \left[\frac{r^2}{2}\right]_0^r \int_{z=0}^{l} \int_{\varphi=0}^{2\pi} d\varphi\, dz = \varepsilon_0 E_r r \int_{z=0}^{l} \int_{\varphi=0}^{2\pi} d\varphi\, dz$$

6.3 Electric Field in Cylindrical Coordinate System

$$\Rightarrow \frac{Q}{\pi a^2 l} \left[\frac{r^2}{2} \right]_0^r = \varepsilon_0 E_r r$$

$$\Rightarrow \frac{Q}{\pi a^2 l} \left(\frac{r^2}{2} - 0 \right) = \varepsilon_0 E_r r$$

$$\Rightarrow \frac{Qr}{2\pi a^2 l} = \varepsilon_0 E_r$$

$$\Rightarrow E_r = \frac{Qr}{2\pi \varepsilon_0 a^2 l}$$

$$\Rightarrow \overrightarrow{E_1} = \frac{Qr}{2\pi \varepsilon_0 a^2 l} \hat{r}$$

By applying Gauss's law for the cylindrical surface ($r > a$) shown in Fig. 6.13, we have:

$$\Rightarrow Q = \int_{z=0}^{l} \int_{\varphi=0}^{2\pi} \varepsilon_0 E_r r d\varphi dz$$

$$\Rightarrow Q = \varepsilon_0 E_r r \int_{z=0}^{l} \int_{\varphi=0}^{2\pi} d\varphi dz$$

$$\Rightarrow Q = \varepsilon_0 E_r r \left[\varphi \right]_0^{2\pi} \left[z \right]_0^l$$

$$\Rightarrow Q = \varepsilon_0 E_r r (2\pi - 0)(l - 0)$$

$$\Rightarrow Q = \varepsilon_0 E_r r 2\pi l$$

$$\Rightarrow E_r = \frac{Q}{2\pi \varepsilon_0 r l}$$

$$\Rightarrow \overrightarrow{E_2} = \frac{Q}{2\pi \varepsilon_0 r l} \hat{r}$$

Therefore:

$$\overrightarrow{E} = \begin{cases} \frac{Qr}{2\pi \varepsilon_0 a^2 l} \hat{r} & r \leq a \\ \frac{Q}{2\pi \varepsilon_0 r l} \hat{r} & r > a \end{cases}$$

Choice (4) is the answer.

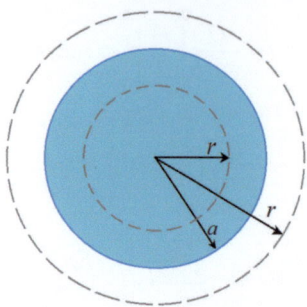

Fig. 6.13 Applying Gauss's law for the cylindrical Gaussian surfaces around the solid cylinder

> **Notes**
> In this problem, the relation below has been used.
> $$\int x^n dx = \frac{x^{n+1}}{n+1}$$

6.24. The problem can be solved by using the superposition principle, as is shown in Fig. 6.14. By applying Gauss's law for the cylindrical surface related to the bottom cylinder shown in Fig. 6.14 (b), we have:

$$Q_{in} = \int \vec{D} \cdot \vec{dS}$$

$$\Rightarrow \int \rho dV = \int \varepsilon_0 \vec{E} \cdot \vec{dS}$$

$$\Rightarrow \int_{z=0}^{l} \int_{\varphi=0}^{2\pi} \int_{r=0}^{r} \rho_0 r dr d\varphi dz = \int_{z=0}^{l} \int_{\varphi=0}^{2\pi} \varepsilon_0 E_r r d\varphi dz$$

$$\Rightarrow \rho_0 \left[\frac{r^2}{2}\right]_0^r \int_{z=0}^{l} \int_{\varphi=0}^{2\pi} d\varphi dz = \varepsilon_0 E_r r \int_{z=0}^{l} \int_{\varphi=0}^{2\pi} d\varphi dz$$

$$\Rightarrow \rho_0 \left(\frac{r^2}{2} - 0\right) = \varepsilon_0 E_r r$$

$$\Rightarrow \rho_0 \frac{r}{2} = \varepsilon_0 E_r$$

$$\Rightarrow E_r = \frac{\rho_0 r}{2\varepsilon_0}$$

6.3 Electric Field in Cylindrical Coordinate System

By considering the direction of electric field, we have:

$$\vec{E} = \frac{\rho_0 r}{2\varepsilon_0} \hat{r}$$

From the previous chapters, we know that $r\hat{r} = \vec{r}$. Therefore:

$$\vec{E} = \frac{\rho_0}{2\varepsilon_0} \vec{r}$$

Now, let us name this electric field as follows.

$$\vec{E_1} = \frac{\rho_0}{2\varepsilon_0} \vec{r_1}$$

Likewise, Gauss's law can be applied for the cylindrical surface related to the top cylinder shown in Fig. 6.14 (c). In this condition, we have:

$$\vec{E_2} = -\frac{\rho_0}{2\varepsilon_0} \vec{r_2}$$

Based on the superposition principle, the total electric field in the common area can be calculated as follows.

$$\vec{E} = \vec{E_1} + \vec{E_2}$$

$$\Rightarrow \vec{E} = \frac{\rho_0}{2\varepsilon_0} \vec{r_1} - \frac{\rho_0}{2\varepsilon_0} \vec{r_2}$$

$$\Rightarrow \vec{E} = \frac{\rho_0}{2\varepsilon_0} \left(\vec{r_1} - \vec{r_2} \right)$$

As can be noticed from Fig. 6.14 (d), $\vec{r_1} - \vec{r_2} = 2d\hat{z}$. Therefore:

$$\Rightarrow \vec{E} = \frac{\rho_0 d}{\varepsilon_0} \hat{z}$$

Choice (4) is the answer.

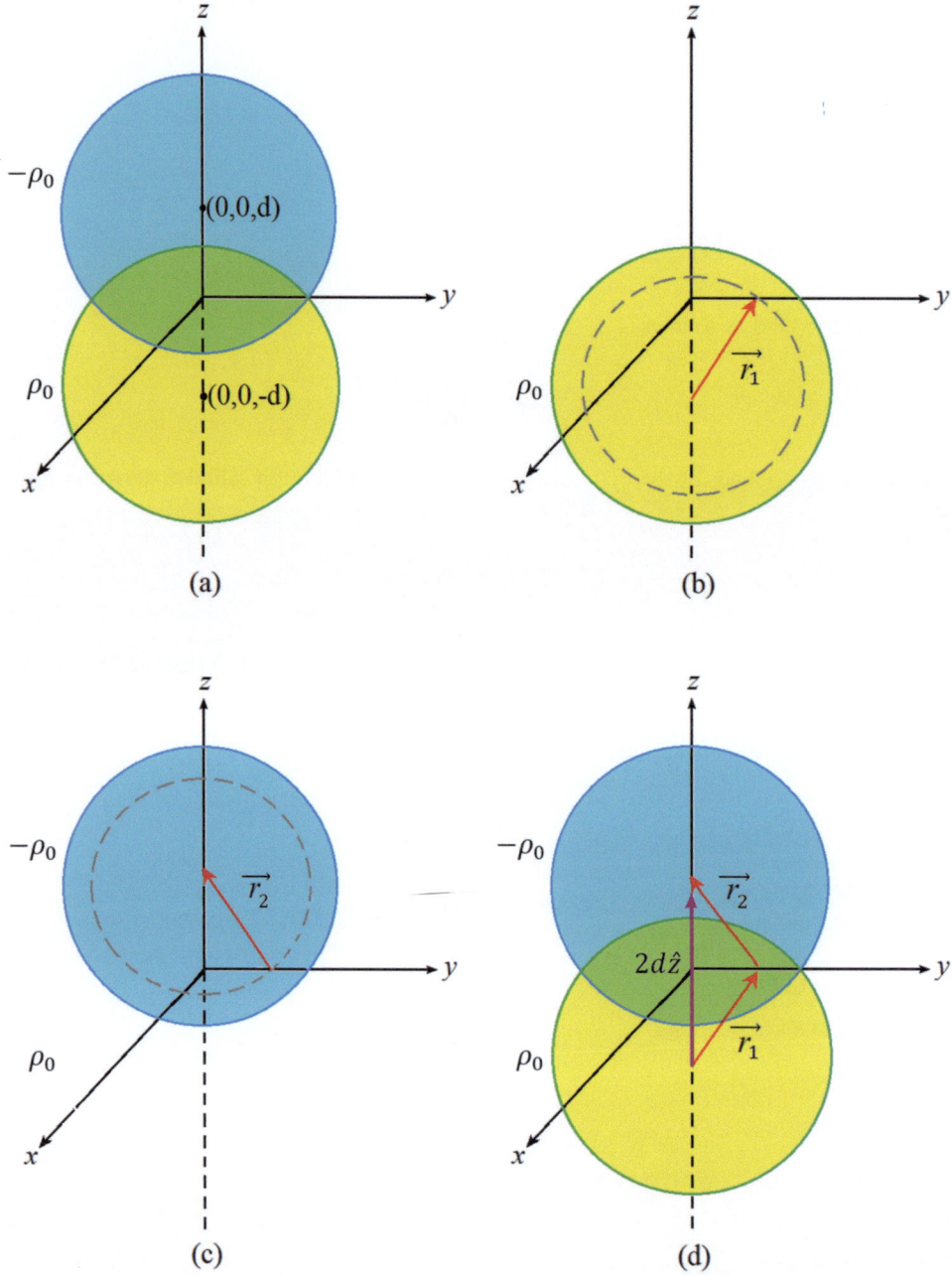

Fig. 6.14 (a) Two cylinders with radius a, length l, uniform volume charge densities ρ_0 and $-\rho_0$, and centers at $(0, 0, -d)$ and $(0, 0, d)$. (b) Applying Gauss's law for the cylindrical surface related to the bottom cylinder. (c) Applying Gauss's law for the cylindrical surface related to the top cylinder. (d) Applying the superposition principle to calculate the total electric field in the common area

6.3 Electric Field in Cylindrical Coordinate System

> **Notes**
>
> In this problem, the relation below has been used.
>
> $$\int x^n dx = \frac{x^{n+1}}{n+1}$$

6.25. The problem can be solved by applying Gauss's law as follows.

$$Q_{in} = \int \vec{D} \cdot \vec{dS}$$

$$\Rightarrow Q_{in} = \int \varepsilon_0 \vec{E} \cdot \vec{dS}$$

Applying Gauss's law for the Gaussian surface ($r \leq a$) shown in Fig. 6.15 will result in zero electric field because electric charge inside a conductor is zero. Therefore:

$$\vec{E} = 0$$

Choice (4) is the answer.

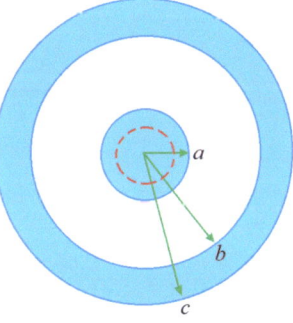

Fig. 6.15 Applying Gauss's law for the cylindrical Gaussian surface

6.26. By applying Gauss's law for the Gaussian surface ($a < r < b$) shown in Fig. 6.16, we have:

$$Q = \int_{z=0}^{l} \int_{\varphi=0}^{2\pi} \varepsilon_0 E_r r d\varphi dz$$

$$\Rightarrow Q = \varepsilon_0 E_r r \left[\varphi\right]_0^{2\pi} \left[z\right]_0^{l}$$

$$\Rightarrow Q = \varepsilon_0 E_r r (2\pi - 0)(l - 0)$$

$$\Rightarrow Q = \varepsilon_0 E_r r 2\pi l$$

$$\Rightarrow E_r = \frac{Q}{2\pi \varepsilon_0 r l}$$

$$\Rightarrow \vec{E} = \frac{Q}{2\pi \varepsilon_0 r l} \hat{r}$$

Choice (2) is the answer.

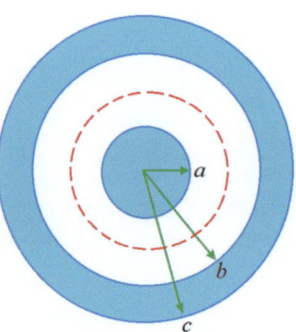

Fig. 6.16 Applying Gauss's law for the cylindrical Gaussian surface

6.27. Applying Gauss's law for the Gaussian surface ($b < r < c$) shown in Fig. 6.17 results in zero electric field because electric charge inside a conductor is zero. Therefore:

$$\vec{E} = 0$$

Choice (1) is the answer.

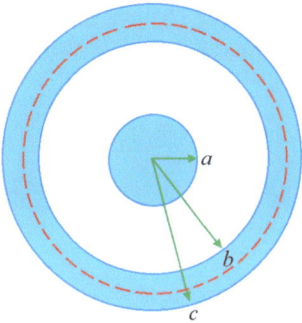

Fig. 6.17 Applying Gauss's law for the cylindrical Gaussian surface

6.28. By applying Gauss's law for the Gaussian surface ($r > c$) shown in Fig. 6.18, we have:

$$Q + (-Q) = \int_{z=0}^{l} \int_{\varphi=0}^{2\pi} \varepsilon_0 E_r r d\varphi dz$$

$$\Rightarrow 0 = \varepsilon_0 E_r r \int_{z=0}^{l} \int_{\varphi=0}^{2\pi} d\varphi dz$$

$$\Rightarrow E_R = 0$$

$$\Rightarrow \vec{E} = 0$$

Choice (3) is the answer.

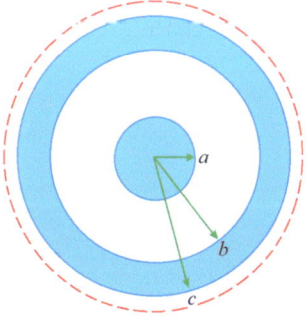

Fig. 6.18 Applying Gauss's law for the cylindrical Gaussian surface

6.29. Based on the information given in the problem, the charges Q and $-Q$ are uniformly distributed across the solid and hollow cylinders, respectively. Hence, the volume charge density of the solid cylinder can be presented as follows.

$$\rho_1 = \frac{Q}{\pi a^2 l}$$

In addition, the volume charge density of the hollow cylinder can be calculated as follows.

$$\rho_2 = \frac{-Q}{\pi (c^2 - b^2) l}$$

By applying Gauss's law for the spherical surface ($r \leq a$) shown in Fig. 6.19, we have:

$$Q_{in} = \int \vec{D} \cdot \vec{dS}$$

$$\Rightarrow \int \rho dV = \int \varepsilon_0 \vec{E} \cdot \vec{dS}$$

$$\Rightarrow \int_{z=0}^{l} \int_{\varphi=0}^{2\pi} \int_{r=0}^{r} \frac{Q}{\pi a^2 l} r dr d\varphi dz = \int_{z=0}^{l} \int_{\varphi=0}^{2\pi} \varepsilon_0 E_r r d\varphi dz$$

$$\Rightarrow \frac{Q}{\pi a^2 l} \left[\frac{r^2}{2}\right]_0^r \int_{z=0}^{l} \int_{\varphi=0}^{2\pi} d\varphi dz = \varepsilon_0 E_r r \int_{z=0}^{l} \int_{\varphi=0}^{2\pi} d\varphi dz$$

$$\Rightarrow \frac{Q}{\pi a^2 l} \left[\frac{r^2}{2}\right]_0^r = \varepsilon_0 E_r r$$

$$\Rightarrow \frac{Q}{\pi a^2 l} \left(\frac{r^2}{2} - 0\right) = \varepsilon_0 E_r r$$

$$\Rightarrow \frac{Qr}{2\pi a^2 l} = \varepsilon_0 E_r$$

$$\Rightarrow E_r = \frac{Qr}{2\pi \varepsilon_0 a^2 l}$$

$$\Rightarrow \vec{E} = \frac{Qr}{2\pi \varepsilon_0 a^2 l} \hat{r}$$

Choice (2) is the answer.

6.3 Electric Field in Cylindrical Coordinate System

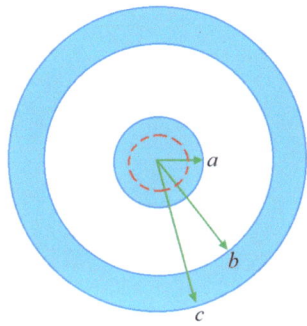

Fig. 6.19 Applying Gauss's law for the cylindrical Gaussian surface

Notes

In this problem, the relation below has been used.

$$\int x^n dx = \frac{x^{n+1}}{n+1}$$

6.30. By applying Gauss's law for the cylindrical surface ($a < r < b$) shown in Fig. 6.20, we have:

$$\Rightarrow Q + \int_{z=0}^{l} \int_{\varphi=0}^{2\pi} \int_{r=a}^{r} 0 = \int_{z=0}^{l} \int_{\varphi=0}^{2\pi} \varepsilon_0 E_r r d\varphi dz$$

$$\Rightarrow Q = \int_{z=0}^{l} \int_{\varphi=0}^{2\pi} \varepsilon_0 E_r r d\varphi dz$$

$$\Rightarrow Q = \varepsilon_0 E_r r \left[\varphi \right]_0^{2\pi} \left[z \right]_0^l$$

$$\Rightarrow Q = \varepsilon_0 E_r r (2\pi - 0)(l - 0)$$

$$\Rightarrow Q = \varepsilon_0 E_r r 2\pi l$$

$$\Rightarrow E_r = \frac{Q}{2\pi \varepsilon_0 r l}$$

$$\Rightarrow \vec{E} = \frac{Q}{2\pi \varepsilon_0 r l} \hat{r}$$

Choice (1) is the answer.

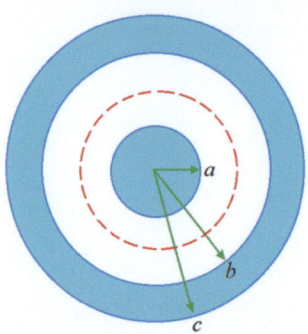

Fig. 6.20 Applying Gauss's law for the cylindrical Gaussian surface

6.31. By applying Gauss's law for the cylindrical surface ($b < r < c$) shown in Fig. 6.21, we have:

$$\int_{z=0}^{l}\int_{\varphi=0}^{2\pi}\int_{r=0}^{a}\frac{Q}{\pi a^2 l}rdrd\varphi dz + \int_{z=0}^{l}\int_{\varphi=0}^{2\pi}\int_{r=a}^{b} 0\times rdrd\varphi dz$$

$$+ \int_{z=0}^{l}\int_{\varphi=0}^{2\pi}\int_{r=b}^{r}\frac{-Q}{\pi(c^2-b^2)l}rdrd\varphi dz = \int_{z=0}^{l}\int_{\varphi=0}^{2\pi}\varepsilon_0 E_r rd\varphi dz$$

$$\Rightarrow \frac{Q}{\pi a^2 l}\left[\frac{r^2}{2}\right]_0^a \int_{z=0}^{l}\int_{\varphi=0}^{2\pi}d\varphi dz + 0 - \frac{Q}{\pi(c^2-b^2)l}\left[\frac{r^2}{2}\right]_b^r \int_{z=0}^{l}\int_{\varphi=0}^{2\pi}d\varphi dz = \varepsilon_0 E_r r \int_{z=0}^{l}\int_{\varphi=0}^{2\pi}d\varphi dz$$

$$\Rightarrow \frac{Q}{\pi a^2 l}\left[\frac{r^2}{2}\right]_0^a - \frac{Q}{\pi(c^2-b^2)l}\left[\frac{r^2}{2}\right]_b^r = \varepsilon_0 E_r r$$

$$\Rightarrow \frac{Q}{\pi a^2 l}\left(\frac{a^2}{2}-0\right) - \frac{Q}{\pi(c^2-b^2)l}\left(\frac{r^2-b^2}{2}\right) = \varepsilon_0 E_r r$$

$$\Rightarrow \frac{Q}{2\pi l} - \frac{Q(r^2-b^2)}{2\pi l(c^2-b^2)} = \varepsilon_0 E_r r$$

$$\Rightarrow E_r = \frac{Q}{2\pi\varepsilon_0 rl} - \frac{Q(r^2-b^2)}{2\pi\varepsilon_0 rl(c^2-b^2)}$$

$$\Rightarrow E_r = \frac{Q}{2\pi\varepsilon_0 rl}\left(1-\frac{r^2-b^2}{c^2-b^2}\right)$$

6.3 Electric Field in Cylindrical Coordinate System

$$\Rightarrow E_r = \frac{Q}{2\pi\varepsilon_0 rl}\left(\frac{c^2-r^2}{c^2-b^2}\right)$$

$$\Rightarrow E_r = \frac{Q}{2\pi\varepsilon_0 rl}\left(\frac{c^2-r^2}{c^2-b^2}\right)$$

$$\Rightarrow \vec{E} = \frac{Q}{2\pi\varepsilon_0 l}\left(\frac{c^2}{r(c^2-b^2)} - \frac{r}{c^2-b^2}\right)\hat{r}$$

Choice (1) is the answer.

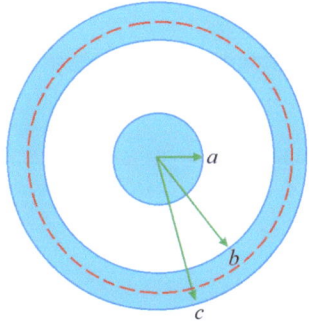

Fig. 6.21 Applying Gauss's law for the cylindrical Gaussian surface

---

Notes

In this problem, the relation below has been used.

$$\int x^n dx = \frac{x^{n+1}}{n+1}$$

6.32. By applying Gauss's law for the cylindrical surface $(r > c)$ shown in Fig. 6.22, we have:

$$Q + (-Q) = \int_{z=0}^{l}\int_{\varphi=0}^{2\pi} \varepsilon_0 E_r r\, d\varphi dz$$

$$\Rightarrow 0 = \varepsilon_0 E_r r \int_{z=0}^{l}\int_{\varphi=0}^{2\pi} d\varphi dz$$

$$\Rightarrow E_r = 0$$

$$\Rightarrow \vec{E} = 0$$

Choice (2) is the answer.

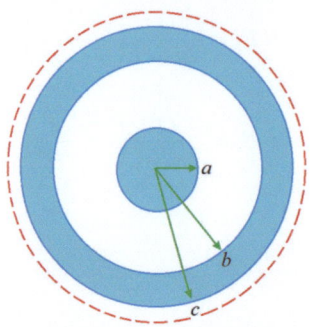

Fig. 6.22 Applying Gauss's law for the cylindrical Gaussian surface

---

6.33. The problem can be solved by using the superposition principle, as is shown in Fig. 6.23. In fact, the large cylinder can be considered a solid cylinder with volume charge density ρ and the cylindrical hole can be considered a small solid cylinder with volume charge density $-\rho$.

Herein, we have:

$$\rho = \frac{Q}{\pi a^2 l}$$

By applying Gauss's law for the large sphere, shown in Fig. 6.23 (b), we have:

$$Q_{in} = \int \vec{D} \cdot \vec{dS}$$

$$\Rightarrow Q = \int \varepsilon_0 \vec{E} \cdot \vec{dS}$$

$$\Rightarrow Q = \int_{z=0}^{l} \int_{\varphi=0}^{2\pi} \varepsilon_0 E_r r \, d\varphi \, dz$$

$$\Rightarrow Q = \varepsilon_0 E_r r \left[\varphi \right]_0^{2\pi} \left[z \right]_0^{l}$$

$$\Rightarrow Q = \varepsilon_0 E_r r (2\pi) l$$

$$\Rightarrow E_r = \frac{Q}{2\pi \varepsilon_0 r l}$$

$$\Rightarrow \vec{E_r} = \frac{Q}{2\pi \varepsilon_0 r l} \hat{r}$$

6.3 Electric Field in Cylindrical Coordinate System

$$\Rightarrow \vec{E_p} = \vec{E_r}\bigg|_{r=4a} = \frac{Q}{8\pi\varepsilon_0 al}\hat{y}$$

By applying Gauss's law for the small sphere, shown in Fig. 6.23 (c), we have:

$$Q_{in} = \int \vec{D}\cdot\vec{dS}$$

$$\Rightarrow \int -\rho dV = \int \varepsilon_0 \vec{E}\cdot\vec{dS}$$

$$\Rightarrow \left(-\frac{Q}{\pi a^2 l}\right)\left(\pi\left(\frac{a}{2}\right)^2 l\right) = \int_{z=0}^{l}\int_{\varphi=0}^{2\pi}\varepsilon_0 E_r r\,d\varphi\,dz$$

$$\Rightarrow -\frac{Q}{4} = \varepsilon_0 E_r r\left[\varphi\right]_0^{2\pi}\left[z\right]_0^{l}$$

$$\Rightarrow -\frac{Q}{4} = \varepsilon_0 E_r r (2\pi) l$$

$$\Rightarrow E_r = -\frac{Q}{8\pi\varepsilon_0 rl}$$

$$\Rightarrow \vec{E} = -\frac{Q}{8\pi\varepsilon_0 rl}\hat{r}$$

$$\Rightarrow \vec{E_p} = \vec{E_r}\bigg|_{r=\frac{7}{2}a} = -\frac{Q}{8\pi\varepsilon_0\left(\frac{7}{2}a\right)l}\hat{y}$$

$$\Rightarrow \vec{E_p} = -\frac{Q}{28\pi\varepsilon_0 al}\hat{y}$$

The total electric field at point p is calculated as follows.

$$\Rightarrow \vec{E_p} = \frac{Q}{8\pi\varepsilon_0 al}\hat{y} - \frac{Q}{28\pi\varepsilon_0 al}\hat{y}$$

$$\Rightarrow \vec{E_p} = \frac{5Q}{56\pi\varepsilon_0 al}\hat{y}$$

Choice (3) is the answer.

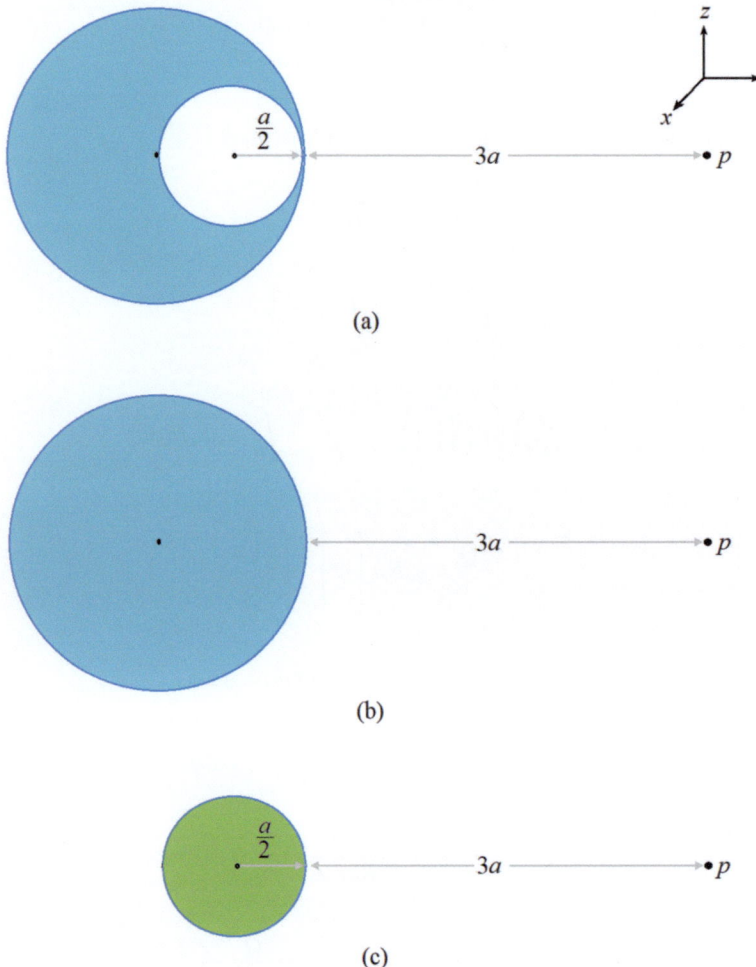

Fig. 6.23 (a) A solid cylinder with length l including a cylindrical hole, where the charge Q has been uniformly distributed across it. (b) Applying Gauss's law for the large sphere. (c) Applying Gauss's law for the small sphere

6.34. The problem can be solved by using the superposition principle, as is shown in Fig 6.24. In fact, the large sphere can be considered a solid cylinder with volume charge density ρ and the cylindrical hole can be considered a small solid cylinder with volume charge density $-\rho$.

By applying Gauss's law for the large sphere, shown in Fig. 6.24 (b), we have:

$$Q_{in} = \int \vec{D} \cdot \vec{dS}$$

$$\Rightarrow \int \rho dV = \int \varepsilon_0 \vec{E} \cdot \vec{dS}$$

6.3 Electric Field in Cylindrical Coordinate System

$$\Rightarrow \int_{z=0}^{l}\int_{\varphi=0}^{2\pi}\int_{r=0}^{r} \rho r\,dr\,d\varphi\,dz = \int_{z=0}^{l}\int_{\varphi=0}^{2\pi} \varepsilon_0 E_r r\,d\varphi\,dz$$

$$\Rightarrow \rho\left[\frac{r^2}{2}\right]_0^r \int_{z=0}^{l}\int_{\varphi=0}^{2\pi} d\varphi\,dz = \varepsilon_0 E_r r \int_{z=0}^{l}\int_{\varphi=0}^{2\pi} d\varphi\,dz$$

$$\Rightarrow \rho\left(\frac{r^2}{2} - 0\right) = \varepsilon_0 E_r r$$

$$\Rightarrow \rho\frac{r}{2} = \varepsilon_0 E_r$$

$$\Rightarrow E_r = \frac{\rho r}{2\varepsilon_0}$$

$$\Rightarrow \overrightarrow{E_r} = \frac{\rho r}{2\varepsilon_0}\widehat{r}$$

$$\Rightarrow \overrightarrow{E_p} = \overrightarrow{E_r}\bigg|_{r=\frac{a}{2}} = \frac{\rho a}{4\varepsilon_0}\widehat{y}$$

For the small cylinder, shown in Fig. 6.24 (c), the Gaussian surface does not include any charge because the electric field at its center is calculated. In other words:

$$\overrightarrow{E_p} = 0$$

The total electric field at point p can be calculated as follows.

$$\Rightarrow \overrightarrow{E_p} = \frac{\rho a}{4\varepsilon_0}\widehat{y} + 0$$

$$\Rightarrow \overrightarrow{E_p} = \frac{\rho a}{4\varepsilon_0}\widehat{y}$$

Choice (1) is the answer.

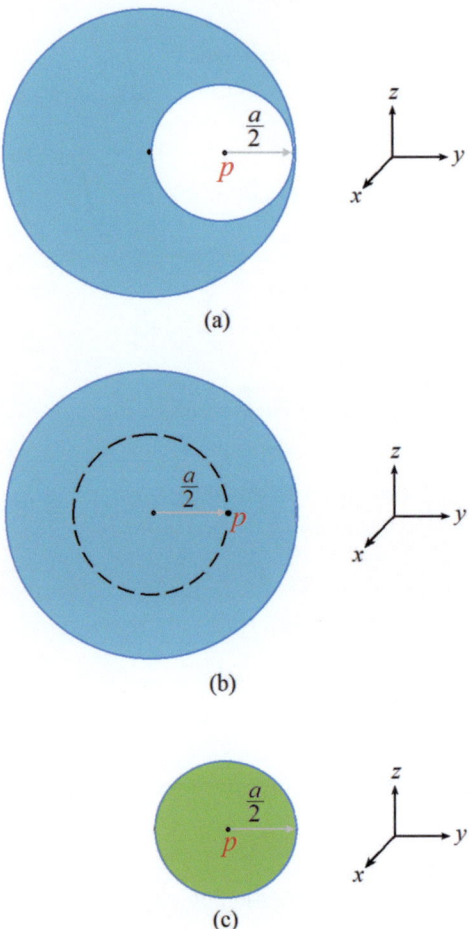

Fig. 6.24 (a) A solid cylinder with length l including a cylindrical hole, where the charge Q has been uniformly distributed across it. (b) Applying Gauss's law for the large sphere. (c) Applying Gauss's law for the small sphere

> **Notes**
>
> In this problem, the relation below has been used.
>
> $$\int x^n dx = \frac{x^{n+1}}{n+1}$$

6.3 Electric Field in Cylindrical Coordinate System

6.35. Based on the information given in the problem, we have:

$$\rho = \frac{1}{r}$$

The problem can be solved by applying Gauss's law for the Gaussian surface shown in Fig. 6.25.

$$Q_{in} = \int \vec{D} \cdot \vec{dS}$$

$$\Rightarrow \int \rho dV = \int \varepsilon_0 \vec{E} \cdot \vec{dS}$$

$$\Rightarrow \int \frac{1}{r} dV = \int \varepsilon_0 \vec{E} \cdot \vec{dS}$$

$$\Rightarrow \int_{z=0}^{l} \int_{\varphi=0}^{2\pi} \int_{r=0}^{r} \frac{1}{r} r dr d\varphi dz = \int_{z=0}^{l} \int_{\varphi=0}^{2\pi} \varepsilon_0 E_r r d\varphi dz$$

$$\Rightarrow \int_{z=0}^{l} \int_{\varphi=0}^{2\pi} \int_{r=0}^{r} dr d\varphi dz = \varepsilon_0 E_r r \int_{z=0}^{l} \int_{\varphi=0}^{2\pi} d\varphi dz$$

$$\Rightarrow \left[r \right]_0^r \int_{z=0}^{l} \int_{\varphi=0}^{2\pi} d\varphi dz = \varepsilon_0 E_r r \int_{z=0}^{l} \int_{\varphi=0}^{2\pi} d\varphi dz$$

$$\Rightarrow r - 0 = \varepsilon_0 E_r r$$

$$\Rightarrow E_r = \frac{1}{\varepsilon_0}$$

$$\vec{E} = \frac{1}{\varepsilon_0} \hat{r}$$

As can be seen, the electric field inside the cylinder is constant although the volume charge density is nonuniform Choice (4) is the answer.

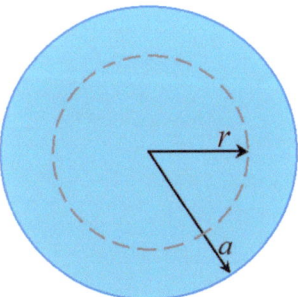

Fig. 6.25 Applying Gauss's law for the Gaussian surface

6.36. The problem can be solved by applying Gauss's law as follows.

$$Q_{in} = \int \vec{D}.\vec{dS}$$

$$\Rightarrow Q_{in} = \int \varepsilon_0 \vec{E}.\vec{dS}$$

By applying Gauss's law for the Gaussian surface ($r > b$), shown in Fig. 6.26, we have:

$$Q + Q_1 + Q_2 = \int_{z=0}^{l} \int_{\varphi=0}^{2\pi} \varepsilon_0 E_r r d\varphi dz$$

$$\Rightarrow Q + \rho_{s1} 2\pi a l + \rho_{s2} 2\pi b l = \varepsilon_0 E_r r \Big[\varphi\Big]_0^{2\pi} \Big[z\Big]_0^{l}$$

$$\Rightarrow Q + \rho_{s1} 2\pi a l + \rho_{s2} 2\pi b l = \varepsilon_0 E_r r (2\pi - 0)(l - 0)$$

$$\Rightarrow Q + \rho_{s1} 2\pi a l + \rho_{s2} 2\pi b l = \varepsilon_0 E_r 2\pi r l$$

$$\Rightarrow E_r = \frac{Q + \rho_{s1} 2\pi a l + \rho_{s2} 2\pi b l}{2\pi \varepsilon_0 r l} = \frac{Q}{2\pi \varepsilon_0 r l} + \frac{\rho_{s1} a}{\varepsilon_0 r} + \frac{\rho_{s2} b}{\varepsilon_0 r}$$

$$\Rightarrow \vec{E} = \left(\frac{Q}{2\pi \varepsilon_0 r l} + \frac{\rho_{s1} a}{\varepsilon_0 r} + \frac{\rho_{s2} b}{\varepsilon_0 r}\right)\hat{r}$$

Choice (1) is the answer.

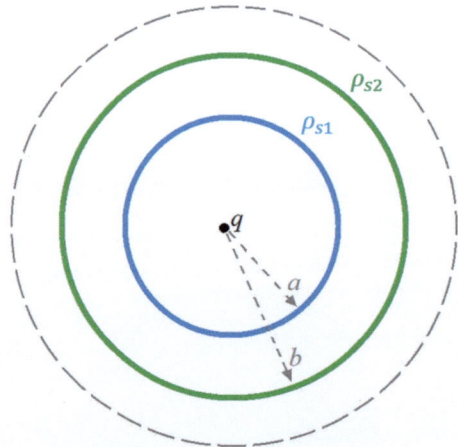

Fig. 6.26 Applying Gauss's law for the Gaussian surface ($r > b$)

6.4 Electric Field in Spherical Coordinate System

6.37. The problem can be solved by applying Gauss's law as follows.

$$Q_{in} = \int \vec{D} \cdot \vec{dS}$$

$$\Rightarrow Q_{in} = \int \varepsilon_0 \vec{E} \cdot \vec{dS}$$

Applying Gauss's law for the Gaussian surface ($R \leq a$) shown in Fig. 6.27 will result in zero electric field because electric charge inside a conductor is zero. Hence:

$$\vec{E_1} = 0$$

By applying Gauss's law for the Gaussian surface ($R > a$) shown in Fig. 6.27, we have:

$$Q = \int_{\varphi=0}^{2\pi} \int_{\theta=0}^{\pi} \varepsilon_0 E_R R^2 \sin\theta d\theta d\varphi$$

$$\Rightarrow Q = \varepsilon_0 E_R R^2 \int_{\varphi=0}^{2\pi} \int_{\theta=0}^{\pi} \sin\theta d\theta d\varphi$$

$$\Rightarrow Q = \varepsilon_0 E_R R^2 \left[-\cos\theta\right]_0^{\pi} \left[\varphi\right]_0^{2\pi}$$

$$\Rightarrow Q = \varepsilon_0 E_R R^2 (-\cos\pi + \cos 0)(2\pi - 0)$$

$$\Rightarrow Q = \varepsilon_0 E_R R^2 (-(-1) + 1)(2\pi)$$

$$\Rightarrow Q = 4\pi\varepsilon_0 E_R R^2$$

$$\Rightarrow E_R = \frac{Q}{4\pi\varepsilon_0 R^2}$$

$$\Rightarrow \vec{E_2} = \frac{Q}{4\pi\varepsilon_0 R^2} \hat{R}$$

Thus:

$$\vec{E} = \begin{cases} 0 & R \leq a \\ \dfrac{Q}{4\pi\varepsilon_0 R^2} \hat{R} & R > a \end{cases}$$

Choice (2) is the answer.

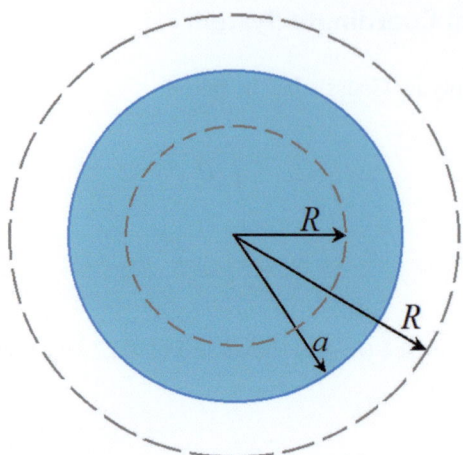

Fig. 6.27 Applying Gauss's law for the spherical Gaussian surfaces around the conducting solid sphere

> **Notes**
> In this problem, the relations below have been used.
>
> $$\int \sin x \, dx = -\cos x$$
>
> $$\cos \pi = -1$$
>
> $$\cos 0 = 1$$

6.39. The volume charge density of the solid sphere can be calculated as follows.

$$\rho = \frac{Q}{\frac{4}{3}\pi a^3} = \frac{3Q}{4\pi a^3}$$

By applying Gauss's law for the spherical surface ($R \leq a$) shown in Fig. 6.28, we have:

$$Q_{in} = \int \vec{D} \cdot \vec{dS}$$

$$\Rightarrow \int \rho dV = \int \varepsilon_0 \vec{E} \cdot \vec{dS}$$

$$\int_{\varphi=0}^{2\pi} \int_{\theta=0}^{\pi} \int_{R=0}^{R} \frac{3Q}{4\pi a^3} R^2 \sin\theta \, dR \, d\theta \, d\varphi = \int_{\varphi=0}^{2\pi} \int_{\theta=0}^{\pi} \varepsilon_0 E_R R^2 \sin\theta \, d\theta \, d\varphi$$

6.4 Electric Field in Spherical Coordinate System

$$\Rightarrow \frac{3Q}{4\pi a^3} \int_{\varphi=0}^{2\pi} \int_{\theta=0}^{\pi} \int_{R=0}^{R} R^2 \sin\theta \, dR \, d\theta \, d\varphi = \varepsilon_0 E_R R^2 \int_{\varphi=0}^{2\pi} \int_{\theta=0}^{\pi} \sin\theta \, d\theta \, d\varphi$$

$$\Rightarrow \frac{3Q}{4\pi a^3} \left[\frac{R^3}{3}\right]_0^R \int_{\varphi=0}^{2\pi} \int_{\theta=0}^{\pi} \sin\theta \, d\theta \, d\varphi = \varepsilon_0 E_R R^2 \int_{\varphi=0}^{2\pi} \int_{\theta=0}^{\pi} \sin\theta \, d\theta \, d\varphi$$

$$\Rightarrow \frac{3Q}{4\pi a^3} \left[\frac{R^3}{3}\right]_0^R = \varepsilon_0 E_R R^2$$

$$\Rightarrow \frac{QR^3}{4\pi a^3} = \varepsilon_0 E_R R^2$$

$$\Rightarrow E_R = \frac{QR}{4\pi\varepsilon_0 a^3}$$

$$\Rightarrow \vec{E_1} = \frac{QR}{4\pi\varepsilon_0 a^3} \hat{R}$$

By applying Gauss's law for the spherical surface ($R > a$) shown in Fig. 6.28, we have:

$$Q = \int_{\varphi=0}^{2\pi} \int_{\theta=0}^{\pi} \varepsilon_0 E_R R^2 \sin\theta \, d\theta \, d\varphi$$

$$\Rightarrow Q = \varepsilon_0 E_R R^2 \int_{\varphi=0}^{2\pi} \int_{\theta=0}^{\pi} \sin\theta \, d\theta \, d\varphi$$

$$\Rightarrow Q = \varepsilon_0 E_R R^2 \left[-\cos\theta\right]_0^{\pi} \left[\varphi\right]_0^{2\pi}$$

$$\Rightarrow Q = \varepsilon_0 E_R R^2 (-\cos\pi + \cos 0)(2\pi - 0)$$

$$\Rightarrow Q = \varepsilon_0 E_R R^2 (-(-1) + 1)(2\pi)$$

$$\Rightarrow E_R = \frac{Q}{4\pi\varepsilon_0 R^2}$$

$$\Rightarrow \vec{E_2} = \frac{Q}{4\pi\varepsilon_0 R^2} \hat{R}$$

Thus:

$$\vec{E} = \begin{cases} \dfrac{QR}{4\pi\varepsilon_0 a^3} \hat{R} & R \leq a \\ \dfrac{Q}{4\pi\varepsilon_0 R^2} \hat{R} & R > a \end{cases}$$

Choice (3) is the answer.

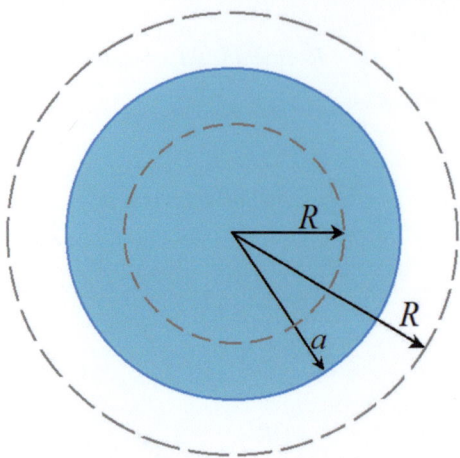

Fig. 6.28 Applying Gauss's law for the spherical Gaussian surfaces around the solid sphere

> **Notes**
> In this problem, the relations below have been used.
>
> $$\int x^n dx = \frac{x^{n+1}}{n+1}$$
>
> $$\int \sin x \, dx = -\cos x$$
>
> $$\cos \pi = -1$$
>
> $$\cos 0 = 1$$

6.41. The problem can be solved by using the superposition principle, as is shown in Fig. 6.29. By applying Gauss's law for the spherical surface related to the bottom sphere shown in Fig. 6.29 (b), we have:

$$Q_{in} = \int \vec{D} \cdot \overrightarrow{dS}$$

$$\Rightarrow \int \rho dV = \int \varepsilon_0 \vec{E} \cdot \overrightarrow{dS}$$

$$\Rightarrow \int_{\varphi=0}^{2\pi} \int_{\theta=0}^{\pi} \int_{R=0}^{R} \rho_0 R^2 \sin\theta \, dR \, d\theta \, d\varphi = \int_{\varphi=0}^{2\pi} \int_{\theta=0}^{\pi} \varepsilon_0 E_R R^2 \sin\theta \, d\theta \, d\varphi$$

6.4 Electric Field in Spherical Coordinate System

$$\Rightarrow \rho_0 \int_{\varphi=0}^{2\pi} \int_{\theta=0}^{\pi} \int_{R=0}^{R} R^2 \sin\theta dR d\theta d\varphi = \varepsilon_0 E_R R^2 \int_{\varphi=0}^{2\pi} \int_{\theta=0}^{\pi} \sin\theta d\theta d\varphi$$

$$\Rightarrow \rho_0 \left[\frac{R^3}{3}\right]_0^R \int_{\varphi=0}^{2\pi} \int_{\theta=0}^{\pi} \sin\theta d\theta d\varphi = \varepsilon_0 E_R R^2 \int_{\varphi=0}^{2\pi} \int_{\theta=0}^{\pi} \sin\theta d\theta d\varphi$$

$$\Rightarrow \rho_0 \left(\frac{R^3}{3} - 0\right) = \varepsilon_0 E_R R^2$$

$$\Rightarrow E_R = \frac{\rho_0 R}{3\varepsilon_0}$$

By considering the direction of electric field, we have:

$$\vec{E} = \frac{\rho_0 R}{3\varepsilon_0} \hat{R}$$

From previous chapters, we know that $R\hat{R} = \vec{R}$. Therefore:

$$\vec{E} = \frac{\rho_0}{3\varepsilon_0} \vec{R}$$

Now, let us name this electric field as follows.

$$\vec{E_1} = \frac{\rho_0}{3\varepsilon_0} \vec{R_1}$$

Likewise, Gauss's law can be applied for the spherical surface related to the top sphere shown in Fig. 6.29 (c). In this condition, we have:

$$\vec{E_2} = -\frac{\rho_0}{3\varepsilon_0} \vec{R_2}$$

Based on the superposition principle, the total electric field in the common area can be calculated as follows.

$$\vec{E} = \vec{E_1} + \vec{E_2}$$

$$\Rightarrow \vec{E} = \frac{\rho_0}{3\varepsilon_0} \vec{R_1} - \frac{\rho_0}{3\varepsilon_0} \vec{R_2}$$

$$\Rightarrow \vec{E} = \frac{\rho_0}{3\varepsilon_0} \left(\vec{R_1} - \vec{R_2}\right)$$

As can be noticed from Fig. 6.29 (d), $\vec{R_1} - \vec{R_2} = 2d\hat{z}$. Therefore:

$$\Rightarrow \vec{E} = \frac{2\rho_0 d}{3\varepsilon_0} \hat{z}$$

Choice (2) is the answer.

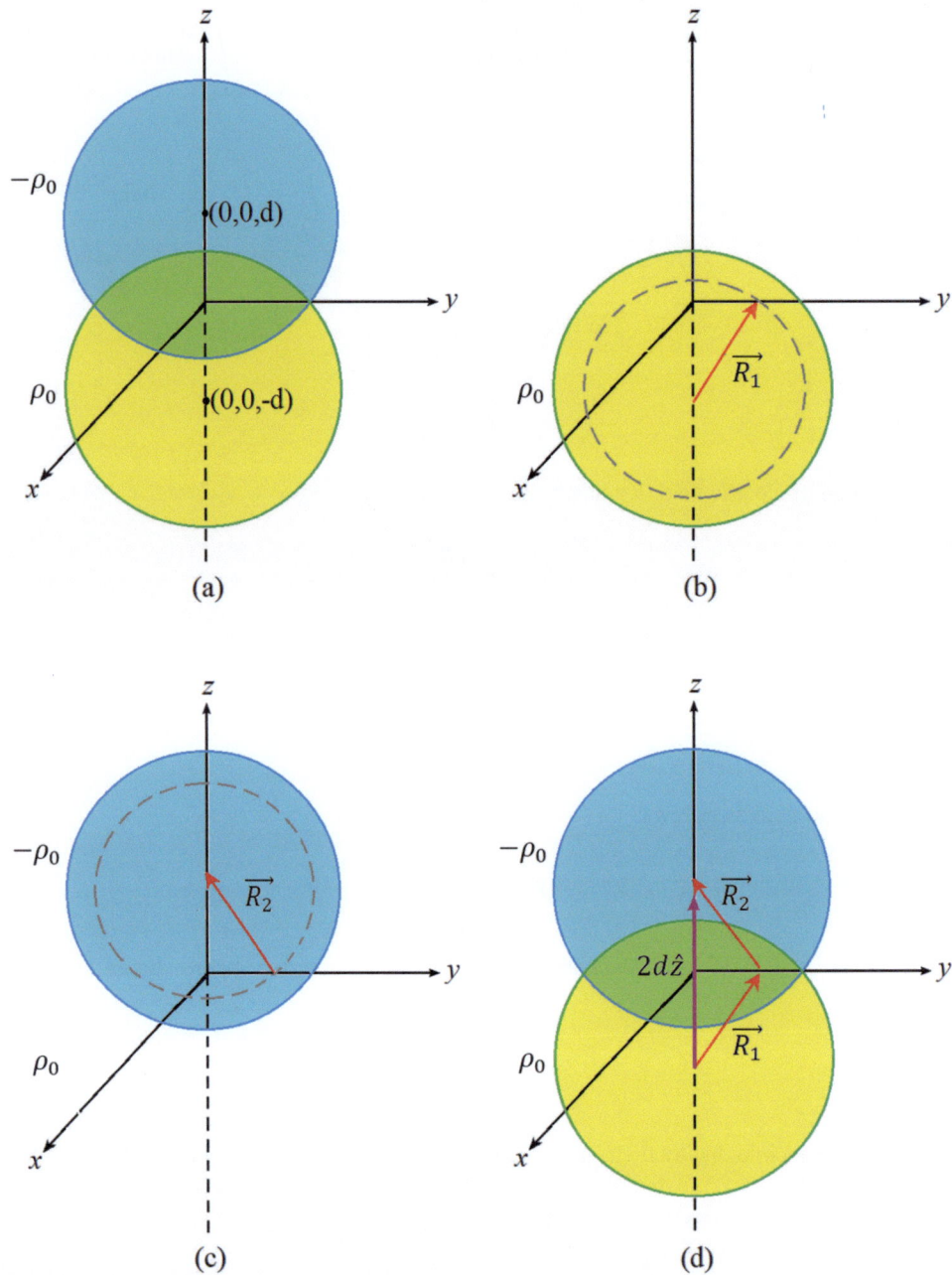

Fig. 6.29 (a) Two spheres with radius a uniform volume charge densities ρ_0 and $-\rho_0$, and centers at $(0, 0, -d)$ and $(0, 0, d)$. (b) Applying Gauss's law for the spherical surface related to the bottom sphere. (c) Applying Gauss's law for the spherical surface related to the top sphere. (d) Applying the superposition principle to calculate the total electric field in the common area

Notes

In this problem, the relation below has been used.

$$\int \sin x\, dx = -\cos x$$

6.4 Electric Field in Spherical Coordinate System

6.42. The problem can be solved by applying Gauss's law as follows.

$$Q_{in} = \int \vec{D} \cdot \vec{dS}$$

$$\Rightarrow Q_{in} = \int \varepsilon_0 \vec{E} \cdot \vec{dS}$$

Applying Gauss's law for the Gaussian surface ($R \leq a$) shown in Fig. 6.30 will result in zero electric field because electric charge inside a conductor is zero. Therefore:

$$\vec{E} = 0$$

Choice (4) is the answer.

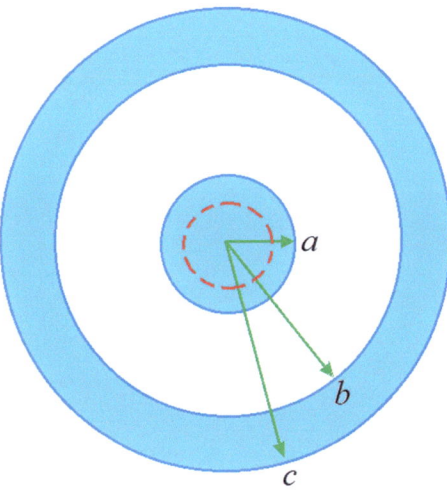

Fig. 6.30 Applying Gauss's law for the spherical Gaussian surface

6.43. By applying Gauss's law for the Gaussian surface ($a < R < b$) shown in Fig. 6.31, we have:

$$Q = \int_{\varphi=0}^{2\pi} \int_{\theta=0}^{\pi} \varepsilon_0 E_R R^2 \sin\theta \, d\theta \, d\varphi$$

$$\Rightarrow Q = \varepsilon_0 E_R R^2 \int_{\varphi=0}^{2\pi} \int_{\theta=0}^{\pi} \sin\theta \, d\theta \, d\varphi$$

$$\Rightarrow Q = \varepsilon_0 E_R R^2 \Big[-\cos\theta\Big]_0^{\pi} \Big[\varphi\Big]_0^{2\pi}$$

$$\Rightarrow Q = \varepsilon_0 E_R R^2 (-\cos\pi + \cos 0)(2\pi - 0)$$

$$\Rightarrow Q = \varepsilon_0 E_R R^2 (-(-1) + 1)(2\pi)$$

$$\Rightarrow Q = 4\pi\varepsilon_0 E_R R^2$$

$$\Rightarrow E_R = \frac{Q}{4\pi\varepsilon_0 R^2}$$

$$\Rightarrow \vec{E} = \frac{Q}{4\pi\varepsilon_0 R^2}\hat{R}$$

Choice (3) is the answer.

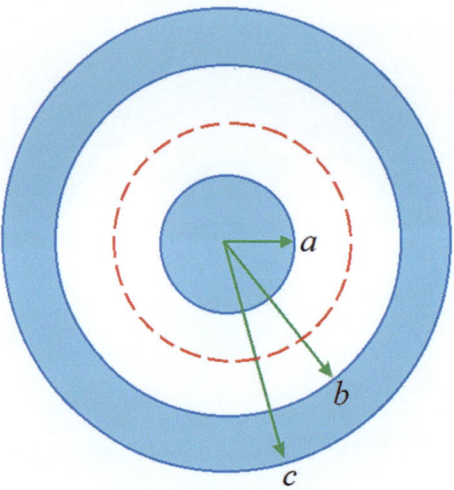

Fig. 6.31 Applying Gauss's law for the spherical Gaussian surface

> **Notes**
>
> In this problem, the relations below have been used.
>
> $$\int \sin x\, dx = -\cos x$$
>
> $$\cos\pi = -1$$
>
> $$\cos 0 = 1$$

6.44. Applying Gauss's law for the Gaussian surface ($b < R < c$) shown in Fig. 6.32 results in zero electric field because electric charge inside a conductor is zero. Therefore:

$$\vec{E} = 0$$

Choice (1) is the answer.

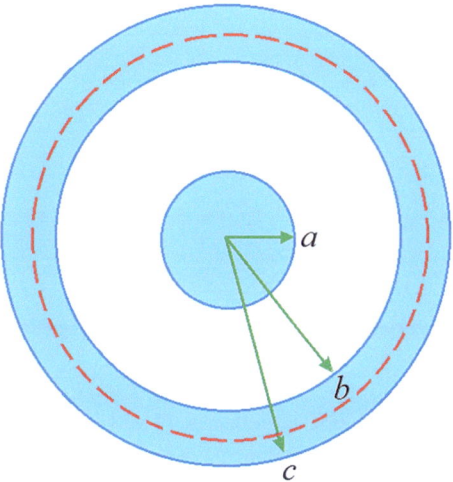

Fig. 6.32 Applying Gauss's law for the spherical Gaussian surface

6.45. By applying Gauss's law for the Gaussian surface ($R > c$) shown in Fig. 6.33, we have:

$$Q + (-Q) = \int_{\varphi=0}^{2\pi} \int_{\theta=0}^{\pi} \varepsilon_0 E_R R^2 \sin\theta d\theta d\varphi$$

$$\Rightarrow 0 = \varepsilon_0 E_R R^2 \int_{\varphi=0}^{2\pi} \int_{\theta=0}^{\pi} \sin\theta d\theta d\varphi$$

$$\Rightarrow E_R = 0$$

$$\Rightarrow \vec{E} = 0$$

Choice (4) is the answer.

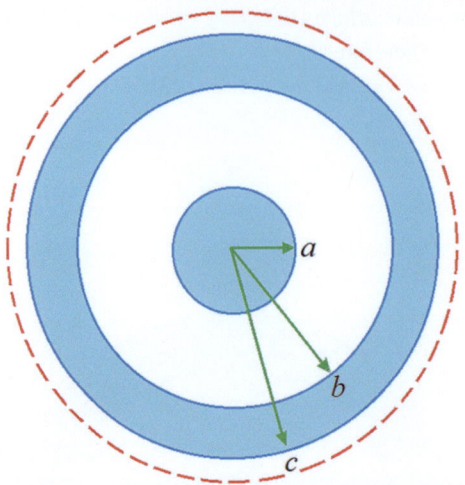

Fig. 6.33 Applying Gauss's law for the spherical Gaussian surface

---

6.46. Based on the information given in the problem, the charges Q and $-Q$ are uniformly distributed across the solid and hollow spheres, respectively. Hence, the volume charge density of the solid sphere can be calculated as follows.

$$\rho_1 = \frac{Q}{\frac{4}{3}\pi a^3} = \frac{3Q}{4\pi a^3}$$

In addition, the volume charge density of the hollow sphere can be calculated as follows.

$$\rho_2 = \frac{-Q}{\frac{4}{3}\pi (c^3 - b^3)} = -\frac{3Q}{4\pi (c^3 - b^3)}$$

By applying Gauss's law for the spherical surface ($R \leq a$) shown in Fig. 6.34, we have:

$$Q_{in} = \int \vec{D} \cdot \vec{dS}$$

$$\Rightarrow \int \rho dV = \int \varepsilon_0 \vec{E} \cdot \vec{dS}$$

$$\int_{\varphi=0}^{2\pi} \int_{\theta=0}^{\pi} \int_{R=0}^{R} \frac{3Q}{4\pi a^3} R^2 \sin\theta dRd\theta d\varphi = \int_{\varphi=0}^{2\pi} \int_{\theta=0}^{\pi} \varepsilon_0 E_R R^2 \sin\theta d\theta d\varphi$$

6.4 Electric Field in Spherical Coordinate System

$$\Rightarrow \frac{3Q}{4\pi a^3} \int_{\varphi=0}^{2\pi} \int_{\theta=0}^{\pi} \int_{R=0}^{R} R^2 \sin\theta \, dR \, d\theta \, d\varphi = \varepsilon_0 E_R R^2 \int_{\varphi=0}^{2\pi} \int_{\theta=0}^{\pi} \sin\theta \, d\theta \, d\varphi$$

$$\Rightarrow \frac{3Q}{4\pi a^3} \left[\frac{R^3}{3}\right]_0^R \int_{\varphi=0}^{2\pi} \int_{\theta=0}^{\pi} \sin\theta \, d\theta \, d\varphi = \varepsilon_0 E_R R^2 \int_{\varphi=0}^{2\pi} \int_{\theta=0}^{\pi} \sin\theta \, d\theta \, d\varphi$$

$$\Rightarrow \frac{3Q}{4\pi a^3} \left[\frac{R^3}{3}\right]_0^R = \varepsilon_0 E_R R^2$$

$$\Rightarrow \frac{QR^3}{4\pi a^3} = \varepsilon_0 E_R R^2$$

$$\Rightarrow E_R = \frac{QR}{4\pi\varepsilon_0 a^3}$$

$$\Rightarrow \vec{E} = \frac{QR}{4\pi\varepsilon_0 a^3} \hat{R}$$

Choice (2) is the answer.

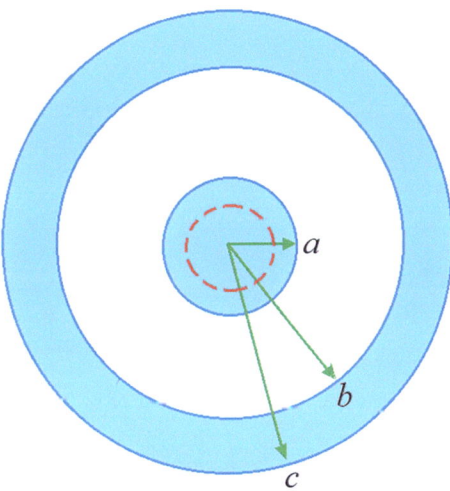

Fig. 6.34 Applying Gauss's law for the spherical Gaussian surface

---

Notes

In this problem, the relation below has been used.

$$\int x^n dx = \frac{x^{n+1}}{n+1}$$

6.47. By applying Gauss's law for the spherical surface ($a < R < b$) shown in Fig. 6.35, we have:

$$Q + \int_{\varphi=0}^{2\pi} \int_{\theta=0}^{\pi} \int_{R=a}^{R} 0 = \int_{\varphi=0}^{2\pi} \int_{\theta=0}^{\pi} \varepsilon_0 E_R R^2 \sin\theta d\theta d\varphi$$

$$\Rightarrow Q + 0 = \varepsilon_0 E_R R^2 \int_{\varphi=0}^{2\pi} \int_{\theta=0}^{\pi} \sin\theta d\theta d\varphi$$

$$\Rightarrow Q = \varepsilon_0 E_R R^2 \Big[-\cos\theta\Big]_0^{\pi} \Big[\varphi\Big]_0^{2\pi}$$

$$\Rightarrow Q = \varepsilon_0 E_R R^2 (-\cos\pi + \cos 0)(2\pi - 0)$$

$$\Rightarrow Q = \varepsilon_0 E_R R^2 (-(-1) + 1)(2\pi)$$

$$\Rightarrow E_R = \frac{Q}{4\pi\varepsilon_0 R^2}$$

$$\Rightarrow \vec{E} = \frac{Q}{4\pi\varepsilon_0 R^2}\hat{R}$$

Choice (1) is the answer.

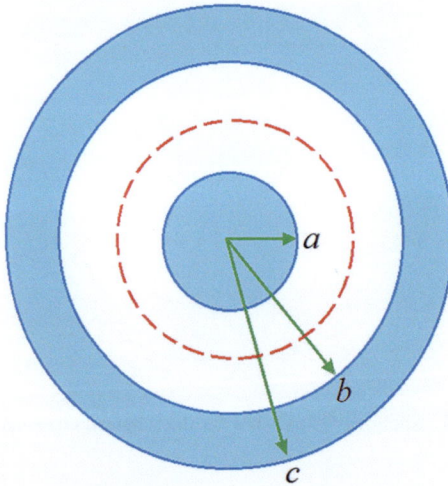

Fig. 6.35 Applying Gauss's law for the spherical Gaussian surface

6.4 Electric Field in Spherical Coordinate System

> **Notes**
> In this problem, the relations below have been used.
> $$\int \sin x \, dx = -\cos x$$
> $$\cos \pi = -1$$
> $$\cos 0 = 1$$

6.48. By applying Gauss's law for the dashed spherical surface ($b < R < c$) shown in Fig. 6.36, we have:

$$\int_{\varphi=0}^{2\pi} \int_{\theta=0}^{\pi} \int_{R=0}^{a} \frac{3Q}{4\pi a^3} R^2 \sin\theta \, dR \, d\theta \, d\varphi + \int_{\varphi=0}^{2\pi} \int_{\theta=0}^{\pi} \int_{R=a}^{b} 0$$

$$+ \int_{\varphi=0}^{2\pi} \int_{\theta=0}^{\pi} \int_{R=b}^{R} \left(-\frac{3Q}{4\pi(c^3-b^3)}\right) R^2 \sin\theta \, dR \, d\theta \, d\varphi = \int_{\varphi=0}^{2\pi} \int_{\theta=0}^{\pi} \varepsilon_0 E_R R^2 \sin\theta \, d\theta \, d\varphi$$

$$\Rightarrow \frac{3Q}{4\pi a^3} \int_{\varphi=0}^{2\pi} \int_{\theta=0}^{\pi} \int_{R=0}^{a} R^2 \sin\theta \, dR \, d\theta \, d\varphi + 0 - \frac{3Q}{4\pi(c^3-b^3)}$$

$$\times \int_{\varphi=0}^{2\pi} \int_{\theta=0}^{\pi} \int_{R=b}^{R} R^2 \sin\theta \, dR \, d\theta \, d\varphi = \varepsilon_0 E_R R^2 \int_{\varphi=0}^{2\pi} \int_{\theta=0}^{\pi} \sin\theta \, d\theta \, d\varphi$$

$$\Rightarrow \frac{3Q}{4\pi a^3} \left[\frac{R^3}{3}\right]_0^a \int_{\varphi=0}^{2\pi} \int_{\theta=0}^{\pi} \sin\theta \, d\theta \, d\varphi - \frac{3Q}{4\pi(c^3-b^3)} \left[\frac{R^3}{3}\right]_b^R \int_{\varphi=0}^{2\pi} \int_{\theta=0}^{\pi} \sin\theta \, d\theta \, d\varphi = \varepsilon_0 E_R R^2 \int_{\varphi=0}^{2\pi} \int_{\theta=0}^{\pi} \sin\theta \, d\theta \, d\varphi$$

$$\Rightarrow \frac{3Q}{4\pi a^3} \left[\frac{R^3}{3}\right]_0^a - \frac{3Q}{4\pi(c^3-b^3)} \left[\frac{R^3}{3}\right]_b^R = \varepsilon_0 E_R R^2$$

$$\Rightarrow \frac{Q}{4\pi} - \frac{Q(R^3-b^3)}{4\pi(c^3-b^3)} = \varepsilon_0 E_R R^2$$

$$\Rightarrow E_R = \frac{Q}{4\pi\varepsilon_0 R^2}\left(1 - \frac{R^3-b^3}{c^3-b^3}\right) = \frac{Q}{4\pi\varepsilon_0 R^2} \frac{c^3-R^3}{c^3-b^3}$$

$$\Rightarrow \vec{E} = \frac{Q}{4\pi\varepsilon_0}\left(\frac{c^3}{R^2(c^3-b^3)} - \frac{R}{c^3-b^3}\right) \hat{R}$$

Choice (4) is the answer.

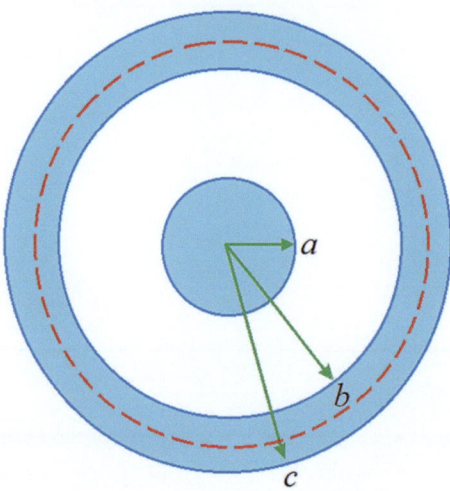

Fig. 6.36 Applying Gauss's law for the spherical Gaussian surface

> **Notes**
> In this problem, the relation below has been used.
>
> $$\int x^n dx = \frac{x^{n+1}}{n+1}$$

6.49. By applying Gauss's law for the spherical surface ($R > c$) shown in Fig. 6.37, we have:

$$Q + (-Q) = \int_{\varphi=0}^{2\pi} \int_{\theta=0}^{\pi} \varepsilon_0 E_R R^2 \sin\theta d\theta d\varphi$$

$$\Rightarrow 0 = \varepsilon_0 E_R R^2 \int_{\varphi=0}^{2\pi} \int_{\theta=0}^{\pi} \sin\theta d\theta d\varphi$$

$$\Rightarrow E_R = 0$$

$$\Rightarrow \vec{E} = 0$$

Choice (1) is the answer.

6.4 Electric Field in Spherical Coordinate System

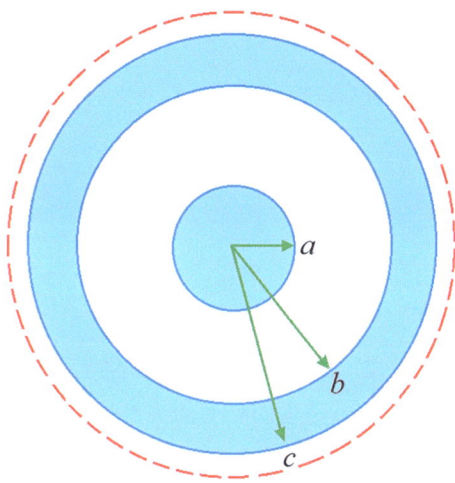

Fig. 6.37 Applying Gauss's law for the spherical Gaussian surface

6.50. The problem can be solved by using the superposition principle, as is shown in Fig. 6.38. In fact, the large sphere can be considered a solid sphere with volume charge density ρ and the spherical hole can be considered as a small solid sphere with volume charge density $-\rho$.

In this regard, we have:

$$\rho = \frac{Q}{\frac{4}{3}\pi a^3}$$

By applying Gauss's law for the spherical surface related to the large sphere, shown in Fig. 6.38 (b), we have:

$$Q_{in} = \int \vec{D} \cdot \vec{dS}$$

$$\Rightarrow Q = \int \varepsilon_0 \vec{E} \cdot \vec{dS}$$

$$\Rightarrow Q = \int \varepsilon_0 \vec{E} \cdot \vec{dS}$$

$$\Rightarrow Q = \int_{\varphi=0}^{2\pi} \int_{\theta=0}^{\pi} \varepsilon_0 E_R R^2 \sin\theta \, d\theta \, d\varphi$$

$$\Rightarrow Q = \varepsilon_0 E_R R^2 \left[-\cos\theta \right]_0^{\pi} \left[\varphi \right]_0^{2\pi}$$

$$\Rightarrow Q = \varepsilon_0 E_R R^2 (-\cos\pi + \cos 0)(2\pi - 0)$$

$$\Rightarrow Q = \varepsilon_0 E_R R^2 (-(-1) + 1)(2\pi)$$

$$\Rightarrow E_R = \frac{Q}{4\pi\varepsilon_0 R^2}$$

$$\Rightarrow \vec{E} = \frac{Q}{4\pi\varepsilon_0 R^2}\hat{R}$$

$$\Rightarrow \vec{E_p} = \vec{E}\bigg|_{R=4a} = \frac{Q}{4\pi\varepsilon_0 (4a)^2}\hat{y}$$

$$\Rightarrow \vec{E_p} = \frac{Q}{16\pi\varepsilon_0 a^2}\hat{y}$$

By applying Gauss's law for the spherical surface related to the small sphere, shown in Fig. 6.38 (c), we have:

$$Q_{in} = \int \vec{D}\cdot\vec{dS}$$

$$\Rightarrow \int \rho dV = \int \varepsilon_0 \vec{E}\cdot\vec{dS}$$

$$\Rightarrow \left(-\frac{Q}{\frac{4}{3}\pi a^3}\right)\left(\frac{4}{3}\pi\left(\frac{a}{2}\right)^3\right) = \int_{\varphi=0}^{2\pi}\int_{\theta=0}^{\pi} \varepsilon_0 E_R R^2 \sin\theta d\theta d\varphi$$

$$\Rightarrow -\frac{Q}{8} = \varepsilon_0 E_R R^2 \left[-\cos\theta\right]_0^\pi \left[\varphi\right]_0^{2\pi}$$

$$\Rightarrow -\frac{Q}{8} = \varepsilon_0 E_R R^2 (-\cos\pi + \cos 0)(2\pi - 0)$$

$$\Rightarrow -\frac{Q}{8} = \varepsilon_0 E_R R^2 (-(-1) + 1)(2\pi)$$

$$\Rightarrow E_R = \frac{-Q}{32\pi\varepsilon_0 R^2}$$

$$\Rightarrow \vec{E} = \frac{-Q}{32\pi\varepsilon_0 R^2}\hat{R}$$

$$\Rightarrow \vec{E_p} = \vec{E}\bigg|_{R=\frac{7}{2}a} = \frac{-Q}{32\pi\varepsilon_0 \left(\frac{7}{2}a\right)^2}\hat{y}$$

$$\Rightarrow \vec{E_p} = -\frac{Q}{392\pi\varepsilon_0 a^2}\hat{y}$$

6.4 Electric Field in Spherical Coordinate System

The total electric field at point p is calculated as follows.

$$\Rightarrow \overrightarrow{E_p} = \frac{Q}{16\pi\varepsilon_0 a^2}\hat{y} - \frac{Q}{392\pi\varepsilon_0 a^2}\hat{y}$$

$$\Rightarrow \overrightarrow{E_p} = \left(\frac{1}{16} - \frac{1}{392}\right)\frac{Q}{\pi\varepsilon_0 a^2}\hat{y} = \left(\frac{49 - 2}{784}\right)\frac{Q}{\pi\varepsilon_0 a^2}\hat{y}$$

$$\Rightarrow \overrightarrow{E_p} = \frac{47Q}{784\pi\varepsilon_0 a^2}\hat{y}$$

Choice (4) is the answer.

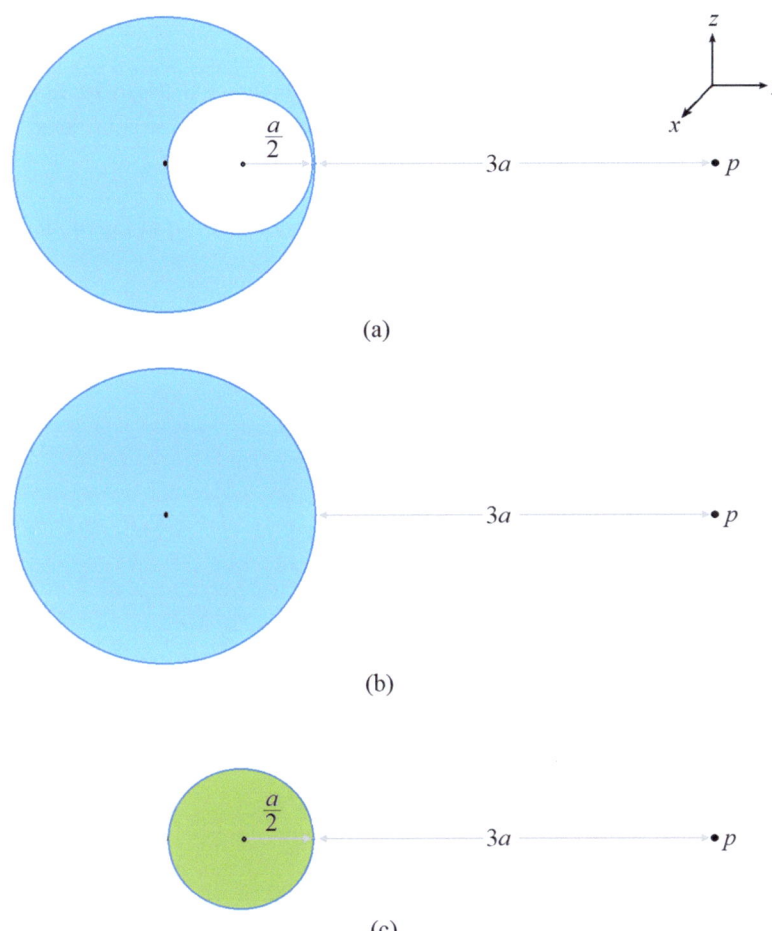

Fig. 6.38 (a) A solid sphere including a spherical hole, where the charge Q has been uniformly distributed across it. (b) Applying Gauss's law for the large sphere. (c) Applying Gauss's law for the small sphere

> **Notes**
>
> In this problem, the relations below have been used.
>
> $$\int \sin x \, dx = -\cos x$$
>
> $$\cos \pi = -1$$
>
> $$\cos 0 = 1$$

6.51. The problem can be solved by using the superposition principle, as is shown in Fig. 6.39. In fact, the large sphere can be considered a solid sphere with volume charge density ρ and the spherical hole can be considered a small solid sphere with volume charge density $-\rho$.

By applying Gauss's law for the spherical surface related to the large sphere, shown in Fig. 6.39 (b), we have:

$$Q_{in} = \int \vec{D} \cdot \vec{dS}$$

$$\Rightarrow \int \rho dV = \int \varepsilon_0 \vec{E} \cdot \vec{dS}$$

$$\Rightarrow \int_{\varphi=0}^{2\pi} \int_{\theta=0}^{\pi} \int_{R=0}^{R} \rho R^2 \sin\theta \, dR \, d\theta \, d\varphi = \int_{\varphi=0}^{2\pi} \int_{\theta=0}^{\pi} \varepsilon_0 E_R R^2 \sin\theta \, d\theta \, d\varphi$$

$$\Rightarrow \rho \left[\frac{R^3}{3}\right]_0^R \int_{\varphi=0}^{2\pi} \int_{\theta=0}^{\pi} \sin\theta \, d\theta \, d\varphi = \varepsilon_0 E_R R^2 \int_{\varphi=0}^{2\pi} \int_{\theta=0}^{\pi} \sin\theta \, d\theta \, d\varphi$$

$$\Rightarrow \rho \left[\frac{R^3}{3}\right]_0^R = \varepsilon_0 E_R R^2$$

$$\Rightarrow \rho \left(\frac{R^3}{3} - 0\right) = \varepsilon_0 E_R R^2$$

$$\Rightarrow \rho \frac{R}{3} = \varepsilon_0 E_R$$

$$\Rightarrow E_R = \frac{\rho R}{3\varepsilon_0}$$

$$\Rightarrow \vec{E} = \frac{\rho R}{3\varepsilon_0} \hat{R}$$

6.4 Electric Field in Spherical Coordinate System

$$\Rightarrow \overrightarrow{E_p} = \left. \vec{E} \right|_{R=\frac{a}{2}} = \frac{\rho a}{6\varepsilon_0}\hat{y}$$

For the small sphere, shown in Fig. 6.39 (c), the Gaussian surface does not include any charge because the electric field at its center needs to be calculated. In other words:

$$\overrightarrow{E_p} = 0$$

The total electric field at point p can be calculated as follows.

$$\Rightarrow \overrightarrow{E_p} = \frac{\rho a}{6\varepsilon_0}\hat{y} + 0$$

$$\Rightarrow \overrightarrow{E_p} = \frac{\rho a}{6\varepsilon_0}\hat{y}$$

Choice (3) is the answer.

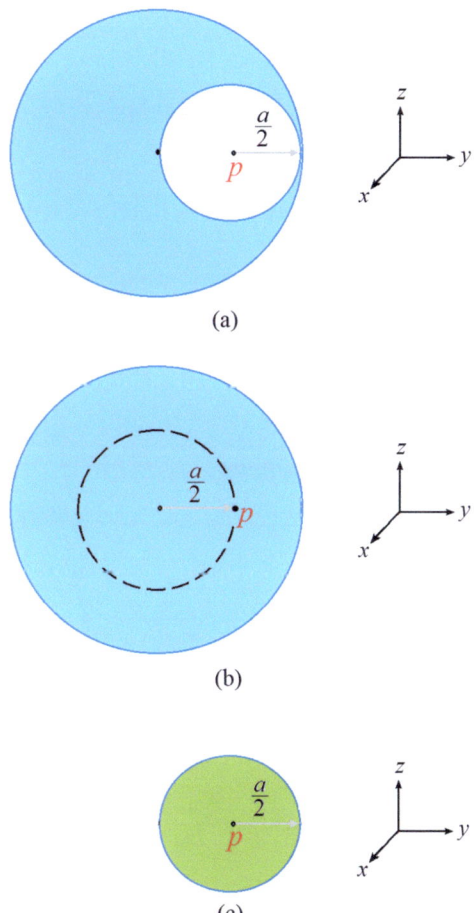

Fig. 6.39 (a) A solid sphere including a spherical hole, where the charge Q has been uniformly distributed across it. (b) Applying Gauss's law for the large sphere. (c) Applying Gauss's law for the small sphere

Notes

In this problem, the relation below has been used.

$$\int x^n dx = \frac{x^{n+1}}{n+1}$$

6.53. Based on the information given in the problem, we have:

$$\rho = \frac{1}{R}$$

The problem can be solved by applying Gauss's law for the Gaussian surface shown in Fig. 6.40.

$$Q_{in} = \int \vec{D} \cdot \vec{dS}$$

$$\Rightarrow \int \rho dV = \int \varepsilon_0 \vec{E} \cdot \vec{dS}$$

$$\Rightarrow \int \frac{1}{R} dV = \int \varepsilon_0 \vec{E} \cdot \vec{dS}$$

$$\Rightarrow \int_{\varphi=0}^{2\pi} \int_{\theta=0}^{\pi} \int_{R=0}^{R} \frac{1}{R} R^2 \sin\theta dR d\theta d\varphi = \int_{\varphi=0}^{2\pi} \int_{\theta=0}^{\pi} \varepsilon_0 E_R R^2 \sin\theta d\theta d\varphi$$

$$\Rightarrow \int_{\varphi=0}^{2\pi} \int_{\theta=0}^{\pi} \int_{R=0}^{R} R \sin\theta dR d\theta d\varphi = \varepsilon_0 E_R R^2 \int_{\varphi=0}^{2\pi} \int_{\theta=0}^{\pi} \sin\theta d\theta d\varphi$$

$$\Rightarrow \left[\frac{R^2}{2}\right]_0^R \int_{\varphi=0}^{2\pi} \int_{\theta=0}^{\pi} \sin\theta d\theta d\varphi = \varepsilon_0 E_R R^2 \int_{\varphi=0}^{2\pi} \int_{\theta=0}^{\pi} \sin\theta d\theta d\varphi$$

$$\Rightarrow \frac{R^2}{2} - 0 = \varepsilon_0 E_R R^2$$

$$\Rightarrow E_R = \frac{1}{2\varepsilon_0}$$

$$\Rightarrow \vec{E} = \frac{1}{2\varepsilon_0} \hat{R}$$

As can be seen, the electric field inside the sphere is constant although the volume charge density is nonuniform. Choice (1) is the answer.

6.4 Electric Field in Spherical Coordinate System

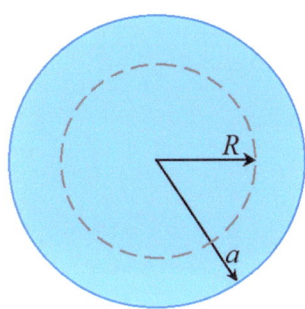

Fig. 6.40 Applying Gauss's law for the spherical Gaussian surface

> **Notes**
> In this problem, the relation below has been used.
> $$\int x^n dx = \frac{x^{n+1}}{n+1}$$

6.54. The problem can be solved by applying Gauss's law as follows.

$$Q_{in} = \int \vec{D} \cdot \vec{dS}$$

$$\Rightarrow Q_{in} = \int \varepsilon_0 \vec{E} \cdot \vec{dS}$$

By applying Gauss's law for the Gaussian surface ($R > b$), shown in Fig. 6.41, we have:

$$Q + Q_1 + Q_2 = \int_{\varphi=0}^{2\pi} \int_{\theta=0}^{\pi} \varepsilon_0 E_R R^2 \sin\theta d\theta d\varphi$$

$$\Rightarrow Q + \rho_{s1} 4\pi a^2 + \rho_{s2} 4\pi b^2 = \varepsilon_0 E_R R^2 \Big[-\cos\theta\Big]_0^{\pi}\Big[\varphi\Big]_0^{2\pi}$$

$$\Rightarrow Q + \rho_{s1} 4\pi a^2 + \rho_{s2} 4\pi b^2 = \varepsilon_0 E_R R^2 (-\cos\pi + \cos 0)(2\pi - 0)$$

$$\Rightarrow Q + \rho_{s1} 4\pi a^2 + \rho_{s2} 4\pi b^2 = \varepsilon_0 E_R R^2 (-(-1) + 1)(2\pi)$$

$$\Rightarrow Q + \rho_{s1} 4\pi a^2 + \rho_{s2} 4\pi b^2 = 4\pi \varepsilon_0 E_R R^2$$

$$\Rightarrow E_R = \frac{Q + \rho_{s1} 4\pi a^2 + \rho_{s2} 4\pi b^2}{4\pi \varepsilon_0 R^2} = \frac{Q}{4\pi \varepsilon_0 R^2} + \frac{\rho_{s1} a^2}{\varepsilon_0 R^2} + \frac{\rho_{s2} b^2}{\varepsilon_0 R^2}$$

$$\Rightarrow \vec{E} = \left(\frac{Q}{4\pi\varepsilon_0 R^2} + \frac{\rho_{s1} a^2}{\varepsilon_0 R^2} + \frac{\rho_{s2} b^2}{\varepsilon_0 R^2} \right) \hat{R}$$

Choice (2) is the answer.

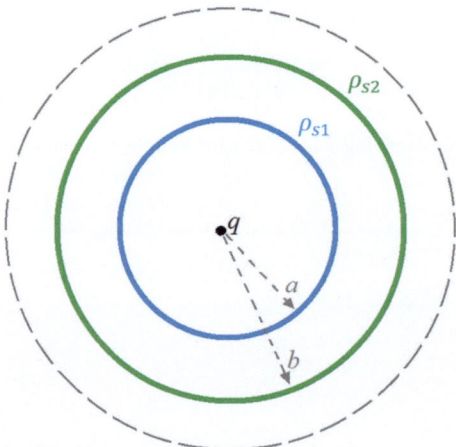

Fig. 6.41 Applying Gauss's law for the Gaussian surface ($R > b$) related to the system including a point charge placed at the origin and two concentric spherical shells centered at the origin

> **Notes**
> In this problem, the relations below have been used.
>
> $$\int \sin x \, dx = -\cos x$$
>
> $$\cos \pi = -1$$
>
> $$\cos 0 = 1$$

6.55. According to the results of Gauss's law, the electric field in a conductor is zero. Therefore, the electric field of the charge q_3 will not be sensed by the charge q_1. Hence, the electric force between them is zero. In other words:

$$F_{13} = 0$$

However, the electric force between the charges q_1 and q_2 can be calculated as follows.

$$F_{12} = \frac{q_1 q_2}{4\pi\varepsilon_0 \left(\frac{b}{2}\right)^2}$$

$$\Rightarrow F_{12} = \frac{q_1 q_2}{\pi \varepsilon_0 b^2}$$

Choice (4) is the answer (Fig. 6.42).

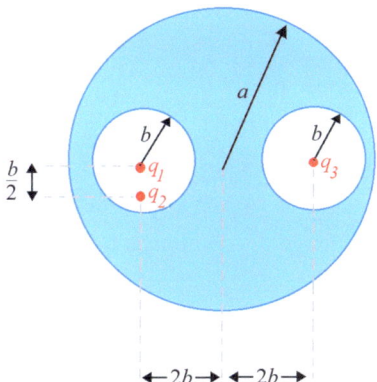

Fig. 6.42 A system including three point charges placed inside the two spherical holes of a conducting sphere

6.56. Based on the information given in the problem, we have:

$$\vec{E} = \frac{1}{R^2} e^{-\frac{R}{R_0}} \hat{R} \tag{6.11}$$

Maxwell's first equation states that:

$$\nabla \cdot \vec{D} = \rho \tag{6.12}$$

$$\Rightarrow \nabla \cdot \left(\varepsilon_0 \varepsilon_r \vec{E} \right) = \rho \tag{6.13}$$

$$\Rightarrow \nabla \cdot \left(\varepsilon_0 (1) \vec{E} \right) = \rho \tag{6.14}$$

$$\Rightarrow \nabla \cdot \vec{E} = \frac{\rho}{\varepsilon_0} \tag{6.15}$$

The relation of divergence for the vector field $\vec{F}(R, \theta, \varphi) = F_R \hat{R} + F_\theta \hat{\theta} + F_\varphi \hat{\varphi}$ in spherical coordinate system is as follows.

$$\nabla \cdot \vec{F} = \frac{1}{R^2} \frac{\partial}{\partial R} \left(R^2 F_R \right) + \frac{1}{R \sin \theta} \frac{\partial}{\partial \theta} (\sin \theta F_\theta) + \frac{1}{R \sin \theta} \frac{\partial}{\partial \varphi} F_\varphi \tag{6.16}$$

Solving (6.11), (6.15), and (6.16):

$$\frac{1}{R^2}\frac{\partial}{\partial R}\left[R^2\left(\frac{1}{R^2}e^{-\frac{R}{R_0}}\right)\right]=\frac{\rho}{\varepsilon_0}$$

$$\Rightarrow \frac{1}{R^2}\frac{\partial}{\partial R}\left[e^{-\frac{R}{R_0}}\right]=\frac{\rho}{\varepsilon_0}$$

$$\Rightarrow \frac{1}{R^2}\left(-\frac{1}{R_0}e^{-\frac{R}{R_0}}\right)=\frac{\rho}{\varepsilon_0}$$

$$\Rightarrow \rho=-\frac{\varepsilon_0}{R_0}\frac{1}{R^2}e^{-\frac{R}{R_0}}$$

Now, we can calculate the total charge of this charge distribution as follows.

$$Q=\int \rho dV$$

$$\Rightarrow Q=\int_{\varphi=0}^{2\pi}\int_{\theta=0}^{\pi}\int_{R=0}^{R}\left(-\frac{\varepsilon_0}{R_0}\frac{1}{R^2}e^{-\frac{R}{R_0}}\right)R^2\sin\theta dRd\theta d\varphi$$

$$\Rightarrow Q=-\frac{\varepsilon_0}{R_0}\int_{\varphi=0}^{2\pi}\int_{\theta=0}^{\pi}\int_{R=0}^{R}\left(e^{-\frac{R}{R_0}}\right)\sin\theta dRd\theta d\varphi$$

$$\Rightarrow Q=-\frac{\varepsilon_0}{R_0}\left[-R_0 e^{-\frac{R}{R_0}}\right]_0^R\left[-\cos\theta\right]_0^\pi\left[\varphi\right]_0^{2\pi}$$

$$\Rightarrow Q=\varepsilon_0\left(e^{-\frac{R}{R_0}}-1\right)(-\cos\pi+\cos 0)(2\pi-0)$$

$$\Rightarrow Q=\varepsilon_0\left(e^{-\frac{R}{R_0}}-1\right)(-(-1)-1)(2\pi)$$

$$\Rightarrow Q=4\pi\varepsilon_0\left(e^{-\frac{R}{R_0}}-1\right)$$

Choice (4) is the answer.

> **Notes**
>
> In this problem, the relations below have been used.
>
> $$\frac{d}{du}e^{u(x)} = e^{u(x)}du$$
>
> $$\int \sin x \, dx = -\cos x$$
>
> $$\cos \pi = -1$$
>
> $$\cos 0 = 1$$
>
> $$\int e^{u(x)} du = e^{u(x)}$$

References

1. Rahmani-Andebili, M., General Physics II – Practice Problems, Methods, and Solutions, Springer Nature, 2025.
2. Rahmani-Andebili, M., General Physics I – Practice Problems, Methods, and Solutions, Springer Nature, 2025.
3. Rahmani-Andebili, M., Mathematics of Engineering and Science – Practice Problems, Methods, and Solutions, Springer Nature, 2024.
4. Rahmani-Andebili, M., Differential Equations – Practice Problems, Methods, and Solutions, Springer Nature, 2022.
5. Rahmani-Andebili, M., Calculus III – Practice Problems, Methods, and Solutions, Springer Nature, 2023.
6. Rahmani-Andebili, M., Calculus II – Practice Problems, Methods, and Solutions, Springer Nature, 2023.
7. Rahmani-Andebili, M., Calculus I (2nd Ed.) – Practice Problems, Methods, and Solutions, Springer Nature, 2023.
8. Rahmani-Andebili, M., Precalculus (2nd Ed.) – Practice Problems, Methods, and Solutions, Springer Nature, 2024.

7

Electric Potential: Part A

Abstract

In this chapter, the basic and advanced problems of electric potential are studied. The subjects include electric potential of uniform and nonuniform charge distributions in the Cartesian, cylindrical, and spherical coordinate systems. Herein, different types of problems and exercises are presented that are categorized as follows.

- **Problems with detailed solution**: They have been designed to teach students the subjects in detail. Moreover, they have been categorized in different levels based on their difficulty levels (easy, normal, and hard) and calculation amounts (small, normal, and large).
- **Partially solved exercises**: They have been designed to encourage students to practice problems while guiding them through the problem-solving procedure and hinting the required formulas.
- **Exercises with final answer**: They have been designed to encourage students to practice more by themselves while hinting them by the final answer as well as to help instructors to give tests or quizzes.

7.1 Electric Potential in the Cartesian Coordinate System

Problem

7.1. Figure 7.1 shows a closed surface that does not include any charge. The electric potential of the surface is V_0. Which of the following choices is correct about the electric potential and field inside the closed surface.

Difficulty level   ○ Easy   ○ Normal   ● Hard
Calculation amount ● Small ○ Normal ○ Large

1) The electric potential inside the closed surface is zero.
2) The electric field inside the closed surface is zero.
3) The electric potential inside the closed surface is nonuniform.
4) The electric field inside the closed surface is constant and nonzero.

Fig. 7.1 A closed surface without any electric charge

Problem

7.2. As is shown in Fig. 7.2, a thin rod with length l and total charge q, which is uniformly distrusted across it, is placed on x-axis in free space. Calculate the electric potential at the origin [1–8].

Difficulty level   ○ Easy   ● Normal   ○ Hard
Calculation amount   ○ Small   ● Normal   ○ Large

1) $V = \dfrac{q}{4\pi\varepsilon_0 l} \ln 2$
2) $V = \dfrac{q}{2\pi\varepsilon_0 l} \ln 2$
3) $V = \dfrac{q}{2\pi\varepsilon_0} \ln 2$
4) $V = \dfrac{q}{4\pi\varepsilon_0} \ln 2$

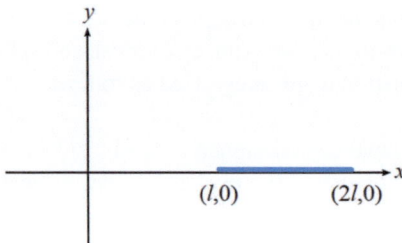

Fig. 7.2 A thin rod with total charge q uniformly distrusted across it and placed in free space

Exercise

7.3. In Problem 7.2, calculate the electric potential as a function of linear charge density ρ_l.

Final Answer

$V = \dfrac{\rho_l}{4\pi\varepsilon_0} \ln 2$

Partially Solved Exercise

7.4. In Problem 7.2, assume that the linear charge density is $\rho_l = x$. Calculate the electric potential at the origin.

Solution

As is shown in Fig. 7.3, a differential length produces a differential electric potential at the origin which is as follows.

$$dV = \frac{dq}{4\pi\varepsilon_0 x}$$

For $dq = \rho_l dx$, we have:

$$dV = \frac{\rho_l dx}{4\pi\varepsilon_0 x}$$

For $\rho_l = x$, we have:

$$dV = \frac{dx}{4\pi\varepsilon_0}$$

7.2 Electric Potential in Cylindrical Coordinate System

Now, the total electric potential can be calculated as follows.

$$V = \int dV$$

$$\Rightarrow V = \int_{x=l}^{x=2l} (\quad)$$

$$\Rightarrow V = \frac{1}{4\pi\varepsilon_0} \int_{x=l}^{x=2l} dx$$

$$\Rightarrow V = \frac{1}{4\pi\varepsilon_0} \left[(\quad) \right]_{l}^{2l}$$

$$\Rightarrow V = \frac{1}{4\pi\varepsilon_0} ((\quad) - (\quad))$$

$$\Rightarrow V = \frac{l}{4\pi\varepsilon_0}$$

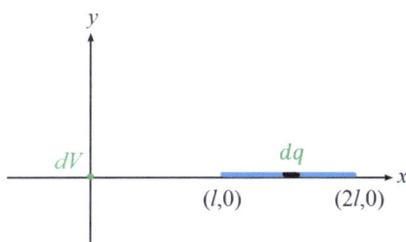

Fig. 7.3 A differential length produces a differential electric potential at the origin

7.2 Electric Potential in Cylindrical Coordinate System

Problem

7.5. As is shown in Fig. 7.4, a thin ring with radius a and total charge q, which is uniformly distributed across it, is placed at the origin in free space. Calculate the electric potential at the center of the ring.
Difficulty level   ○ Easy   ● Normal   ○ Hard
Calculation amount   ○ Small   ● Normal   ○ Large
1) $V = 0$
2) $V = \dfrac{q}{4\pi\varepsilon_0 a}$
3) $V = \dfrac{q}{8\pi^2 \varepsilon_0 a}$
4) $V = \dfrac{q}{2\pi\varepsilon_0 a}$

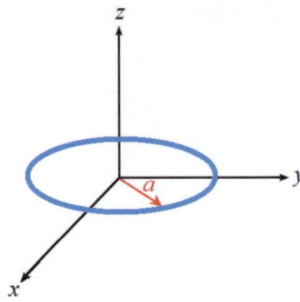

Fig. 7.4 A thin ring with charge q uniformly distributed across it and placed in free space

Exercise

7.6. In Problem 7.5, calculate the electric potential as a function of linear charge density ρ_l.

Final Answer

$$V = \frac{\rho_l}{2\varepsilon_0}$$

Problem

7.7. As is shown in Fig. 7.5, a very long thin rod with linear charge density ρ_l is available. Calculate the electric potential difference between the two points at $r = r_1$ and $r = r_2$.

Difficulty level — ○ Easy ● Normal ○ Hard
Calculation amount — ○ Small ● Normal ○ Large

1) $V_2 - V_1 = \dfrac{\rho_l(r_2 - r_1)}{2\pi\varepsilon_0}$
2) $V_2 - V_1 = \dfrac{\rho_l}{2\pi\varepsilon_0} \ln \dfrac{r_1}{r_2}$
3) $V_2 - V_1 = \dfrac{\rho_l}{2\pi\varepsilon_0} \ln \dfrac{r_2}{r_1}$
4) $V_2 - V_1 = 0$

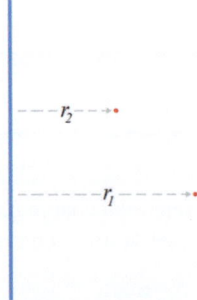

Fig. 7.5 A long thin rod with linear charge density

Problem

7.8. The linear charge density of the semicircle ring, which is placed at the origin and shown in Fig. 7.6, is $\rho_l(x) = 6x^2$ in the SI system. Calculate the electric potential at the origin.

Difficulty level — ○ Easy ○ Normal ● Hard
Calculation amount — ○ Small ● Normal ○ Large

7.2 Electric Potential in Cylindrical Coordinate System

1) $V = 0$
2) $V = \dfrac{3a}{2\varepsilon_0}$
3) $V = \dfrac{3a}{4\varepsilon_0}$
4) $V = \dfrac{3a^2}{4\varepsilon_0}$

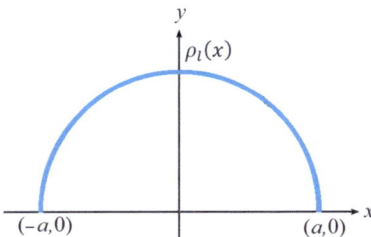

Fig. 7.6 A semicircle ring with linear charge density

Partially Solved Exercise

7.9. Solve Problem 7.8 by assuming that $\rho_l = \rho_0$.

Solution

As is shown in Fig. 7.7, a differential length produces a differential electric potential at the origin that can be calculated as follows.

$$dV = \frac{dq}{4\pi\varepsilon_0 a}$$

Herein, the differential charge can be calculated as follows.

$$dq = \rho_l dl = \rho_0 a d\varphi$$

The total electric potential at the origin can be calculated as follows.

$$V = \int dV$$

$$\Rightarrow V = \int_{\varphi=0}^{\varphi=\pi} (\quad)$$

$$\Rightarrow V = \frac{\rho_0}{4\pi\varepsilon_0} \int_{\varphi=0}^{\varphi=\pi} (\quad)$$

$$\Rightarrow V = \frac{\rho_0}{4\pi\varepsilon_0} \left[(\quad) \right]_0^{\pi}$$

$$\Rightarrow V = \frac{\rho_0}{4\pi\varepsilon_0}((\quad) - (\quad))$$

$$\Rightarrow V = \frac{\rho_0}{4\varepsilon_0}$$

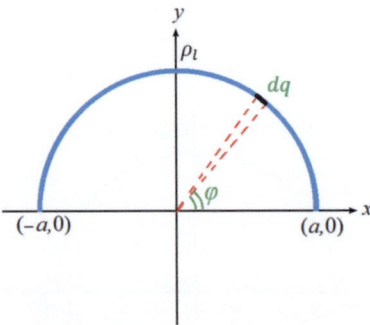

Fig. 7.7 A differential length produces a differential electric potential at the origin

Problem

7.10. Figure 7.8 shows a solid cylinder with radius a, very long length l, and total charge Q which is uniformly distributed across it. This solid cylinder is placed inside a hollow cylinder with inner and outer radiuses $2a$ and $4a$, respectively, and charge $-Q$ that is uniformly distributed across it. The solid and hollow cylinders are concentric. Which one of the following choices is correct about the electric potential in the region $r > 4a$?

Difficulty level ● Easy ○ Normal ○ Hard
Calculation amount ● Small ○ Normal ○ Large

1) $V = 0$
2) $V = \dfrac{Qr}{2\pi\varepsilon_0 a^2 l}$
3) $V = \dfrac{Q}{2\pi\varepsilon_0 l} \ln\left(\dfrac{4a}{r}\right)$
4) $V = \dfrac{Q}{2\pi\varepsilon_0 l} \left(\ln\left(\dfrac{2a}{r}\right) - \dfrac{1}{2}\right)$

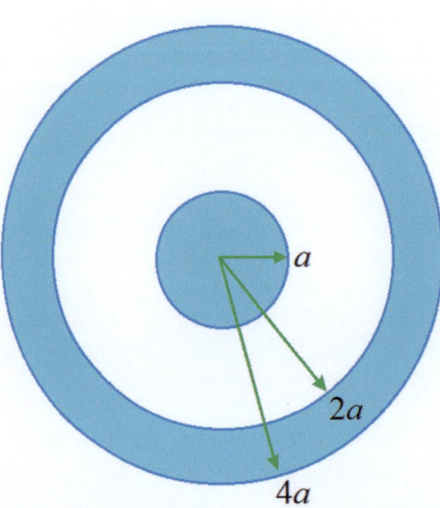

Fig. 7.8 A solid cylinder with radius a, length l, and total charge Q, which is uniformly distributed, placed inside a hollow cylinder with charge $-Q$ which is uniformly distributed across it

7.3 Electric Potential in Spherical Coordinate System

Problem

7.11. In Problem 7.10, which one of the following choices is correct about the electric potential in the region $2a < r < 4a$?

Difficulty level ○ Easy ○ Normal ● Hard
Calculation amount ○ Small ○ Normal ● Large

1) $V = \dfrac{Q}{2\pi\varepsilon_0 l}\left[\dfrac{4}{3}\ln\left(\dfrac{4a}{r}\right) + \dfrac{2}{3}\right]$

2) $V = \dfrac{Q}{2\pi\varepsilon_0 l}\left[\dfrac{4}{3}\ln\left(\dfrac{4a}{r}\right) + \dfrac{r^2}{24a^2}\right]$

3) $V = \dfrac{Q}{2\pi\varepsilon_0 l}\left[\dfrac{4}{3}\ln\left(\dfrac{4a}{r}\right) + \dfrac{2}{3} - \dfrac{r^2}{24a^2}\right]$

4) $V = \dfrac{Q}{2\pi\varepsilon_0 l}\left[\dfrac{4}{3}\ln\left(\dfrac{4a}{r}\right) - \dfrac{2}{3} + \dfrac{r^2}{24a^2}\right]$

Problem

7.12. In Problem 7.10, which one of the following choices is correct about the electric potential in the region $a < r < 2a$?

Difficulty level ○ Easy ○ Normal ● Hard
Calculation amount ○ Small ○ Normal ● Large

1) $V = 0$

2) $V = \dfrac{Q}{2\pi\varepsilon_0 l}\left(\ln\left(\dfrac{2a}{r}\right) + \dfrac{4}{3}\ln 2 - \dfrac{1}{2}\right)$

3) $V = \dfrac{Q}{2\pi\varepsilon_0 l}\left(\ln\left(\dfrac{2a}{r}\right) - \dfrac{4}{3}\ln 2 - \dfrac{1}{2}\right)$

4) $V = \dfrac{Q}{2\pi\varepsilon_0 l}\left(\ln\left(\dfrac{2a}{r}\right) + \dfrac{4}{3}\ln 2 + \dfrac{1}{2}\right)$

Problem

7.13. In Problem 7.10, which one of the following choices is correct about the electric potential in the region $r \leq a$?

Difficulty level ○ Easy ○ Normal ● Hard
Calculation amount ○ Small ○ Normal ● Large

1) $V = \dfrac{Q}{2\pi\varepsilon_0 l}\left(-\dfrac{r^2}{2a^2} + \ln 2\right)$

2) $V = 0$

3) $V = \dfrac{Q}{2\pi\varepsilon_0 l}\left(-\dfrac{r^2}{2a^2} + \dfrac{7}{3}\ln 2\right)$

4) $V = \dfrac{Q}{2\pi\varepsilon_0 l}\left(\dfrac{r^2}{2a^2} + \dfrac{7}{3}\ln 2\right)$

7.3 Electric Potential in Spherical Coordinate System

Problem

7.14. In Fig. 7.9, when the switch is closed the current flows from the first conducting sphere to the second conducting one. Which one of the following choices is correct about them?

Difficulty level   ● Easy   ○ Normal   ○ Hard
Calculation amount   ● Small   ○ Normal   ○ Large
1) $Q_1 > Q_2$
2) $Q_1 < Q_2$
3) $V_1 > V_2$
4) $V_1 < V_2$

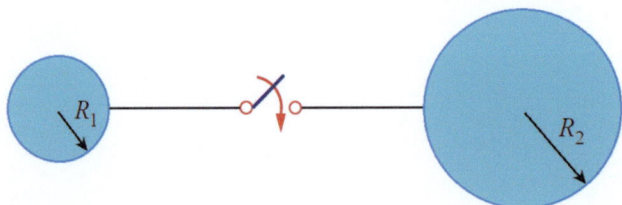

Fig. 7.9 Two spherical conductors connected to each other by a switch

Problem

7.15. In Fig. 7.10, before closing the switch, the initial charge of the conducting spheres is Q_1 and Q_2, and after closing the switch, the final charge of the conducting spheres is Q'_1 and Q'_2. Which one of the following choices is correct about them?

(a) $\dfrac{Q_1}{Q_2} = \dfrac{R_1}{R_2}$

(b) $\dfrac{Q'_1}{Q'_2} = \dfrac{R_1}{R_2}$

(c) $\dfrac{Q_1}{Q_2} = \dfrac{R_2}{R_1}$

(d) $\dfrac{Q'_1}{Q'_2} = \dfrac{R_2}{R_1}$

(e) $Q_1 + Q_2 = Q'_1 + Q'_2$

Difficulty level   ○ Easy   ● Normal   ○ Hard
Calculation amount   ○ Small   ● Normal   ○ Large
1) (*a*) and (*e*)
2) (*d*) and (*e*)
3) (*b*) and (*e*)
4) (*a*) and (*b*)

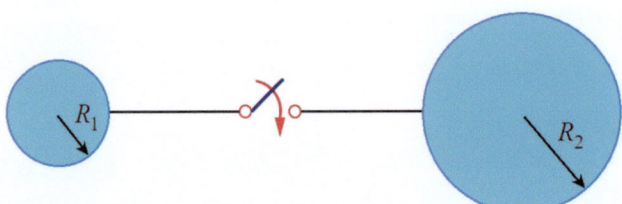

Fig. 7.10 Two spherical conductors connected to each other by a switch

7.3 Electric Potential in Spherical Coordinate System

Partially Solved Exercise

7.16. In Fig. 7.11, before closing the switch, the initial charge of the conducting spheres is Q_1 and Q_2, and after closing the switch, the final charge of the conducting spheres is Q'_1 and Q'_2. Which one of the following choices is correct about their charges? Herein, assume that $R_2 = 2R_1$.

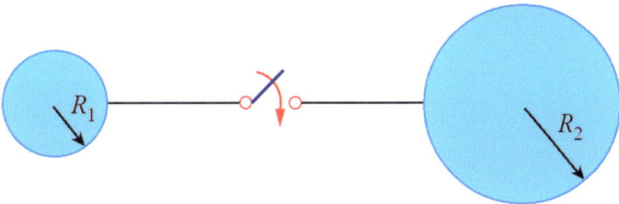

Fig. 7.11 Two spherical conductors connected to each other by a switch

Solution

After closing the switch, the electric potential of the conducting spheres will be equal. In other words:

$$V'_1 = V'_2$$

On the other hand, we know, the electric potential of a spherical conductor with total charge Q and radius R can be calculated as follows.

$$V = \frac{Q}{4\pi\varepsilon_0 R}$$

Hence:

$$\frac{Q'_1}{4\pi\varepsilon_0 R_1} = \frac{Q'_2}{4\pi\varepsilon_0 R_2}$$

$$\Rightarrow \frac{Q'_1}{R_1} = \frac{Q'_2}{()R_1}$$

$$\Rightarrow Q'_2 = ()Q'_1 \tag{7.1}$$

On the other hand, based on the conservation charge principle, we have:

$$Q_1 + Q_2 = Q'_1 + Q'_2 \tag{7.1}$$

Solving (7.1) and (7.2):

$$Q_1 + Q_2 = Q'_1 + ()Q'_1 = ()Q'_1$$

$$\Rightarrow Q'_1 = \frac{1}{3}(Q_1 + Q_2) \tag{7.1}$$

Solving (7.1) and (7.3):

$$Q'_2 = \frac{2}{3}(Q_1 + Q_2)$$

Problem

7.17. Figure 7.12 shows a single conducting solid sphere with radius a and total charge Q. Calculate the electric potential in the regions $R \leq a$ and $R > a$?

Difficulty level — ○ Easy ○ Normal ● Hard
Calculation amount — ○ Small ○ Normal ● Large

1) $V = \begin{cases} \dfrac{Q}{4\pi\varepsilon_0 a} & R \leq a \\ \dfrac{Q}{4\pi\varepsilon_0 R} & R > a \end{cases}$

2) $V = \begin{cases} \dfrac{Q}{4\pi\varepsilon_0 R} & R \leq a \\ \dfrac{Q}{4\pi\varepsilon_0 R} & R > a \end{cases}$

3) $V = \begin{cases} \dfrac{Q}{4\pi\varepsilon_0 R^2} & R \leq a \\ \dfrac{Q}{4\pi\varepsilon_0 R} & R > a \end{cases}$

4) $V = \begin{cases} \dfrac{Q}{4\pi\varepsilon_0 a} & R \leq a \\ \dfrac{Q}{4\pi\varepsilon_0 R^2} & R > a \end{cases}$

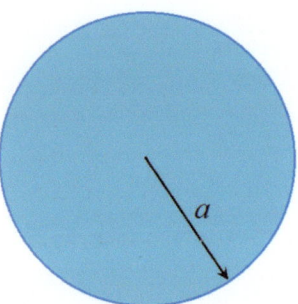

Fig. 7.12 A conducting solid sphere with total charge Q

Exercise

7.18. In Problem 7.17, calculate the electric potential as a function of surface charge density ρ_s.

Final Answer

$V = \begin{cases} \dfrac{\rho_s a}{\varepsilon_0} & R \leq a \\ \dfrac{\rho_s a^2}{\varepsilon_0 R} & R > a \end{cases}$

7.3 Electric Potential in Spherical Coordinate System

Problem

7.19. Figure 7.13 shows a single solid sphere with radius a and total charge Q which is uniformly distributed across it. Calculate the electric potential in the regions $R \leq a$ and $R > a$?

Difficulty level ○ Easy ○ Normal ● Hard
Calculation amount ○ Small ○ Normal ● Large

1) $V = \begin{cases} \dfrac{1}{4\pi\varepsilon_0}\dfrac{Q}{2a}\left(3 - \left(\dfrac{R}{a}\right)^3\right) & R \leq a \\ \dfrac{1}{4\pi\varepsilon_0}\dfrac{Q}{R} & R > a \end{cases}$

2) $V = \begin{cases} \dfrac{1}{4\pi\varepsilon_0}\dfrac{Q}{2a}\left(3 - \left(\dfrac{R}{a}\right)^2\right) & R \leq a \\ \dfrac{1}{4\pi\varepsilon_0}\dfrac{Q}{R} & R > a \end{cases}$

3) $V = \begin{cases} \dfrac{1}{4\pi\varepsilon_0}\dfrac{Q}{2a}\left(1 - \left(\dfrac{R}{a}\right)^2\right) & R \leq a \\ \dfrac{1}{4\pi\varepsilon_0}\dfrac{Q}{R} & R > a \end{cases}$

4) $V = \begin{cases} \dfrac{1}{4\pi\varepsilon_0}\dfrac{Q}{2a}\left(3 - \left(\dfrac{R}{a}\right)^2\right) & R \leq a \\ \dfrac{1}{4\pi\varepsilon_0}\dfrac{Q}{R^2} & R > a \end{cases}$

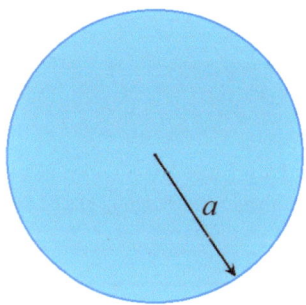

Fig. 7.13 A solid sphere with total charge Q which is uniformly distributed across it

Exercise

7.20. In Problem 7.19, calculate the electric potential as a function of volume charge density ρ_v.

Final Answer

$V = \begin{cases} \dfrac{\rho_v}{6\varepsilon_0}(3a^2 - R^2) & R \leq a \\ \dfrac{\rho_v a^3}{3\varepsilon_0 R} & R > a \end{cases}$

Problem

7.21. Figure 7.14 shows a conducting solid sphere with radius a and total charge Q which is inside a conducting hollow sphere with charge $-Q$ and inner and outer radiuses b and c, respectively. The solid and hollow spheres are concentric. Which one of the following choices is correct about the electric potential in the region $R > 4a$?

Difficulty level ● Easy ○ Normal ○ Hard
Calculation amount ● Small ○ Normal ○ Large

1) $V = \dfrac{Q}{4\pi\varepsilon_0 R}$
2) $V = 0$
3) $V = \dfrac{Q}{4\pi\varepsilon_0 R^2}$
4) $V = \dfrac{Q}{2\pi\varepsilon_0 R^2}$

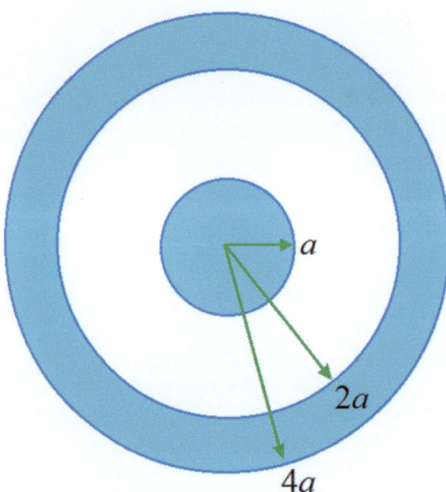

Fig. 7.14 A conducting solid sphere with radius a and total charge Q placed inside a conducting hollow sphere with charge $-Q$

Problem

7.22. In Problem 7.21, which one of the following choices is correct about the electric potential in the region $2a < R < 4a$?

Difficulty level ○ Easy ● Normal ○ Hard
Calculation amount ● Small ○ Normal ○ Large

1) $V = \dfrac{Q}{4\pi\varepsilon_0 R^2}$
2) $V = \dfrac{Q}{2\pi\varepsilon_0 R^2}$
3) $V = \dfrac{Q}{4\pi\varepsilon_0 R}$
4) $V = 0$

Problem

7.23. In Problem 7.21, which one of the following choices is correct about the electric potential in the region $a < R < 2a$?
Difficulty level ○ Easy ○ Normal ● Hard
Calculation amount ○ Small ● Normal ○ Large

1) $V = \dfrac{Q}{4\pi\varepsilon_0}\left(\dfrac{1}{R} - \dfrac{1}{2a}\right)$
2) 0
3) $V = \dfrac{Q}{4\pi\varepsilon_0}\left(\dfrac{1}{R} - \dfrac{1}{a}\right)$
4) $V = -\dfrac{Q}{4\pi\varepsilon_0}\left(\dfrac{1}{2R} - \dfrac{1}{2a}\right)$

Problem

7.24. In Problem 7.21, which one of the following choices is correct about the electric potential in the region $R \leq a$?
Difficulty level ○ Easy ● Normal ○ Hard
Calculation amount ○ Small ● Normal ○ Large

1) 0
2) $V = \dfrac{Q}{8\pi\varepsilon_0 a}$
3) $V = \dfrac{Q}{4\pi\varepsilon_0 a}$
4) $V = \dfrac{Q}{2\pi\varepsilon_0 a}$

Problem

7.25. Figure 7.15 shows a solid sphere with radius a and total charge Q which is uniformly distributed across it. This solid sphere is placed inside a hollow sphere with inner and outer radiuses $2a$ and $4a$, respectively, and charge $-Q$ that is uniformly distributed across it. The solid and hollow spheres are concentric. Which one of the following choices is correct about the electric potential in the region $R > 4a$?
Difficulty level ● Easy ○ Normal ○ Hard
Calculation amount ● Small ○ Normal ○ Large

1) $V = \dfrac{Q}{4\pi\varepsilon_0 R^2}$
2) $V = \dfrac{Q}{4\pi\varepsilon_0}\left[\dfrac{33}{28a} - \dfrac{R^2}{2a^3}\right]$
3) $V = 0$
4) $V = \dfrac{QR}{4\pi\varepsilon_0 a^3}$

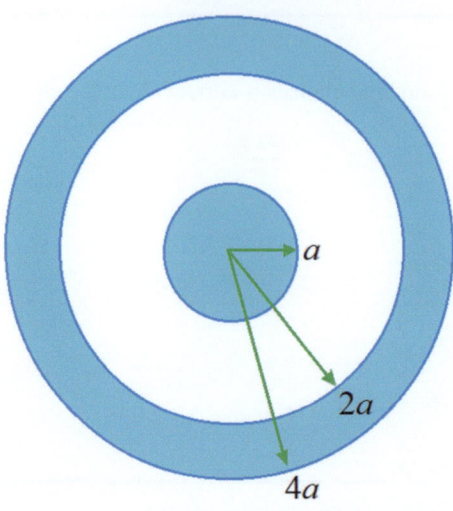

Fig. 7.15 A solid sphere with radius a and total charge Q, which is uniformly distributed, placed inside a hollow sphere with charge $-Q$ that is uniformly distributed across it

Problem

7.26. In Problem 7.25, which one of the following choices is correct about the electric potential in the region $2a < R < 4a$?
Difficulty level ○ Easy ○ Normal ● Hard
Calculation amount ○ Small ● Normal ○ Large

1) $V = \dfrac{Q}{4\pi\varepsilon_0}\left(-\dfrac{3}{7a} + \dfrac{8}{7R} + \dfrac{R^2}{112a^3}\right)$

2) $V = 0$

3) $V = \dfrac{Q}{4\pi\varepsilon_0}\left(-\dfrac{3}{7a} + \dfrac{8}{7R}\right)$

4) $V = \dfrac{Q}{4\pi\varepsilon_0}\left(\dfrac{3}{7a} + \dfrac{8}{7R} + \dfrac{R^2}{112a^3}\right)$

Problem

7.27. In Problem 7.25, which one of the following choices is correct about the electric potential in the region $a < R < 2a$?
Difficulty level ○ Easy ○ Normal ● Hard
Calculation amount ○ Small ○ Normal ● Large

1) $V = 0$

2) $V = \dfrac{Q}{4\pi\varepsilon_0}\left(\dfrac{3}{7a} + \dfrac{8}{7R} + \dfrac{R^2}{112a^3}\right)$

3) $V = \dfrac{Q}{4\pi\varepsilon_0 R}$

4) $V = \dfrac{Q}{4\pi\varepsilon_0}\left(\dfrac{1}{R} - \dfrac{9}{28a}\right)$

Problem

7.28. In Problem 7.25, which one of the following choices is correct about the electric potential in the region $R \leq a$?

Difficulty level ○ Easy ○ Normal ● Hard
Calculation amount ○ Small ○ Normal ● Large

1) $V = \dfrac{Q}{4\pi\varepsilon_0} \left(\dfrac{33}{28a} - \dfrac{R^2}{2a^3} \right)$

2) $V = \dfrac{Q}{4\pi\varepsilon_0} \left(\dfrac{1}{R} - \dfrac{9}{28a} \right)$

3) $V = \dfrac{Q}{4\pi\varepsilon_0} \left(\dfrac{1}{R} + \dfrac{9}{28a} \right)$

4) $V = 0$

References

1. Rahmani-Andebili, M., General Physics II – Practice Problems, Methods, and Solutions, Springer Nature, 2025.
2. Rahmani-Andebili, M., General Physics I – Practice Problems, Methods, and Solutions, Springer Nature, 2025.
3. Rahmani-Andebili, M., Mathematics of Engineering and Science – Practice Problems, Methods, and Solutions, Springer Nature, 2024.
4. Rahmani-Andebili, M., Differential Equations – Practice Problems, Methods, and Solutions, Springer Nature, 2022.
5. Rahmani-Andebili, M., Calculus III – Practice Problems, Methods, and Solutions, Springer Nature, 2023.
6. Rahmani-Andebili, M., Calculus II – Practice Problems, Methods, and Solutions, Springer Nature, 2023.
7. Rahmani-Andebili, M., Calculus I (2nd Ed.) – Practice Problems, Methods, and Solutions, Springer Nature, 2023.
8. Rahmani-Andebili, M., Precalculus (2nd Ed.) – Practice Problems, Methods, and Solutions, Springer Nature, 2024.

Electric Potential: Part B

Abstract

In this chapter, the problems of the seventh chapter are fully solved, in detail, step-by-step, and with different methods.

8.1 Electric Potential in the Cartesian Coordinate System

8.1. Since the electric potential of the surface is V_0, based on Gauss's law, the electric potential inside the closed surface is uniform and the same as V_0. Thus, Choices (1) and (3) are wrong. Moreover, based on Gauss's law, the electric field inside a closed surface is zero if there is no net electric charge inside the closed surface. Hence, Choice (4) is wrong, and Choice (2) is correct.

Choice (2) is the answer (Fig. 8.1).

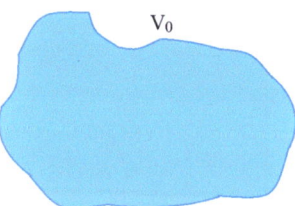

Fig. 8.1 A closed surface without any electric charge

8.2. The linear charge density of the rod can be calculated as follows [1–8].

$$\rho_l = \frac{q}{l}$$

As is shown in Fig. 8.2, a differential length produces a differential electric potential at the origin.

$$dV = \frac{dq}{4\pi\varepsilon_0 x}$$

Herein, the differential charge can be calculated as follows.

$$dq = \rho_l dx = \frac{q}{l}dx$$

Thus:

$$V = \int dV$$

$$\Rightarrow V = \int_{x=l}^{x=2l} \frac{\frac{q}{l}dx}{4\pi\varepsilon_0 x}$$

$$\Rightarrow V = \frac{q}{4\pi\varepsilon_0 l} \int_{x=l}^{x=2l} \frac{dx}{x}$$

$$\Rightarrow V = \frac{q}{4\pi\varepsilon_0 l} \Big[\ln x\Big]_{l}^{2l}$$

$$\Rightarrow V = \frac{q}{4\pi\varepsilon_0 l} (\ln 2l - \ln l)$$

$$\Rightarrow V = \frac{q}{4\pi\varepsilon_0 l} \ln 2$$

Choice (1) is the answer.

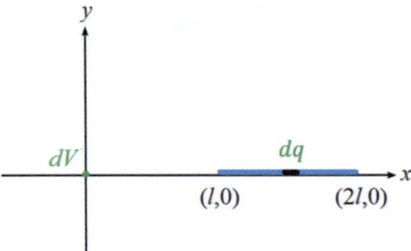

Fig. 8.2 A thin rod with total charge q uniformly distrusted across it and placed in free space

> **Notes**
>
> In this problem, the relations below have been used.
>
> $$\int \frac{dx}{x} = \ln x + c$$
>
> $$\ln a - \ln b = \ln\left(\frac{a}{b}\right)$$

8.2 Electric Potential in Cylindrical Coordinate System

8.5. The linear charge density of the rod can be calculated as follows.

$$\rho_l = \frac{q}{2\pi a}$$

As is shown in Fig. 8.3, a differential length produces a differential electric potential at the origin.

$$dV = \frac{dq}{4\pi\varepsilon_0 a}$$

Herein, the differential charge can be calculated as follows.

$$dq = \rho_l dl = \rho_l a d\varphi = \frac{q}{2\pi} d\varphi$$

Thus:

$$V = \int dV$$

$$\Rightarrow V = \int_{\varphi=0}^{\varphi=2\pi} \frac{\frac{q}{2\pi} d\varphi}{4\pi\varepsilon_0 a}$$

$$\Rightarrow V = \left(\frac{q}{2\pi}\right)\left(\frac{1}{4\pi\varepsilon_0 a}\right) \int_{\varphi=0}^{\varphi=2\pi} d\varphi$$

$$\Rightarrow V = \left(\frac{q}{2\pi}\right)\left(\frac{1}{4\pi\varepsilon_0 a}\right)(2\pi)$$

$$\Rightarrow V = \frac{q}{4\pi\varepsilon_0 a}$$

Choice (2) is the answer.

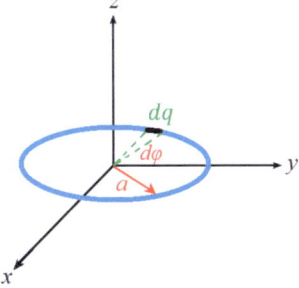

Fig. 8.3 A thin ring with charge q uniformly distributed across it and placed in free space

8.7. First, we need to calculate the electric field in the area. However, we have already calculated it in the previous chapter as follows.

$$\vec{E} = \frac{\rho_l}{2\pi\varepsilon_0 r}\hat{r}$$

Now, we can calculate the electric potential difference as follows.

$$V_2 - V_1 = -\int_{r_2}^{r_1} \vec{E}\cdot\vec{dr} = \int_{r_1}^{r_2} \vec{E}\cdot\vec{dr}$$

$$\Rightarrow V_2 - V_1 = \int_{r_1}^{r_2} \frac{\rho_l}{2\pi\varepsilon_0 r}\hat{r}\cdot dr\hat{r}$$

$$\Rightarrow V_2 - V_1 = \frac{\rho_l}{2\pi\varepsilon_0}\int_{r_1}^{r_2} \frac{1}{r}dr$$

$$\Rightarrow V_2 - V_1 = \frac{\rho_l}{2\pi\varepsilon_0}\left[\ln r\right]_{r_1}^{r_2}$$

$$\Rightarrow V_2 - V_1 = \frac{\rho_l}{2\pi\varepsilon_0}\ln\frac{r_2}{r_1}$$

Choice (3) is the answer (Fig. 8.4).

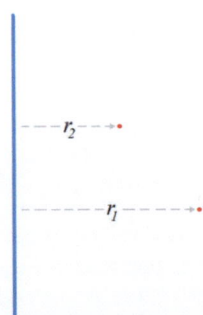

Fig. 8.4 A long thin rod with linear charge density

Notes

In this problem, the relations below have been used.

$$\int \frac{dx}{x} = \ln x + c$$

$$\ln a - \ln b = \ln\left(\frac{a}{b}\right)$$

8.8. Based on the information given in the problem, we have:

$$\rho_l(x) = 6x^2$$

As is shown in Fig. 8.5, a differential length produces a differential electric potential at the origin that can be calculated as follows.

$$dV = \frac{dq}{4\pi\varepsilon_0 a}$$

Herein, the differential charge can be calculated as follows.

$$dq = \rho_l dl = 6x^2 a\, d\varphi$$

From Fig. 8.5, we have:

$$x = a\cos\varphi$$

Therefore:

$$dq = \rho_l dl = 6(a\cos\varphi)^2 a\, d\varphi = 6a^3 \cos^2\varphi\, d\varphi$$

$$V = \int dV$$

$$\Rightarrow V = \int_{\varphi=0}^{\varphi=\pi} \frac{6a^3 \cos^2\varphi\, d\varphi}{4\pi\varepsilon_0 a}$$

$$\Rightarrow V = \frac{3a^2}{2\pi\varepsilon_0} \int_{\varphi=0}^{\varphi=\pi} \cos^2\varphi\, d\varphi$$

$$\Rightarrow V = \frac{3a^2}{2\pi\varepsilon_0} \int_{\varphi=0}^{\varphi=\pi} \left(\frac{1}{2} + \frac{\cos 2\varphi}{2}\right) d\varphi$$

$$\Rightarrow V = \frac{3a^2}{2\pi\varepsilon_0} \left[\frac{\varphi}{2} + \frac{\sin 2\varphi}{4}\right]_0^{\pi}$$

$$\Rightarrow V = \frac{3a^2}{2\pi\varepsilon_0} \left(\frac{\pi - 0}{2} + \frac{\sin 2\pi - \sin 0}{4}\right)$$

$$\Rightarrow V = \frac{3a^2}{4\varepsilon_0}$$

Choice (4) is the answer.

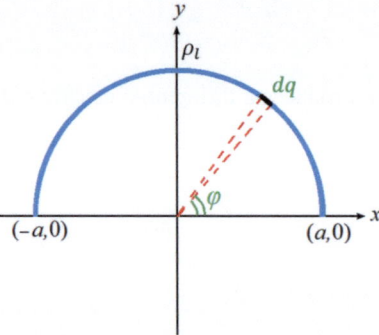

Fig. 8.5 A differential length produces a differential electric potential at the origin

Notes

In this problem, the relations below have been used.

$$\cos^2\varphi = \frac{1 + \cos 2\varphi}{2}$$

$$\int \cos ax\, dx = \frac{1}{a} \sin ax$$

$$\sin 2\pi = 0$$

$$\sin 0 = 0$$

8.10. First, we need to calculate the electric field in all the regions, that is, $r \leq a$, $a < r < 2a$, $2a < r < 4a$, and $r > 4a$. However, we have already calculated them in the previous chapter for $2a = b$ and $4a = c$ as follows.

$$\begin{cases} \overrightarrow{E_1} = \dfrac{Qr}{2\pi\varepsilon_0 a^2 l}\widehat{r} & r \leq a \\ \overrightarrow{E_2} = \dfrac{Q}{2\pi\varepsilon_0 rl}\widehat{r} & a < r < b \\ \overrightarrow{E_3} = \dfrac{Q}{2\pi\varepsilon_0 l}\left(\dfrac{c^2}{r(c^2-b^2)} - \dfrac{r}{c^2-b^2}\right)\widehat{r} & b < r < c \\ \overrightarrow{E_4} = 0 & r > c \end{cases}$$

Therefore, for this problem, the electric fields in different regions are as follows.

$$\begin{cases} \overrightarrow{E_1} = \dfrac{Qr}{2\pi\varepsilon_0 a^2 l}\widehat{r} & r \leq a \\ \overrightarrow{E_2} = \dfrac{Q}{2\pi\varepsilon_0 rl}\widehat{r} & a < r < 2a \\ \overrightarrow{E_3} = \dfrac{Q}{2\pi\varepsilon_0 l}\left(\dfrac{4}{3r} - \dfrac{r}{12a^2}\right)\widehat{r} & 2a < r < 4a \\ \overrightarrow{E_4} = 0 & ar > 4a \end{cases}$$

Now, we can calculate the electric potential in the region $r > 4a$ as follows.

$$V = -\int_{\infty}^{r} \overrightarrow{E} . \overrightarrow{dr} = \int_{r}^{\infty} \overrightarrow{E} . \overrightarrow{dr} = \int_{r}^{\infty} \overrightarrow{E_4} . \overrightarrow{dr} = \int_{r}^{\infty} 0 . dr\widehat{r}$$

$$\Rightarrow V = 0$$

Choice (1) is the answer (Fig. 8.6).

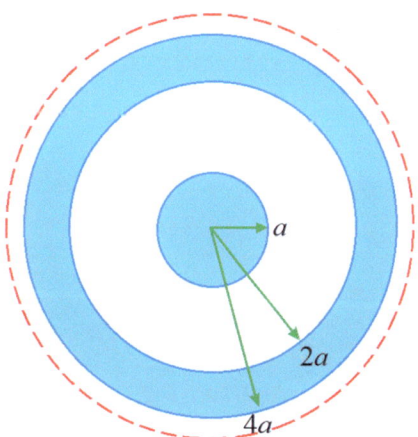

Fig. 8.6 The Gaussian surface used to calculate the electric potential

8.11. The electric potential in the region $2a < r < 4a$ can be calculated as follows.

$$V = -\int_{\infty}^{r} \vec{E} \cdot \vec{dr} = \int_{r}^{\infty} \vec{E} \cdot \vec{dr}$$

$$\Rightarrow V = \int_{r}^{4a} \vec{E_3} \cdot \vec{dr} + \int_{4a}^{\infty} \vec{E_4} \cdot \vec{dr}$$

$$\Rightarrow V = \int_{r}^{4a} \frac{Q}{2\pi\varepsilon_0 l}\left(\frac{4}{3r} - \frac{r}{12a^2}\right)\hat{r} \cdot dr\hat{r} + \int_{4a}^{\infty} 0 \cdot dr\hat{r}$$

$$\Rightarrow V = \int_{r}^{4a} \frac{Q}{2\pi\varepsilon_0 l}\left(\frac{4}{3r} - \frac{r}{12a^2}\right) dr$$

$$\Rightarrow V = \frac{Q}{2\pi\varepsilon_0 l}\left[\frac{4}{3}\ln r - \frac{r^2}{24a^2}\right]_{r}^{4a}$$

$$\Rightarrow V = \frac{Q}{2\pi\varepsilon_0 l}\left[\frac{4}{3}(\ln 4a - \ln r) - \frac{1}{24a^2}\left((4a)^2 - r^2\right)\right]$$

$$\Rightarrow V = \frac{Q}{2\pi\varepsilon_0 l}\left[\frac{4}{3}\ln\left(\frac{4a}{r}\right) - \frac{2}{3} + \frac{r^2}{24a^2}\right]$$

Choice (4) is the answer (Fig. 8.7).

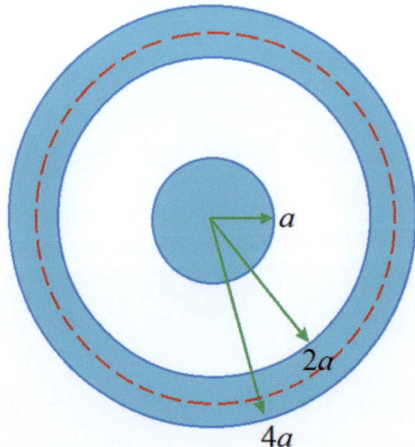

Fig. 8.7 The Gaussian surface used to calculate the electric potential

8.2 Electric Potential in Cylindrical Coordinate System

> **Notes**
>
> In this problem, the relations below have been used.
>
> $$\int x^n dx = \frac{x^{n+1}}{n+1}$$
>
> $$\int \frac{dx}{x} = \ln x + c$$
>
> $$\ln a - \ln b = \ln\left(\frac{a}{b}\right)$$

8.12. The electric potential in the region $a < r < 2a$ can be calculated as follows.

$$V = -\int_{\infty}^{r} \vec{E}.\vec{dr}$$

$$\Rightarrow V = \int_{r}^{\infty} \vec{E}.\vec{dr}$$

$$\Rightarrow V = \int_{r}^{2a} \vec{E_2}.\vec{dr} + \int_{2a}^{4a} \vec{E_3}.\vec{dr} + \int_{4a}^{\infty} \vec{E_4}.\vec{dr}$$

$$\Rightarrow V = \int_{r}^{2a} \frac{Q}{2\pi\varepsilon_0 r l}\hat{r}.dr\hat{r} + \int_{2a}^{4a} \frac{Q}{2\pi\varepsilon_0 l}\left(\frac{4}{3r} - \frac{r}{12a^2}\right)\hat{r}.dr\hat{r} + \int_{4a}^{\infty} 0.dr\hat{r}$$

$$\Rightarrow V = \int_{r}^{2a} \frac{Q}{2\pi\varepsilon_0 r l}dr + \int_{2a}^{4a} \frac{Q}{2\pi\varepsilon_0 l}\left(\frac{4}{3r} - \frac{r}{12a^2}\right)dr + 0$$

$$\Rightarrow V = \frac{Q}{2\pi\varepsilon_0 l}\Big[\ln r\Big]_{r}^{2a} + \frac{Q}{2\pi\varepsilon_0 l}\left[\frac{4}{3}\ln r - \frac{r^2}{24a^2}\right]_{2a}^{4a}$$

$$\Rightarrow V = \frac{Q}{2\pi\varepsilon_0 l}(\ln 2a - \ln r) + \frac{Q}{2\pi\varepsilon_0 l}\left(\frac{4}{3}(\ln 4a - \ln 2a) - \frac{1}{24a^2}\left((4a)^2 - (2a)^2\right)\right)$$

$$\Rightarrow V = \frac{Q}{2\pi\varepsilon_0 l}\ln\left(\frac{2a}{r}\right) + \frac{Q}{2\pi\varepsilon_0 l}\left(\frac{4}{3}\ln 2 - \frac{1}{2}\right)$$

$$\Rightarrow V = \frac{Q}{2\pi\varepsilon_0 l}\left(\ln\left(\frac{2a}{r}\right) + \frac{4}{3}\ln 2 - \frac{1}{2}\right)$$

Choice (2) is the answer (Fig. 8.8).

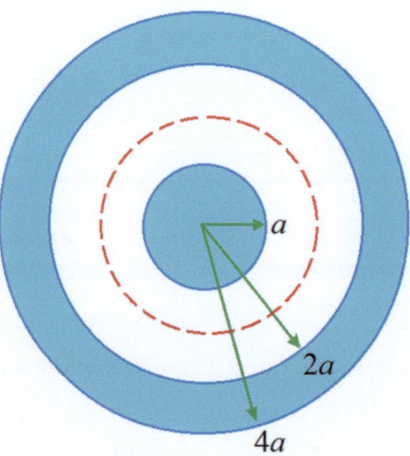

Fig. 8.8 The Gaussian surface used to calculate the electric potential

> **Notes**
> In this problem, the relations below have been used.
>
> $$\int x^n dx = \frac{x^{n+1}}{n+1}$$
>
> $$\int \frac{dx}{x} = \ln x + c$$
>
> $$\ln a - \ln b = \ln\left(\frac{a}{b}\right)$$

8.13. The electric potential in the region $r < a$ can be calculated as follows.

$$V = -\int_{\infty}^{r} \vec{E} \cdot \vec{dr}$$

$$\Rightarrow V = \int_{r}^{\infty} \vec{E} \cdot \vec{dr}$$

8.2 Electric Potential in Cylindrical Coordinate System

$$\Rightarrow V = \int_r^a \vec{E_1}.\vec{dr} + \int_a^{2a} \vec{E_2}.\vec{dr} + \int_{2a}^{4a} \vec{E_3}.\vec{dr} + \int_{4a}^{\infty} \vec{E_4}.\vec{dr}$$

$$\Rightarrow V = \int_r^a \frac{Qr}{2\pi\varepsilon_0 a^2 l}\hat{r}.dr\hat{r} + \int_a^{2a} \frac{Q}{2\pi\varepsilon_0 rl}\hat{r}.dr\hat{r} + \int_{2a}^{4a} \frac{Q}{2\pi\varepsilon_0 l}\left(\frac{4}{3r} - \frac{r}{12a^2}\right)\hat{r}.dr\hat{r} + \int_{4a}^{\infty} 0.dr\hat{r}$$

$$\Rightarrow V = \int_r^a \frac{Qr}{2\pi\varepsilon_0 a^2 l}dr + \int_a^{2a} \frac{Q}{2\pi\varepsilon_0 rl}dr + \int_{2a}^{4a} \frac{Q}{2\pi\varepsilon_0 l}\left(\frac{4}{3r} - \frac{r}{12a^2}\right)dr + 0$$

$$\Rightarrow V = \frac{Q}{2\pi\varepsilon_0 a^2 l}\left[\frac{r^2}{2}\right]_r^a + \frac{Q}{2\pi\varepsilon_0 l}\left[\ln r\right]_a^{2a} + \frac{Q}{2\pi\varepsilon_0 l}\left[\frac{4}{3}\ln r - \frac{r^2}{24a^2}\right]_{2a}^{4a}$$

$$\Rightarrow V = \frac{Q}{2\pi\varepsilon_0 a^2 l}\left(\frac{a^2}{2} - \frac{r^2}{2}\right) + \frac{Q}{2\pi\varepsilon_0 l}(\ln 2a - \ln a) + \frac{Q}{2\pi\varepsilon_0 l}\left(\frac{4}{3}(\ln 4a - \ln 2a) - \frac{(4a)^2 - (2a)^2}{24a^2}\right)$$

$$\Rightarrow V = \frac{Q}{4\pi\varepsilon_0 l} - \frac{Qr^2}{4\pi\varepsilon_0 a^2 l} + \frac{Q}{2\pi\varepsilon_0 l}\ln 2 + \frac{Q}{2\pi\varepsilon_0 l}\left(\frac{4}{3}\ln 2 - \frac{1}{2}\right)$$

$$\Rightarrow V = \frac{Q}{4\pi\varepsilon_0 l}\left(1 - \frac{r^2}{a^2} + 2\ln 2 + \frac{8}{3}\ln 2 - 1\right)$$

$$\Rightarrow V = \frac{Q}{2\pi\varepsilon_0 l}\left(-\frac{r^2}{2a^2} + \frac{7}{3}\ln 2\right)$$

Choice (3) is the answer (Fig. 8.9).

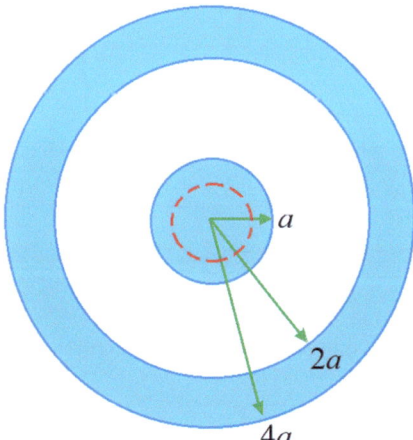

Fig. 8.9 The Gaussian surface used to calculate the electric potential

Notes

In this problem, the relations below have been used.

$$\int x^n dx = \frac{x^{n+1}}{n+1}$$

$$\int \frac{dx}{x} = \ln x + c$$

$$\ln a - \ln b = \ln\left(\frac{a}{b}\right)$$

8.3 Electric Potential in Spherical Coordinate System

8.14. The current always flows from one conductor with higher electric potential to the other conductor that has lower electric potential. Hence, it is concluded that the first conducting sphere has higher electric potential with respect to the second one. In other words:

$$V_1 > V_2$$

Choice (3) is the answer (Fig. 8.10).

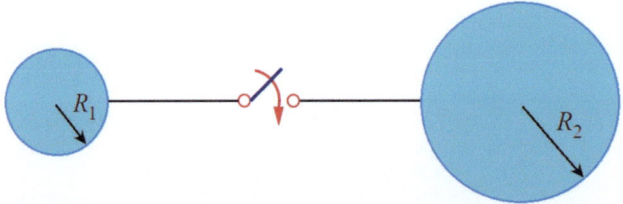

Fig. 8.10 Two spherical conductors connected to each other by a switch

8.15. After closing the switch, the conducting spheres become equipotential. In other words:

$$V'_1 = V'_2$$

As we know, the electric potential of a spherical conductor with total charge Q and radius R can be calculated as follows.

$$V = \frac{Q}{4\pi\varepsilon_0 R}$$

Therefore:

8.3 Electric Potential in Spherical Coordinate System

$$\frac{Q'_1}{4\pi\varepsilon_0 R_1} = \frac{Q'_2}{4\pi\varepsilon_0 R_2}$$

$$\Rightarrow \frac{Q'_1}{Q'_2} = \frac{R_1}{R_2}$$

Moreover, based on the charge conservation principle, we have:

$$Q_1 + Q_2 = Q'_1 + Q'_2$$

Therefore, (b) and (e) are correct. Choice (3) is the answer (Fig. 8.11).

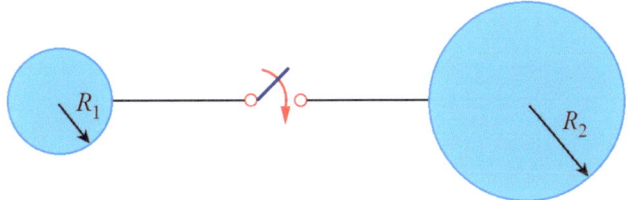

Fig. 8.11 Two spherical conductors connected to each other by a switch

---

8.17. First, we need to calculate the electric field in all the regions, that is, $R \leq a$ and $R > a$. However, we have already calculated them in the previous chapter as follows.

$$\vec{E} = \begin{cases} 0 & R \leq a \\ \dfrac{Q}{4\pi\varepsilon_0 R^2}\hat{R} & R > a \end{cases}$$

The electric potential in the region $R \leq a$ can be calculated as follows.

$$V = -\int_{\infty}^{R} \vec{E}\cdot\overrightarrow{dR} = \int_{R}^{\infty} \vec{E}\cdot\overrightarrow{dR}$$

$$\Rightarrow V = \int_{R}^{a} 0\cdot\overrightarrow{dR} + \int_{a}^{\infty} \frac{Q}{4\pi\varepsilon_0 R^2}\hat{R}\cdot\overrightarrow{dR}$$

$$\Rightarrow V = 0 + \frac{Q}{4\pi\varepsilon_0} \int_{a}^{\infty} \frac{1}{R^2}\hat{R}\cdot dR\hat{R}$$

$$\Rightarrow V = \frac{Q}{4\pi\varepsilon_0} \int_{a}^{\infty} \frac{1}{R^2} dR$$

$$\Rightarrow V = -\frac{Q}{4\pi\varepsilon_0}\left[\frac{1}{R}\right]_a^\infty$$

$$\Rightarrow V = -\frac{Q}{4\pi\varepsilon_0}\left(\frac{1}{\infty} - \frac{1}{a}\right)$$

$$\Rightarrow V_1 = \frac{Q}{4\pi\varepsilon_0 a}$$

Also, the electric potential in the region $R > a$ can be calculated as follows.

$$V = -\int_\infty^R \vec{E}\cdot\vec{dR} = \int_R^\infty \vec{E}\cdot\vec{dR}$$

$$\Rightarrow V = \int_R^\infty \frac{Q}{4\pi\varepsilon_0 R^2}\hat{R}\cdot\vec{dR}$$

$$\Rightarrow V = \int_R^\infty \frac{Q}{4\pi\varepsilon_0 R^2}\hat{R}\cdot dR\hat{R}$$

$$\Rightarrow V = \frac{Q}{4\pi\varepsilon_0}\int_R^\infty \frac{1}{R^2}dR$$

$$\Rightarrow V = -\frac{Q}{4\pi\varepsilon_0}\left[\frac{1}{R}\right]_R^\infty$$

$$\Rightarrow V = -\frac{Q}{4\pi\varepsilon_0}\left(\frac{1}{\infty} - \frac{1}{R}\right)$$

$$\Rightarrow V_2 = \frac{Q}{4\pi\varepsilon_0 R}$$

Hence:

$$V = \begin{cases} \dfrac{Q}{4\pi\varepsilon_0 a} & R \leq a \\ \dfrac{Q}{4\pi\varepsilon_0 R} & R > a \end{cases}$$

Choice (1) is the answer (Fig. 8.12).

8.3 Electric Potential in Spherical Coordinate System

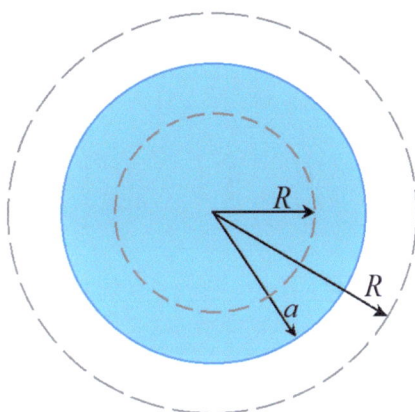

Fig. 8.12 The Gaussian surfaces used to calculate the electric potential

Notes

In this problem, the relation below has been used.

$$\int x^n dx = \frac{x^{n+1}}{n+1}$$

8.19. First, we need to calculate the electric field in all the regions, that is, $R \leq a$ and $R > a$. However, we have already calculated them in the previous chapter as follows.

$$\vec{E} = \begin{cases} \dfrac{QR}{4\pi\varepsilon_0 a^3}\hat{R} & R \leq a \\[6pt] \dfrac{Q}{4\pi\varepsilon_0 R^2}\hat{R} & R > a \end{cases}$$

The electric potential in the region $R \leq a$ can be calculated as follows.

$$V = -\int\limits_{\infty}^{R} \vec{E}.\overrightarrow{dR} = \int\limits_{R}^{\infty} \vec{E}.\overrightarrow{dR}$$

$$\Rightarrow V = \int\limits_{R}^{a} \frac{QR}{4\pi\varepsilon_0 a^3}\hat{R}.\overrightarrow{dR} + \int\limits_{a}^{\infty} \frac{Q}{4\pi\varepsilon_0 R^2}\hat{R}.\overrightarrow{dR}$$

$$\Rightarrow V = \int\limits_{R}^{a} \frac{QR}{4\pi\varepsilon_0 a^3}\hat{R}.dR\hat{R} + \frac{Q}{4\pi\varepsilon_0}\int\limits_{a}^{\infty} \frac{1}{R^2}\hat{R}.dR\hat{R}$$

$$\Rightarrow V = \frac{Q}{4\pi\varepsilon_0 a^3} \int_R^a R\,dR + \frac{Q}{4\pi\varepsilon_0} \int_a^\infty \frac{1}{R^2}\,dR$$

$$\Rightarrow V = \frac{Q}{4\pi\varepsilon_0 a^3} \left[\frac{R^2}{2}\right]_R^a - \frac{Q}{4\pi\varepsilon_0} \left[\frac{1}{R}\right]_a^\infty$$

$$\Rightarrow V = \frac{Q}{4\pi\varepsilon_0 a^3} \left(\frac{a^2}{2} - \frac{R^2}{2}\right) - \frac{Q}{4\pi\varepsilon_0}\left(\frac{1}{\infty} - \frac{1}{a}\right)$$

$$\Rightarrow V = \frac{Q}{8\pi\varepsilon_0 a} - \frac{QR^2}{8\pi\varepsilon_0 a^3} + \frac{Q}{4\pi\varepsilon_0 a}$$

$$\Rightarrow V = \frac{3Q}{8\pi\varepsilon_0 a} - \frac{QR^2}{8\pi\varepsilon_0 a^3}$$

$$\Rightarrow V_1 = \frac{1}{4\pi\varepsilon_0} \frac{Q}{2a}\left(3 - \left(\frac{R}{a}\right)^2\right)$$

Also, the electric potential in the region $R > a$ can be calculated as follows.

$$V = - \int_\infty^R \vec{E}\cdot\overrightarrow{dR} = \int_R^\infty \vec{E}\cdot\overrightarrow{dR}$$

$$\Rightarrow V = \int_R^\infty \frac{Q}{4\pi\varepsilon_0 R^2} \hat{R}\cdot\overrightarrow{dR}$$

$$\Rightarrow V = \int_R^\infty \frac{Q}{4\pi\varepsilon_0 R^2} \hat{R}\cdot dR\hat{R}$$

$$\Rightarrow V = \frac{Q}{4\pi\varepsilon_0} \int_R^\infty \frac{1}{R^2}\,dR$$

$$\Rightarrow V = -\frac{Q}{4\pi\varepsilon_0}\left[\frac{1}{R}\right]_R^\infty$$

$$\Rightarrow V = -\frac{Q}{4\pi\varepsilon_0}\left(\frac{1}{\infty} - \frac{1}{R}\right)$$

$$\Rightarrow V_2 = \frac{1}{4\pi\varepsilon_0} \frac{Q}{R}$$

Hence:

$$V = \begin{cases} \dfrac{1}{4\pi\varepsilon_0}\dfrac{Q}{2a}\left(3-\left(\dfrac{R}{a}\right)^2\right) & R \leq a \\ \dfrac{1}{4\pi\varepsilon_0}\dfrac{Q}{R} & R > a \end{cases}$$

Choice (2) is the answer (Fig. 8.13).

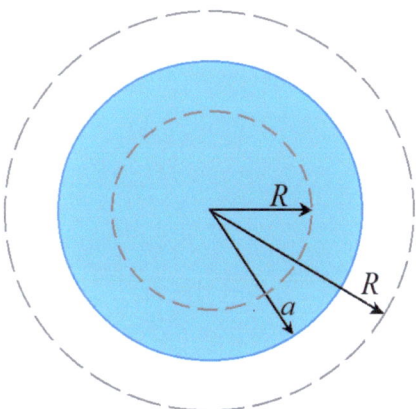

Fig. 8.13 The Gaussian surfaces used to calculate the electric potential

> **Notes**
> In this problem, the relation below has been used.
>
> $$\int x^n dx = \frac{x^{n+1}}{n+1}$$

8.21. First, we need to calculate the electric field in all the regions, that is, $R \leq a$, $a < R < 2a$, $2a < R < 4a$, and $R > 4a$. However, we have already calculated them in the previous chapter for $2a = b$ and $4a = c$ as follows.

$$\begin{cases} \overrightarrow{E_1} = 0 & R \leq a \\ \overrightarrow{E_2} = \dfrac{Q}{4\pi\varepsilon_0 R^2}\hat{R} & a < R < b \\ \overrightarrow{E_3} = 0 & b < R < c \\ \overrightarrow{E_4} = 0 & R > c \end{cases}$$

Therefore, for this problem, the electric field in different regions are as follows.

$$\begin{cases} \overrightarrow{E_1} = 0 & R \leq a \\ \overrightarrow{E_2} = \dfrac{Q}{4\pi\varepsilon_0 R^2}\widehat{R} & a < R < 2a \\ \overrightarrow{E_3} = 0 & 2a < R < 4a \\ \overrightarrow{E_4} = 0 & R > 4a \end{cases}$$

Now, we can calculate the electric potential in the region $R > 4a$ as follows.

$$V = -\int_{\infty}^{R} \vec{E} \cdot \overrightarrow{dR} = \int_{R}^{\infty} \vec{E} \cdot \overrightarrow{dR} = \int_{R}^{\infty} \overrightarrow{E_4} \cdot \overrightarrow{dR} = \int_{R}^{\infty} 0 \cdot dR\widehat{R}$$

$$\Rightarrow V = 0$$

Choice (2) is the answer (Fig. 8.14).

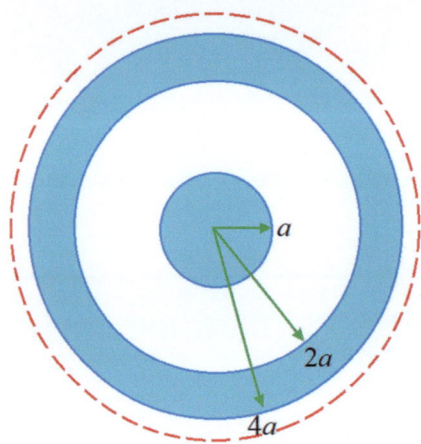

Fig. 8.14 The Gaussian surface used to calculate the electric potential

8.22. The electric potential in the region $2a < R < 4a$ can be calculated as follows.

$$V = -\int_{\infty}^{R} \vec{E} \cdot \overrightarrow{dR} = \int_{R}^{\infty} \vec{E} \cdot \overrightarrow{dR}$$

$$\Rightarrow V = \int_{R}^{4a} \overrightarrow{E_3} \cdot \overrightarrow{dR} + \int_{4a}^{\infty} \overrightarrow{E_4} \cdot \overrightarrow{dR}$$

8.3 Electric Potential in Spherical Coordinate System

$$\Rightarrow V = \int_{R}^{4a} 0.\overrightarrow{dR} + \int_{4a}^{\infty} 0.\overrightarrow{dR}$$

$$\Rightarrow V = 0$$

Choice (4) is the answer (Fig. 8.15).

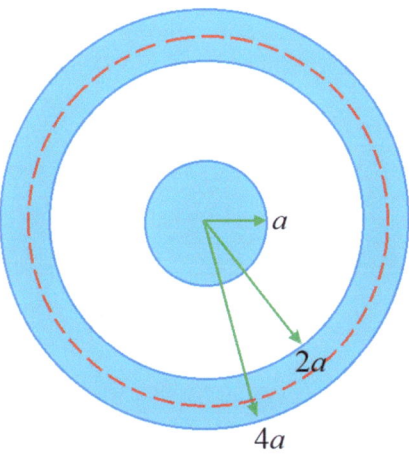

Fig. 8.15 The Gaussian surface used to calculate the electric potential

8.23. The electric potential in the region $a < R < 2a$ can be calculated as follows.

$$V = -\int_{\infty}^{R} \vec{E}.\overrightarrow{dR} = \int_{R}^{\infty} \vec{E}.\overrightarrow{dR}$$

$$\Rightarrow V = \int_{R}^{2a} \overrightarrow{E_2}.\overrightarrow{dR} + \int_{2a}^{4a} \overrightarrow{E_3}.\overrightarrow{dR} + \int_{4a}^{\infty} \overrightarrow{E_4}.\overrightarrow{dR}$$

$$\Rightarrow V = \int_{R}^{2a} \frac{Q}{4\pi\varepsilon_0 R^2}\widehat{R}.dR\widehat{R} + \int_{2a}^{4a} 0.dR\widehat{R} + \int_{4a}^{\infty} 0.dR\widehat{R}$$

$$\Rightarrow V = \frac{Q}{4\pi\varepsilon_0}\int_{R}^{2a}\frac{1}{R^2}dR + 0 + 0$$

$$\Rightarrow V = -\frac{Q}{4\pi\varepsilon_0}\left[\frac{1}{R}\right]_R^{2a}$$

$$\Rightarrow V = -\frac{Q}{4\pi\varepsilon_0}\left(\frac{1}{2a} - \frac{1}{R}\right)$$

$$\Rightarrow V = \frac{Q}{4\pi\varepsilon_0}\left(\frac{1}{R} - \frac{1}{2a}\right)$$

Choice (1) is the answer (Fig. 8.16).

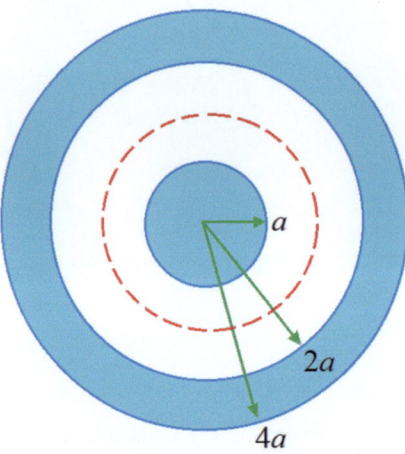

Fig. 8.16 The Gaussian surface used to calculate the electric potential

Notes

In this problem, the relation below has been used.

$$\int x^n dx = \frac{x^{n+1}}{n+1}$$

8.24. The electric potential in the region $R \leq a$ can be calculated as follows.

$$V = -\int_\infty^R \vec{E} \cdot \vec{dR} = \int_R^\infty \vec{E} \cdot \vec{dR}$$

$$\Rightarrow V = \int_R^a \vec{E_1} \cdot \vec{dR} + \int_a^{2a} \vec{E_2} \cdot \vec{dR} + \int_{2a}^{4a} \vec{E_3} \cdot \vec{dR} + \int_{4a}^\infty \vec{E_4} \cdot \vec{dR}$$

8.3 Electric Potential in Spherical Coordinate System

$$\Rightarrow V = \int_R^a 0.\overrightarrow{dR} + \int_a^{2a} \frac{Q}{4\pi\varepsilon_0 R^2}\widehat{R}.dR\widehat{R} + \int_{2a}^{4a} 0.dR\widehat{R} + \int_{4a}^{\infty} 0.dR\widehat{R}$$

$$\Rightarrow V = 0 + \frac{Q}{4\pi\varepsilon_0} \int_a^{2a} \frac{1}{R^2} dR + 0 + 0$$

$$\Rightarrow V = -\frac{Q}{4\pi\varepsilon_0} \left[\frac{1}{R}\right]_a^{2a}$$

$$\Rightarrow V = -\frac{Q}{4\pi\varepsilon_0} \left(\frac{1}{2a} - \frac{1}{a}\right)$$

$$\Rightarrow V = \frac{Q}{8\pi\varepsilon_0 a}$$

Choice (2) is the answer (Fig. 8.17).

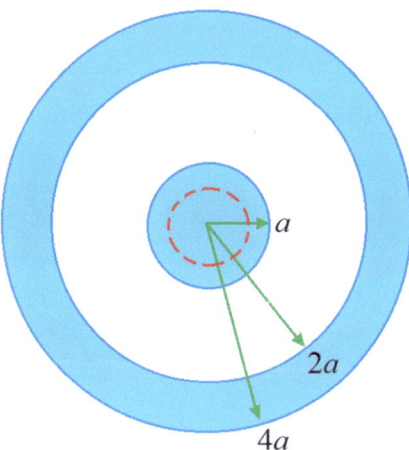

Fig. 8.17 The Gaussian surface used to calculate the electric potential

---

Notes

In this problem, the relation below has been used.

$$\int x^n dx = \frac{x^{n+1}}{n+1}$$

8.25. First, we need to calculate the electric field in all the regions, that is, $R \leq a$, $a < R < 2a$, $2a < R < 4a$, and $R > 4a$. However, we have already calculated them in the previous chapter for $2a = b$ and $4a = c$ as follows.

$$\begin{cases} \overrightarrow{E_1} = \dfrac{QR}{4\pi\varepsilon_0 a^3}\widehat{R} & R \leq a \\ \overrightarrow{E_2} = \dfrac{Q}{4\pi\varepsilon_0 R^2}\widehat{R} & a < R < b \\ \overrightarrow{E_3} = \dfrac{Q}{4\pi\varepsilon_0}\left(\dfrac{c^3}{R^2(c^3 - b^3)} - \dfrac{R}{c^3 - b^3}\right)\widehat{R} & b < R < c \\ \overrightarrow{E_4} = 0 & R > c \end{cases}$$

Therefore, for this problem, the electric field in different regions are as follows.

$$\begin{cases} \overrightarrow{E_1} = \dfrac{QR}{4\pi\varepsilon_0 a^3}\widehat{R} & R \leq a \\ \overrightarrow{E_2} = \dfrac{Q}{4\pi\varepsilon_0 R^2}\widehat{R} & a < R < 2a \\ \overrightarrow{E_3} = \dfrac{Q}{4\pi\varepsilon_0}\left(\dfrac{8}{7R^2} - \dfrac{R}{56a^3}\right)\widehat{R} & 2a < R < 4a \\ \overrightarrow{E_4} = 0 & R > 4a \end{cases}$$

Now, we can calculate the electric potential in the region $R > 4a$ as follows.

$$V = -\int_{\infty}^{R} \overrightarrow{E}.\overrightarrow{dR} = \int_{R}^{\infty} \overrightarrow{E}.\overrightarrow{dR} = \int_{R}^{\infty} \overrightarrow{E_4}.\overrightarrow{dR} = \int_{R}^{\infty} 0.dR\widehat{R}$$

$$\Rightarrow V = 0$$

Choice (3) is the answer (Fig. 8.18).

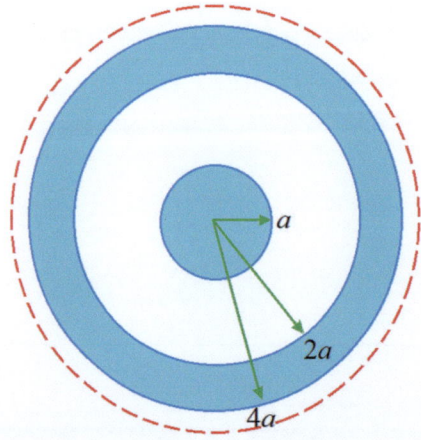

Fig. 8.18 The Gaussian surface used to calculate the electric potential

8.26. The electric potential in the region $2a < R < 4a$ can be calculated as follows.

$$V = -\int_{\infty}^{R} \vec{E} \cdot \overrightarrow{dR} = \int_{R}^{\infty} \vec{E} \cdot \overrightarrow{dR}$$

$$\Rightarrow V = \int_{R}^{4a} \vec{E_3} \cdot \overrightarrow{dR} + \int_{4a}^{\infty} \vec{E_4} \cdot \overrightarrow{dR}$$

$$\Rightarrow V = \int_{R}^{4a} \frac{Q}{4\pi\varepsilon_0}\left(\frac{8}{7R^2} - \frac{R}{56a^3}\right)\hat{R} \cdot dR\hat{R} + \int_{4a}^{\infty} 0 \cdot dR\hat{R}$$

$$\Rightarrow V = \int_{R}^{4a} \frac{Q}{4\pi\varepsilon_0}\left(\frac{8}{7R^2} - \frac{R}{56a^3}\right)dR + 0$$

$$\Rightarrow V = \left[\frac{Q}{4\pi\varepsilon_0}\left(-\frac{8}{7R} - \frac{R^2}{112a^3}\right)\right]_{R}^{4a}$$

$$\Rightarrow V = \frac{Q}{4\pi\varepsilon_0}\left(-\frac{8}{7}\left(\frac{1}{4a} - \frac{1}{R}\right) - \frac{(4a)^2 - R^2}{112a^3}\right)$$

$$\Rightarrow V = \frac{Q}{4\pi\varepsilon_0}\left(-\frac{2}{7a} + \frac{8}{7R} - \frac{1}{7a} + \frac{R^2}{112a^3}\right)$$

$$\Rightarrow V = \frac{Q}{4\pi\varepsilon_0}\left(-\frac{3}{7a} + \frac{8}{7R} + \frac{R^2}{112a^3}\right)$$

Choice (1) is the answer (Fig. 8.19).

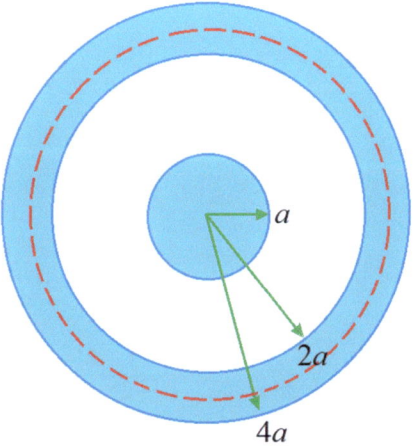

Fig. 8.19 The Gaussian surface used to calculate the electric potential

Notes

In this problem, the relation below has been used.

$$\int x^n dx = \frac{x^{n+1}}{n+1}$$

8.27. The electric potential in the region $a < R < 2a$ can be calculated as follows.

$$V = -\int_{\infty}^{R} \vec{E} \cdot \vec{dR}$$

$$\Rightarrow V = \int_{R}^{\infty} \vec{E} \cdot \vec{dR}$$

$$\Rightarrow V = \int_{R}^{2a} \vec{E_2} \cdot \vec{dR} + \int_{2a}^{4a} \vec{E_3} \cdot \vec{dR} + \int_{4a}^{\infty} \vec{E_4} \cdot \vec{dR}$$

$$\Rightarrow V = \int_{R}^{2a} \frac{Q}{4\pi\varepsilon_0 R^2} \hat{R} \cdot dR\hat{R} + \int_{2a}^{4a} \frac{Q}{4\pi\varepsilon_0}\left(\frac{8}{7R^2} - \frac{R}{56a^3}\right)\hat{R} \cdot dR\hat{R} + \int_{4a}^{\infty} 0 \cdot dR\hat{R}$$

$$\Rightarrow V = \int_{R}^{2a} \frac{Q}{4\pi\varepsilon_0 R^2} dR + \int_{2a}^{4a} \frac{Q}{4\pi\varepsilon_0}\left(\frac{8}{7R^2} - \frac{R}{56a^3}\right) dR + 0$$

$$\Rightarrow V = \left[-\frac{Q}{4\pi\varepsilon_0 R}\right]_{R}^{2a} + \left[\frac{Q}{4\pi\varepsilon_0}\left(-\frac{8}{7R} - \frac{R^2}{112a^3}\right)\right]_{2a}^{4a}$$

$$\Rightarrow V = -\frac{Q}{4\pi\varepsilon_0}\left(\frac{1}{2a} - \frac{1}{R}\right) - \frac{Q}{4\pi\varepsilon_0}\left(\frac{8}{7}\left(\frac{1}{4a} - \frac{1}{2a}\right) + \frac{(4a)^2 - (2a)^2}{112a^3}\right)$$

$$\Rightarrow V = -\frac{Q}{4\pi\varepsilon_0}\left(\frac{1}{2a} - \frac{1}{R}\right) - \frac{Q}{4\pi\varepsilon_0}\left(-\frac{2}{7a} + \frac{12}{112a}\right)$$

$$\Rightarrow V = \frac{Q}{4\pi\varepsilon_0}\left(-\frac{1}{2a} + \frac{1}{R} + \frac{2}{7a} - \frac{3}{28a}\right)$$

$$\Rightarrow V = \frac{Q}{4\pi\varepsilon_0}\left(\frac{1}{R} + \frac{-14 + 8 - 3}{28a}\right)$$

8.3 Electric Potential in Spherical Coordinate System

$$\Rightarrow V = \frac{Q}{4\pi\varepsilon_0}\left(\frac{1}{R} - \frac{9}{28a}\right)$$

Choice (4) is the answer (Fig. 8.20).

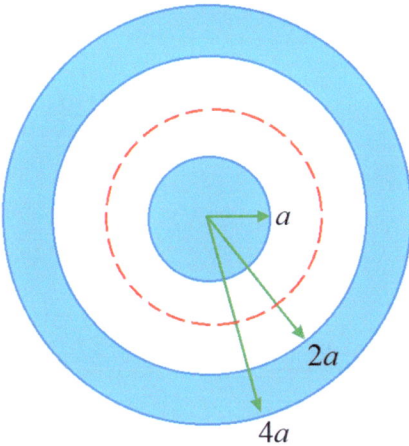

Fig. 8.20 The Gaussian surface used to calculate the electric potential

---

Notes

In this problem, the relation below has been used.

$$\int x^n dx = \frac{x^{n+1}}{n+1}$$

8.28. The electric potential in the region $R \leq a$ can be calculated as follows.

$$V = -\int_{\infty}^{R} \vec{E} \cdot \vec{dR}$$

$$\Rightarrow V = \int_{R}^{\infty} \vec{E} \cdot \vec{dR}$$

$$\Rightarrow V = \int_{R}^{a} \vec{E_1} \cdot \vec{dR} + \int_{a}^{2a} \vec{E_2} \cdot \vec{dR} + \int_{2a}^{4a} \vec{E_3} \cdot \vec{dR} + \int_{4a}^{\infty} \vec{E_4} \cdot \vec{dR}$$

$$\Rightarrow V = \int_R^a \frac{QR}{4\pi\varepsilon_0 a^3}\widehat{R} \cdot dR\widehat{R} + \int_a^{2a} \frac{Q}{4\pi\varepsilon_0 R^2}\widehat{R} \cdot dR\widehat{R} + \int_{2a}^{4a} \frac{Q}{4\pi\varepsilon_0}\left(\frac{8}{7R^2} - \frac{R}{56a^3}\right)\widehat{R} \cdot dR\widehat{R} + \int_{4a}^{\infty} 0 \cdot dR\widehat{R}$$

$$\Rightarrow V = \int_R^a \frac{QR}{4\pi\varepsilon_0 a^3} dR + \int_a^{2a} \frac{Q}{4\pi\varepsilon_0 R^2} dR + \int_{2a}^{4a} \frac{Q}{4\pi\varepsilon_0}\left(\frac{8}{7R^2} - \frac{R}{56a^3}\right) dR + 0$$

$$\Rightarrow V = \left[\frac{QR^2}{8\pi\varepsilon_0 a^3}\right]_R^a + \left[-\frac{Q}{4\pi\varepsilon_0 R}\right]_a^{2a} + \left[\frac{Q}{4\pi\varepsilon_0}\left(-\frac{8}{7R} - \frac{R^2}{112a^3}\right)\right]_{2a}^{4a}$$

$$\Rightarrow V = \left(\frac{Q(a^2 - R^2)}{8\pi\varepsilon_0 a^3}\right) - \left(\frac{Q}{4\pi\varepsilon_0}\left(\frac{1}{2a} - \frac{1}{a}\right)\right) - \frac{Q}{4\pi\varepsilon_0}\left(\frac{8}{7}\left(\frac{1}{4a} - \frac{1}{2a}\right) + \frac{(4a)^2 - (2a)^2}{112a^3}\right)$$

$$\Rightarrow V = \frac{Q}{4\pi\varepsilon_0}\left(\frac{a^2 - R^2}{2a^3}\right) - \frac{Q}{4\pi\varepsilon_0}\left(-\frac{1}{2a}\right) - \frac{Q}{4\pi\varepsilon_0}\left(-\frac{2}{7a} + \frac{12}{112a}\right)$$

$$\Rightarrow V = \frac{Q}{4\pi\varepsilon_0}\left(\frac{1}{2a} - \frac{R^2}{2a^3} + \frac{1}{2a} + \frac{2}{7a} - \frac{3}{28a}\right)$$

$$\Rightarrow V = \frac{Q}{4\pi\varepsilon_0}\left(\frac{14 + 14 + 8 - 3}{28a} - \frac{R^2}{2a^3}\right)$$

$$\Rightarrow V = \frac{Q}{4\pi\varepsilon_0}\left(\frac{33}{28a} - \frac{R^2}{2a^3}\right)$$

Choice (1) is the answer (Fig. 8.21).

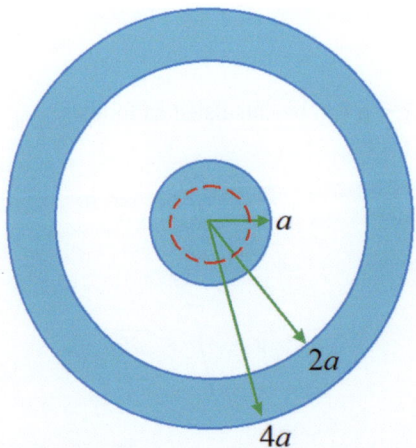

Fig. 8.21 The Gaussian surface used to calculate the electric potential

> **Notes**
>
> In this problem, the relation below has been used.
>
> $$\int x^n dx = \frac{x^{n+1}}{n+1}$$

References

1. Rahmani-Andebili, M., General Physics II – Practice Problems, Methods, and Solutions, Springer Nature, 2025.
2. Rahmani-Andebili, M., General Physics I – Practice Problems, Methods, and Solutions, Springer Nature, 2025.
3. Rahmani-Andebili, M., Mathematics of Engineering and Science – Practice Problems, Methods, and Solutions, Springer Nature, 2024.
4. Rahmani-Andebili, M., Differential Equations – Practice Problems, Methods, and Solutions, Springer Nature, 2022.
5. Rahmani-Andebili, M., Calculus III – Practice Problems, Methods, and Solutions, Springer Nature, 2023.
6. Rahmani-Andebili, M., Calculus II – Practice Problems, Methods, and Solutions, Springer Nature, 2023.
7. Rahmani-Andebili, M., Calculus I (2nd Ed.) – Practice Problems, Methods, and Solutions, Springer Nature, 2023.
8. Rahmani-Andebili, M., Precalculus (2nd Ed.) – Practice Problems, Methods, and Solutions, Springer Nature, 2024.

9. Electric Potential Energy: Part A

Abstract

In this chapter, the basic and advanced problems of electric potential energy are studied. The subjects include electric potential energy of discrete and continuous uniform and nonuniform charge distributions in the Cartesian, cylindrical, and spherical coordinate systems. Herein, different types of problems and exercises are presented that are categorized as follows.

- *Problems with detailed solution*: They have been designed to teach students the subjects in detail. Moreover, they have been categorized in different levels based on their difficulty levels (easy, normal, and hard) and calculation amounts (small, normal, and large).
- *Partially solved exercises*: They have been designed to encourage students to practice problems while guiding them through the problem-solving procedure and hinting the required formulas.
- *Exercises with final answer*: They have been designed to encourage students to practice more by themselves while hinting them by the final answer as well as to help instructors to give tests or quizzes.

9.1 Electric Potential Energy Due to Discrete Charge Distribution

Problem

9.1. From 10 km distance, a proton with the initial velocity of 3 km/s is shot toward a stationary proton which is fixed at its place. Calculate the final distance between these two protons [1–8]. Herein, assume that $m_p = 2 \times 10^{-27}$ kg, $q_p = 1.6 \times 10^{-19}$ C, and $\varepsilon_0 = \frac{10^{-9}}{36\pi} \frac{C^2}{N.m^2}$.

Difficulty level ○ Easy ● Normal ○ Hard
Calculation amount ○ Small ● Normal ○ Large
1) 5 mm
2) 2.56 μm
3) 25.6 nm
4) 5 cm

Problem

9.2. Calculate the electric potential energy of the system shown in Fig. 9.1. Herein, we have:

$$d = 12 \; cm$$

$$q_1 = 15 \, \mu C$$

$$q_2 = -60 \, \mu C$$

$$q_3 = 30 \, \mu C$$

$$\varepsilon_0 = \frac{10^{-9}}{36\pi} \frac{C^2}{N.m^2}$$

Difficulty level ○ Easy ○ Normal ● Hard
Calculation amount ○ Small ● Normal ○ Large

1) $U = -168.75 \, J$
2) $U = 168.75 \, J$
3) $U = 168.75 \, mJ$
4) $U = -168.75 \, mJ$

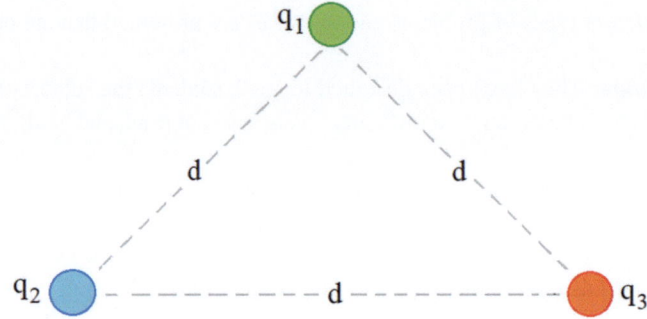

Fig. 9.1 Three point charges on the corners of an equilateral triangle

Partially Solved Exercise

9.3. Three point charges are placed at the corners of an equilateral triangle with one meter side length shown in Fig. 9.2. Calculate the amount of work needed to move them to the other equilateral triangle with 25 *cm* side length shown in Fig. 9.2. Herein, $q_1 = 1 \, \mu C$, $q_2 = 2 \, \mu C$, and $q_3 = 3 \, \mu C$.

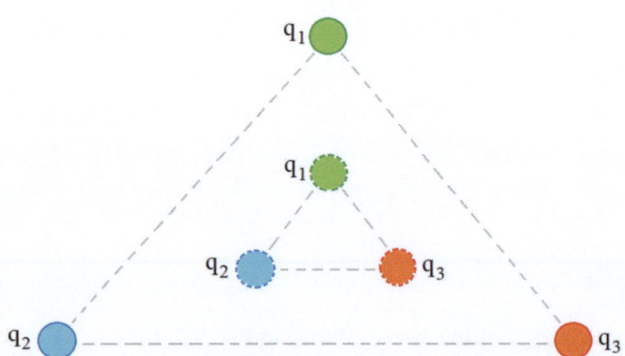

Fig. 9.2 Three point charges move from the corners of an equilateral triangle to the corners of the smaller one

9.1 Electric Potential Energy Due to Discrete Charge Distribution

Solution

Based on the information given in the problem, we have:

$$q_1 = 1\ \mu C$$

$$q_2 = 2\ \mu C$$

$$q_3 = 3\ \mu C$$

$$R = 1\ m$$

$$R' = 0.25\ m$$

The amount of work needed to move the charges can be calculated as follows.

$$W = \Delta U = U' - U$$

The electric potential energy of a system of N charges can be calculated as follows.

$$U = \frac{1}{2}\sum_{i=1}^{N} q_i V_i$$

Herein, V_i is the total electric potential of the other charges at the place of the charge q_i.

For the initial state of the system, we can write:

$$U = \frac{1}{2}\sum_{i=1}^{3} q_i V_i$$

$$\Rightarrow U = \frac{1}{2}\left[q_1\left(\frac{q_2}{4\pi\varepsilon_0 R} + \frac{q_3}{4\pi\varepsilon_0 R}\right) + q_2\left(\frac{q_3}{4\pi\varepsilon_0 R} + \frac{q_1}{4\pi\varepsilon_0 R}\right) + q_3\left(\frac{q_1}{4\pi\varepsilon_0 R} + \frac{q_2}{4\pi\varepsilon_0 R}\right)\right]$$

$$\Rightarrow U = \frac{1}{8\pi\varepsilon_0 R}[q_1(q_2+q_3) + q_2(q_3+q_1) + q_3(q_1+q_2)]$$

$$\Rightarrow U = \frac{1}{8\pi\varepsilon_0(\quad)}[(\quad)((\quad)+(\quad))+(\quad)((\quad)+(\quad))+(\quad)((\quad)+(\quad))]$$

$$U = (\quad)[(\quad)+(\quad)+(\quad)]$$

$$\Rightarrow U = (\quad)[\quad]$$

$$\Rightarrow U = \frac{(\quad)}{\pi\varepsilon_0}\ J$$

For the final state of the system, we have:

$$U' = \frac{1}{2}\sum_{i=1}^{3} q_i V_i$$

$$\Rightarrow U' = \frac{1}{2}\left[q_1\left(\frac{q_2}{4\pi\varepsilon_0 R'} + \frac{q_3}{4\pi\varepsilon_0 R'}\right) + q_2\left(\frac{q_3}{4\pi\varepsilon_0 R'} + \frac{q_1}{4\pi\varepsilon_0 R'}\right) + q_3\left(\frac{q_1}{4\pi\varepsilon_0 R'} + \frac{q_2}{4\pi\varepsilon_0 R'}\right)\right]$$

$$\Rightarrow U' = \frac{1}{8\pi\varepsilon_0 R'}[q_1(q_2+q_3) + q_2(q_3+q_1) + q_3(q_1+q_2)]$$

$$\Rightarrow U' = \frac{1}{8\pi\varepsilon_0(\quad)}[(\quad)((\quad) + (\quad)) + (\quad)((\quad) + (\quad)) + (\quad)((\quad) + (\quad))]$$

$$\Rightarrow U' = (\quad)[(\quad) + (\quad) + (\quad)]$$

$$\Rightarrow U' = (\quad)[\quad]$$

$$\Rightarrow U' = \frac{(\quad)}{\pi\varepsilon_0} \, J$$

The amount of work needed to move the charges is calculated as follows.

$$W = U' - U = \frac{(\quad)}{\pi\varepsilon_0} - \frac{(\quad)}{\pi\varepsilon_0}$$

$$\Rightarrow W = \frac{33 \times 10^{-12}}{4\pi\varepsilon_0} \, J$$

Problem

9.4. As is illustrated in Fig. 9.3, three particles with mass m and charge q are placed and kept at the corners of an equilateral triangle with side length a. The particles are freed to move. Calculate the final velocity of the particles.

Difficulty level   ○ Easy   ○ Normal   ● Hard
Calculation amount   ○ Small   ● Normal   ○ Large

1) $v = \dfrac{\sqrt{3}q}{\sqrt{2\pi\varepsilon_0 ma}}$

2) $v = \dfrac{q}{\sqrt{3\pi\varepsilon_0 ma}}$

3) $v = \dfrac{q}{\sqrt{\pi\varepsilon_0 ma}}$

4) $v = \dfrac{q}{\sqrt{2\pi\varepsilon_0 ma}}$

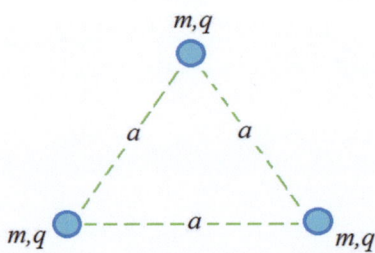

Fig. 9.3 Three particles with mass m and charge q placed on the corners of an equilateral triangle

9.1 Electric Potential Energy Due to Discrete Charge Distribution

Problem

9.5. In the system shown in Fig. 9.4, calculate the amount of work needed to move the point charge q' from point A to point B.
Difficulty level ○ Easy ○ Normal ● Hard
Calculation amount ○ Small ● Normal ○ Large

1) $W = 0$
2) $W = \dfrac{kq'q}{a}\left(1 + \dfrac{1}{\sqrt{2}}\right)$
3) $W = \dfrac{kq'q}{a}\left(1 - \dfrac{1}{\sqrt{2}}\right)$
4) $W = \dfrac{2kq'q}{a}$

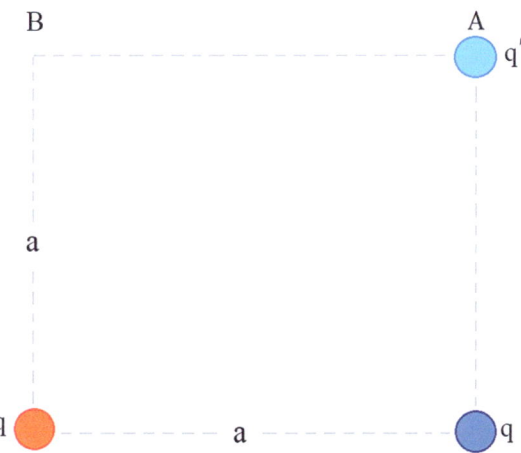

Fig. 9.4 Three point charges on the corners of a square

Exercise

9.6. As is shown in Fig. 9.5, an external force is applied to bring the dipole from infinity to the neighborhood of the point charge q. Calculate the amount of work done by the external force. Herein, $q = 1\ \mu C$ and $\varepsilon_0 = \dfrac{10^{-9}}{36\pi}\ F/m$.

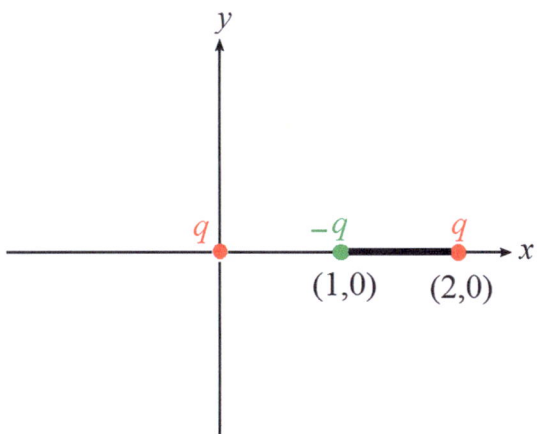

Fig. 9.5 Applying an external force to bring the dipole from infinity to the neighborhood of the point charge q

Final Answer
$W = -4.5 \times 10^{-3} J$

9.2 Electric Potential Energy Due to Continuous Charge Distribution

Problem
9.7. Calculate the total amount of electric potential energy stored in the whole space due to a conducting solid sphere with radius a and total charge Q (Fig. 9.6).

Difficulty level ○ Easy ○ Normal ● Hard
Calculation amount ○ Small ● Normal ○ Large

1) $U = \dfrac{Q}{4\pi\varepsilon_0 R^2}$
2) $U = \dfrac{Q^2}{4\pi\varepsilon_0}$
3) $U = \dfrac{Q^2}{8\pi\varepsilon_0 a}$
4) $U = \dfrac{Q^2}{8\pi\varepsilon_0 a^2}$

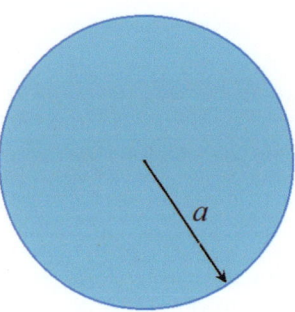

Fig. 9.6 A conducting solid sphere with total charge Q

Exercise
9.8. In Problem 9.7, calculate the total electric energy stored in the whole space as a function of surface charge density ρ_s.

Final Answer
$U = \dfrac{2\rho_s^2 \pi a^3}{\varepsilon_0}$

Exercise
9.9. In Problem 9.7, calculate the amount of electric energy stored inside the conducting solid sphere.

Final Answer
$U = 0$

9.2 Electric Potential Energy Due to Continuous Charge Distribution

Exercise

9.10. In Problem 9.7, calculate the amount of electric energy stored outside the conducting solid sphere as a function of Q.

Final Answer

$$U = \frac{Q^2}{8\pi\varepsilon_0 a}$$

Exercise

9.11. In Problem 9.7, calculate the amount of electric energy stored outside the conducting solid sphere as a function of ρ_s.

Final Answer

$$U = \frac{2\rho_s^2 \pi a^3}{\varepsilon_0}$$

Problem

9.12. Calculate the total amount of electric energy stored in the whole space due to a solid sphere with radius a and total charge Q which is uniformly distributed across it (Fig. 9.7).

Difficulty level — ○ Easy ○ Normal ● Hard
Calculation amount — ○ Small ● Normal ○ Large

1) $U = \dfrac{3Q^2}{20\pi\varepsilon_0 R}$

2) $U = \dfrac{Q^2}{20\pi\varepsilon_0 R^2}$

3) $U = \dfrac{Q^2}{20\pi\varepsilon_0 a^2}$

4) $U = \dfrac{3Q^2}{20\pi\varepsilon_0 a}$

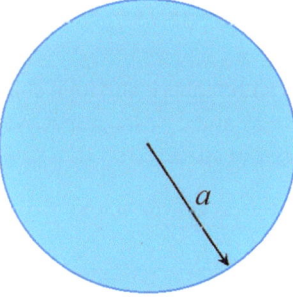

Fig. 9.7 A solid sphere with total charge Q which is uniformly distributed across it

Exercise

9.13. In Problem 9.12, calculate the total electric energy stored in the whole space as a function of volume charge density ρ_v.

Final Answer

$$U = \frac{4\pi\rho_v^2 a^5}{15\varepsilon_0}$$

Exercise

9.14. In Problem 9.12, calculate the amount of electric energy stored inside the solid sphere as a function of Q.

Final Answer

$$U = \frac{Q^2}{40\pi\varepsilon_0 a}$$

Exercise

9.15. In Problem 9.12, calculate the amount of electric energy stored inside the solid sphere as a function of ρ_v.

Final Answer

$$U = \frac{2\pi\rho_v^2 a^5}{45\varepsilon_0}$$

Exercise

9.16. In Problem 9.12, calculate the amount of electric energy stored outside the solid sphere as a function of Q.

Final Answer

$$U = \frac{Q^2}{8\pi\varepsilon_0 a}$$

Exercise

9.17. In Problem 9.12, calculate the amount of electric energy stored outside the solid sphere as a function of Q.

Final Answer

$$U = \frac{2\pi\rho_v^2 a^5}{9\varepsilon_0}$$

Problem

9.18. Figure 9.8 shows a conducting solid sphere with radius a and total charge Q which is inside a conducting hollow sphere with charge $-Q$ and inner and outer radiuses $b = 2a$ and $c = 4a$, respectively. The solid and hollow spheres are concentric. Calculate the total amount of electric energy stored in space.
Difficulty level ○ Easy ○ Normal ● Hard
Calculation amount ○ Small ● Normal ○ Large

1) $U = \dfrac{Q^2}{16\pi\varepsilon_0 a}$

2) $U = \dfrac{Q^2}{8\pi\varepsilon_0 a}$

3) $U = \dfrac{Q^2}{16\pi\varepsilon_0 a^2}$

4) $U = \dfrac{Q^2}{16\pi\varepsilon_0 R}$

9.2 Electric Potential Energy Due to Continuous Charge Distribution

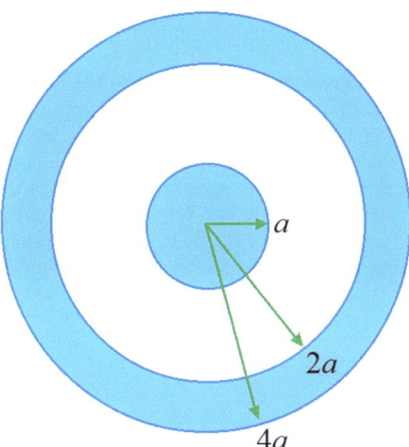

Fig. 9.8 A conducting solid sphere with total charge Q placed inside a conducting hollow sphere with charge $-Q$

Exercise

9.19. Calculate the total amount of electric energy stored in a spherical layer of electric charge in the area $a < R < b$ with uniform volume charge density ρ_v (Fig. 9.9).

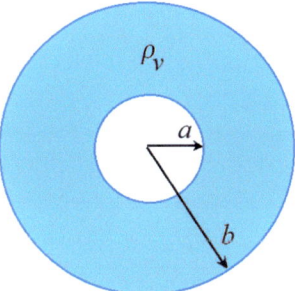

Fig. 9.9 A spherical layer of uniform electric charge

Final Answer

$$U = \frac{2\pi\rho_v^2}{15\varepsilon_0}\left(3a^5 + 2b^5 - 5a^3b^2\right)$$

Exercise

9.20. In Problem 9.19, how can we use the result to calculate the total amount of electric energy stored in a solid sphere of uniform electric charge with radius R_0 and volume charge density ρ_v.

Final Answer

By considering $a = 0$ and $b = R_0$, the spherical layer is converted to a solid sphere with radius R_0. In this regard, the total amount of electric energy is as follows which is equal to the one calculated in this chapter.

$$U = \frac{4\pi\rho_v^2 R_0^5}{15\varepsilon_0}$$

References

1. Rahmani-Andebili, M., General Physics II – Practice Problems, Methods, and Solutions, Springer Nature, 2025.
2. Rahmani-Andebili, M., General Physics I – Practice Problems, Methods, and Solutions, Springer Nature, 2025.
3. Rahmani-Andebili, M., Mathematics of Engineering and Science – Practice Problems, Methods, and Solutions, Springer Nature, 2024.
4. Rahmani-Andebili, M., Differential Equations – Practice Problems, Methods, and Solutions, Springer Nature, 2022.
5. Rahmani-Andebili, M., Calculus III – Practice Problems, Methods, and Solutions, Springer Nature, 2023.
6. Rahmani-Andebili, M., Calculus II – Practice Problems, Methods, and Solutions, Springer Nature, 2023.
7. Rahmani-Andebili, M., Calculus I (2nd Ed.) – Practice Problems, Methods, and Solutions, Springer Nature, 2023.
8. Rahmani-Andebili, M., Precalculus (2nd Ed.) – Practice Problems, Methods, and Solutions, Springer Nature, 2024.

Electric Potential Energy: Part B

Abstract
In this chapter, the problems of the ninth chapter are fully solved, in detail, step-by-step, and with different methods.

10.1 Electric Potential Energy Due to Discrete Charge Distribution

10.1. Based on the information given in the problem, we have [1–8]:

$$R = 10 \text{ km}$$

$$v = 3 \text{ km/s}$$

$$m_p = 2 \times 10^{-27} \text{ kg}$$

$$q_p = 1.6 \times 10^{-19} \text{ C}$$

$$\varepsilon_0 = \frac{10^{-9}}{36\pi} \frac{C^2}{N.m^2}$$

The problem can be solved by applying the principle of conservation of mechanical energy as follows.

$$E_1 = E_2$$

$$U_1 + K_1 = U_2 + K_2$$

$$\Rightarrow 0 + \frac{1}{2} mv^2 = \frac{q_1 q_2}{4\pi \varepsilon_0 R} + 0$$

$$\Rightarrow \frac{1}{2} \times 2 \times 10^{-27} \times (3 \times 10^3)^2 = \frac{(1.6 \times 10^{-19})^2}{4\pi \times \frac{10^{-9}}{36\pi} \times R}$$

$$\Rightarrow 9 \times 10^{-21} = 2.56 \times 9 \times 10^{-29} \frac{1}{R}$$

$$\Rightarrow R = 2.56 \times 10^{-8} \, m = 25.6 \, nm$$

Choice (3) is the answer.

10.2. Based on the information given in the problem, we have:

$$d = 12 \, cm$$

$$q_1 = 15 \, \mu C$$

$$q_2 = -60 \, \mu C$$

$$q_3 = 30 \, \mu C$$

$$\varepsilon_0 = \frac{10^{-9}}{36\pi} \, \frac{C^2}{N.m^2}$$

The electric potential energy of a system of N charges can be calculated as follows.

$$U = \frac{1}{2} \sum_{i=1}^{N} q_i V_i$$

Herein, V_i is the total electric potential of the other charges at place of the charge q_i.

For this problem, we can write:

$$U = \frac{1}{2} \sum_{i=1}^{3} q_i V_i$$

$$\Rightarrow U = \frac{1}{2} \left[q_1 \left(\frac{q_2}{4\pi\varepsilon_0 d} + \frac{q_3}{4\pi\varepsilon_0 d} \right) + q_2 \left(\frac{q_3}{4\pi\varepsilon_0 d} + \frac{q_1}{4\pi\varepsilon_0 d} \right) + q_3 \left(\frac{q_1}{4\pi\varepsilon_0 d} + \frac{q_2}{4\pi\varepsilon_0 d} \right) \right]$$

$$\Rightarrow U = \frac{1}{8\pi\varepsilon_0 d} [q_1(q_2 + q_3) + q_2(q_3 + q_1) + q_3(q_1 + q_2)]$$

$$\Rightarrow U = \frac{1}{8\pi \left(\frac{10^{-9}}{36\pi} \right)(0.12)} [15 \times 10^{-6}(-60 \times 10^{-6} + 30 \times 10^{-6}) - 60 \times 10^{-6}(30 \times 10^{-6} + 15 \times 10^{-6}) + 30 \times 10^{-6}(15 \times 10^{-6} - 60 \times 10^{-6})]$$

$$\Rightarrow U = \frac{3 \times 10^{11}}{8} [15 \times 10^{-6}(-30 \times 10^{-6}) - 60 \times 10^{-6}(45 \times 10^{-6}) + 30 \times 10^{-6}(-45 \times 10^{-6})]$$

$$\Rightarrow U = \frac{3 \times 10^{11}}{8} [-45 \times 10^{-11} - 270 \times 10^{-11} - 135 \times 10^{-11}]$$

$$\Rightarrow U = \frac{3 \times 10^{11}}{8} \left[-450 \times 10^{-11} \right]$$

$$\Rightarrow U = -168.75 \, J$$

Choice (1) is the answer (Fig. 10.1).

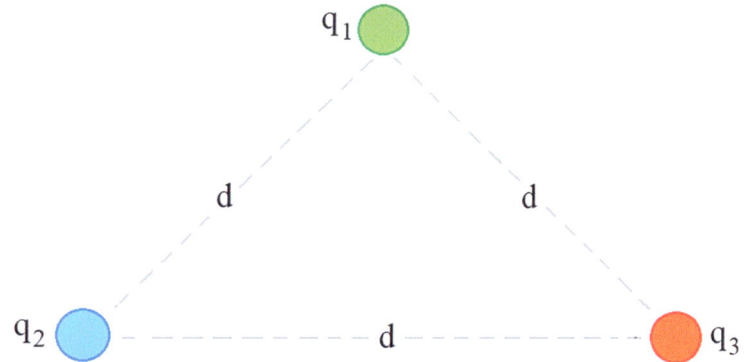

Fig. 10.1 Three point charges on the corners of an equilateral triangle

10.4. The problem can be solved by applying the principle of conservation of mechanical energy as follows.

$$E = E'$$

$$U + K = U' + K'$$

$$\Rightarrow U + 0 = 0 + K'$$

$$\Rightarrow U = K' \tag{10.1}$$

The electric potential energy of a system of N charges can be calculated as follows.

$$U = \frac{1}{2} \sum_{i=1}^{N} q_i V_i$$

Herein, V_i is the total electric potential of the other charges at the place of the charge q_i.

For this problem, we can write:

$$U = \frac{1}{2} \sum_{i=1}^{3} q_i V_i$$

$$\Rightarrow U = \frac{1}{2} \left[q \left(\frac{q}{4\pi\varepsilon_0 a} + \frac{q}{4\pi\varepsilon_0 a} \right) + q \left(\frac{q}{4\pi\varepsilon_0 a} + \frac{q}{4\pi\varepsilon_0 a} \right) + q \left(\frac{q}{4\pi\varepsilon_0 a} + \frac{q}{4\pi\varepsilon_0 a} \right) \right]$$

$$\Rightarrow U = \frac{1}{8\pi\varepsilon_0 a}\left[2q^2 + 2q^2 + 2q^2\right]$$

$$\Rightarrow U = \frac{3q^2}{4\pi\varepsilon_0 a} \qquad (10.2)$$

Solving (10.1) and (10.2):

$$\frac{3q^2}{4\pi\varepsilon_0 a} = \frac{1}{2}mv^2 + \frac{1}{2}mv^2 + \frac{1}{2}mv^2$$

$$\Rightarrow \frac{3q^2}{4\pi\varepsilon_0 a} = \frac{3}{2}mv^2$$

$$\Rightarrow \frac{q^2}{2\pi\varepsilon_0 a} = mv^2$$

$$\Rightarrow v = \frac{q}{\sqrt{2\pi\varepsilon_0 ma}}$$

Choice (4) is the answer (Fig. 10.2).

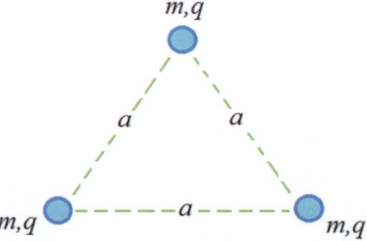

Fig. 10.2 Three particles with mass m and charge q placed on the corners of an equilateral triangle

---

10.5. The amount of work is irrespective of the path. Hence, the amount of work needed to move the point charge q' from point A to point B can be calculated as follows.

$$W = \Delta U$$

$$\Rightarrow W = q'\Delta V = q'(V_B - V_A)$$

The electric potential at points A and B can be calculated as follows.

$$V_A = q'\left(\frac{q}{4\pi\varepsilon_0 a} + \frac{q}{4\pi\varepsilon_0 a\sqrt{2}}\right)$$

$$V_B = q'\left(\frac{q}{4\pi\varepsilon_0 a\sqrt{2}} + \frac{q}{4\pi\varepsilon_0 a}\right)$$

As can be seen, $V_A = V_B$. Therefore:

$$W = 0$$

Choice (1) is the answer (Fig. 10.3).

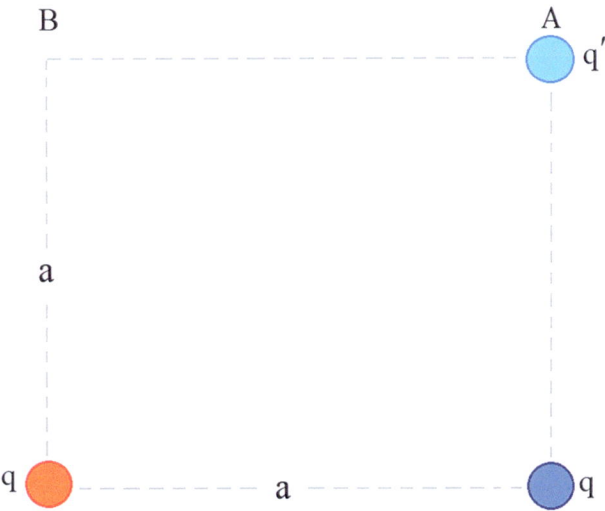

Fig. 10.3 Three point charges on the corners of a square

10.2 Electric Potential Energy Due to Continuous Charge Distribution

10.7. The following relation is used to calculate electric potential energy resulting from the continuous charge distribution. Herein, the integration is applied to the whole space which is from the center of the charge distribution to infinity.

$$U = \frac{1}{2}\varepsilon_0 \int E^2 dV$$

As can be noticed, the electric field in every region of space must be determined first. Hence, for this problem, we need to calculate the electric field in the regions $R \leq a$ and $R > a$. However, we have already calculated them in the previous chapters as follows.

$$\vec{E} = \begin{cases} 0 & R \leq a \\ \dfrac{Q}{4\pi\varepsilon_0 R^2}\hat{R} & R > a \end{cases}$$

Therefore:

$$U = \frac{1}{2}\varepsilon_0 \left(\int_{\varphi=0}^{2\pi} \int_{\theta=0}^{\pi} \int_{R=0}^{a} (0)^2 \times R^2 \sin\theta \, dRd\theta d\varphi + \int_{\varphi=0}^{2\pi} \int_{\theta=0}^{\pi} \int_{R=a}^{\infty} \left(\frac{Q}{4\pi\varepsilon_0 R^2}\right)^2 R^2 \sin\theta \, dRd\theta d\varphi \right)$$

$$\Rightarrow U = \frac{1}{2}\varepsilon_0 \left(\frac{Q}{4\pi\varepsilon_0}\right)^2 \int_{\varphi=0}^{2\pi} \int_{\theta=0}^{\pi} \int_{R=a}^{\infty} \frac{1}{R^2} \sin\theta \, dRd\theta d\varphi$$

$$\Rightarrow U = \frac{1}{2}\varepsilon_0 \left(\frac{Q}{4\pi\varepsilon_0}\right)^2 (4\pi) \left[-\frac{1}{R}\right]_a^{\infty}$$

$$\Rightarrow U = \frac{1}{2}\varepsilon_0 \left(\frac{Q}{4\pi\varepsilon_0}\right)^2 (4\pi) \left(\frac{1}{a} - \frac{1}{\infty}\right)$$

$$\Rightarrow U = \frac{Q^2}{8\pi\varepsilon_0 a}$$

Choice (3) is the answer (Fig. 10.4).

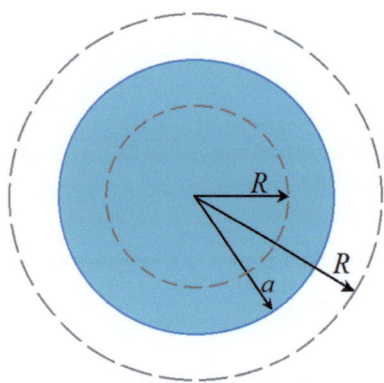

Fig. 10.4 The Gaussian surfaces to calculate the electric energy stored in the whole space

---

Notes

In this problem, the relations below have been used.

$$\int_{\varphi=0}^{2\pi} \int_{\theta=0}^{\pi} \sin\theta \, d\theta d\varphi = [-\cos\theta]_0^{\pi}[\varphi]_0^{2\pi} = (-\cos\pi + \cos 0)(2\pi - 0) = (-(-1)+1)(2\pi) = 4\pi$$

$$\int x^n dx = \frac{x^{n+1}}{n+1}$$

10.2 Electric Potential Energy Due to Continuous Charge Distribution

10.12. The following relation is used to calculate electric energy resulting from the continuous charge distribution. Herein, the integration is applied to the whole space which is from the center of the charge distribution to infinity.

$$U = \frac{1}{2}\varepsilon_0 \int E^2 dV$$

As can be noticed, the electric field in every region of space must be determined first. Hence, for this problem, we need to calculate the electric field in the regions $R \leq a$ and $R > a$. However, we have already calculated them in the previous chapters as follows.

$$\vec{E} = \begin{cases} \dfrac{QR}{4\pi\varepsilon_0 a^3}\hat{R} & R \leq a \\ \dfrac{Q}{4\pi\varepsilon_0 R^2}\hat{R} & R > a \end{cases}$$

Therefore:

$$U = \frac{1}{2}\varepsilon_0 \left(\int_{\varphi=0}^{2\pi}\int_{\theta=0}^{\pi}\int_{R=0}^{a} \left(\frac{QR}{4\pi\varepsilon_0 a^3}\right)^2 R^2 \sin\theta \, dRd\theta d\varphi + \int_{\varphi=0}^{2\pi}\int_{\theta=0}^{\pi}\int_{R=a}^{\infty} \left(\frac{Q}{4\pi\varepsilon_0 R^2}\right)^2 R^2 \sin\theta \, dRd\theta d\varphi \right)$$

$$\Rightarrow U = \frac{1}{2}\varepsilon_0 \left(\left(\frac{Q}{4\pi\varepsilon_0 a^3}\right)^2 \int_{\varphi=0}^{2\pi}\int_{\theta=0}^{\pi}\int_{R=0}^{a} R^4 \sin\theta \, dRd\theta d\varphi + \left(\frac{Q}{4\pi\varepsilon_0}\right)^2 \int_{\varphi=0}^{2\pi}\int_{\theta=0}^{\pi}\int_{R=a}^{\infty} \frac{1}{R^2} \sin\theta \, dRd\theta d\varphi \right)$$

$$\Rightarrow U = \frac{1}{2}\varepsilon_0 \left(\left(\frac{Q}{4\pi\varepsilon_0 a^3}\right)^2 (4\pi) \left[\frac{R^5}{5}\right]_0^a + \left(\frac{Q}{4\pi\varepsilon_0}\right)^2 (4\pi) \left[-\frac{1}{R}\right]_a^{\infty} \right)$$

$$\Rightarrow U = \frac{1}{2}\varepsilon_0 \left(\frac{Q}{4\pi\varepsilon_0 a^3}\right)^2 (4\pi)\left(\frac{a^5}{5} - 0\right) + \frac{1}{2}\varepsilon_0 \left(\frac{Q}{4\pi\varepsilon_0}\right)^2 (4\pi)\left(\frac{1}{a} - \frac{1}{\infty}\right)$$

$$\Rightarrow U = \frac{Q^2}{40\pi\varepsilon_0 a} + \frac{Q^2}{8\pi\varepsilon_0 a}$$

$$\Rightarrow U = \frac{3Q^2}{20\pi\varepsilon_0 a}$$

Choice (4) is the answer (Fig. 10.5).

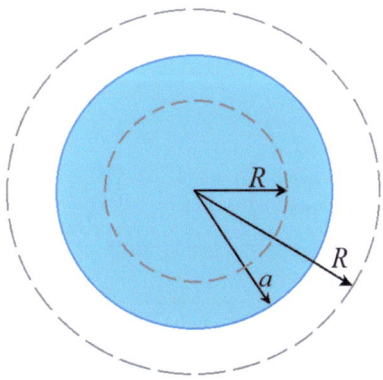

Fig. 10.5 The Gaussian surfaces to calculate the electric energy stored in the whole space

Notes

In this problem, the relations below have been used.

$$\int_{\varphi=0}^{2\pi}\int_{\theta=0}^{\pi} \sin\theta\, d\theta d\varphi = [-\cos\theta]_0^\pi [\varphi]_0^{2\pi} = (-\cos\pi + \cos 0)(2\pi - 0) = (-(-1)+1)(2\pi) = 4\pi$$

$$\int x^n dx = \frac{x^{n+1}}{n+1}$$

10.18. The following relation is used to calculate electric energy resulting from the continuous charge distribution. Herein, the integration is applied to the whole space which is from the center of the charge distribution to infinity.

$$U = \frac{1}{2}\varepsilon_0 \int E^2 dV$$

As can be noticed, the electric field in every region of space must be determined first. Hence, for this problem, we need to calculate the electric field in the regions $R \leq a$, $a < R < 2a$, $2a < R < 4a$, and $R > 4a$. However, we have already calculated them in the previous chapters as follows.

$$\begin{cases} \vec{E_1} = 0 & R \leq a \\ \vec{E_2} = \dfrac{Q}{4\pi\varepsilon_0 R^2}\hat{R} & a < R < 2a \\ \vec{E_3} = 0 & 2a < R < 4a \\ \vec{E_4} = 0 & R > 4a \end{cases}$$

Therefore:

$$U = \frac{1}{2}\varepsilon_0 \left(\int_{\varphi=0}^{2\pi}\int_{\theta=0}^{\pi}\int_{R=0}^{a} (0)^2 R^2 \sin\theta\, dRd\theta d\varphi + \int_{\varphi=0}^{2\pi}\int_{\theta=0}^{\pi}\int_{R=a}^{2a} \left(\frac{Q}{4\pi\varepsilon_0 R^2}\right)^2 R^2 \sin\theta\, dRd\theta d\varphi \right.$$

$$\left. + \int_{\varphi=0}^{2\pi}\int_{\theta=0}^{\pi}\int_{R=2a}^{4a} (0)^2 R^2 \sin\theta\, dRd\theta d\varphi + \int_{\varphi=0}^{2\pi}\int_{\theta=0}^{\pi}\int_{R=4a}^{\infty} (0)^2 R^2 \sin\theta\, dRd\theta d\varphi \right)$$

$$\Rightarrow U = \frac{1}{2}\varepsilon_0 \left(0 + \left(\frac{Q}{4\pi\varepsilon_0}\right)^2 \int_{\varphi=0}^{2\pi}\int_{\theta=0}^{\pi}\int_{R=a}^{2a} \frac{1}{R^2} \sin\theta\, dRd\theta d\varphi + 0 + 0 \right)$$

$$\Rightarrow U = \frac{1}{2}\varepsilon_0 \left(\frac{Q}{4\pi\varepsilon_0}\right)^2 (4\pi) \left[-\frac{1}{R}\right]_a^{2a}$$

$$\Rightarrow U = \frac{1}{2}\varepsilon_0 \left(\frac{Q}{4\pi\varepsilon_0}\right)^2 (4\pi)\left(\frac{1}{a} - \frac{1}{2a}\right)$$

$$\Rightarrow U = \frac{1}{2}\varepsilon_0 \left(\frac{Q}{4\pi\varepsilon_0}\right)^2 (4\pi)\left(\frac{1}{2a}\right)$$

$$\Rightarrow U = \frac{Q^2}{16\pi\varepsilon_0 a}$$

Choice (1) is the answer.

> **Notes**
>
> In this problem, the relations below have been used.
>
> $$\int_{\varphi=0}^{2\pi}\int_{\theta=0}^{\pi} \sin\theta\, d\theta d\varphi = [-\cos\theta]_0^{\pi}[\varphi]_0^{2\pi} = (-\cos\pi + \cos 0)(2\pi - 0) = (-(-1)+1)(2\pi) = 4\pi$$
>
> $$\int x^n dx = \frac{x^{n+1}}{n+1}$$

References

1. Rahmani-Andebili, M., General Physics II – Practice Problems, Methods, and Solutions, Springer Nature, 2025.
2. Rahmani-Andebili, M., General Physics I – Practice Problems, Methods, and Solutions, Springer Nature, 2025.
3. Rahmani-Andebili, M., Mathematics of Engineering and Science – Practice Problems, Methods, and Solutions, Springer Nature, 2024.
4. Rahmani-Andebili, M., Differential Equations – Practice Problems, Methods, and Solutions, Springer Nature, 2022.
5. Rahmani-Andebili, M., Calculus III – Practice Problems, Methods, and Solutions, Springer Nature, 2023.
6. Rahmani-Andebili, M., Calculus II – Practice Problems, Methods, and Solutions, Springer Nature, 2023.
7. Rahmani-Andebili, M., Calculus I (2nd Ed.) – Practice Problems, Methods, and Solutions, Springer Nature, 2023.
8. Rahmani-Andebili, M., Precalculus (2nd Ed.) – Practice Problems, Methods, and Solutions, Springer Nature, 2024.

11 Polarization and Electric Field in Dielectrics and Boundary Conditions: Part A

Abstract

In this chapter, the basic and advanced problems of boundary conditions, polarization, and electric field in dielectrics in the Cartesian, cylindrical, and spherical coordinate systems are studied. Herein, different types of problems and exercises are presented that are categorized as follows.

- **Problems with detailed solution**: They have been designed to teach students the subjects in detail. Moreover, they have been categorized in different levels based on their difficulty levels (easy, normal, and hard) and calculation amounts (small, normal, and large).
- **Partially solved exercises**: They have been designed to encourage students to practice problems while guiding them through the problem-solving procedure and hinting the required formulas.
- **Exercises with final answer**: They have been designed to encourage students to practice more by themselves while hinting them by the final answer as well as to help instructors to give tests or quizzes.

11.1 Polarization and Electric Field in Dielectrics and Boundary Conditions in the Cartesian Coordinate System

Problem

11.1. In Fig. 11.1, calculate the electric flux density in the second region if $\vec{D}_1 = 3\hat{x} - 4\hat{y}$. Herein, on the boundary of the environments, there is no free charge [1–8].

Difficulty level   ○ Easy   ● Normal   ○ Hard
Calculation amount   ○ Small   ● Normal   ○ Large

1) $\vec{D}_2 = 12\hat{x} + 3\hat{y}$
2) $\vec{D}_2 = 4\hat{x} - 3\hat{y}$
3) $\vec{D}_2 = 3\hat{x} - 12\hat{y}$
4) $\vec{D}_2 = 9\hat{x} - 12\hat{y}$

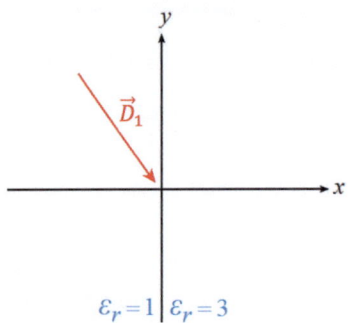

Fig. 11.1 Studying the boundary conditions for the electric flux density

Problem

11.2. In the area $z > 0$, $\varepsilon_{r1} = 2$ and $\vec{E_1} = \hat{x} - 3\hat{y} + 5\hat{z}$, and in the area $z \leq 0$, $\varepsilon_{r2} = 5$ as is illustrated in Fig 11.2. Calculate the refracted angle of electric field in the area $z \leq 0$.

Difficulty level ○ Easy ● Normal ○ Hard
Calculation amount ○ Small ● Normal ○ Large
1) $\alpha_2 \approx 57.7°$
2) $\alpha_2 \approx 32.3°$
3) $\alpha_2 \approx 90°$
4) $\alpha_2 \approx 0°$

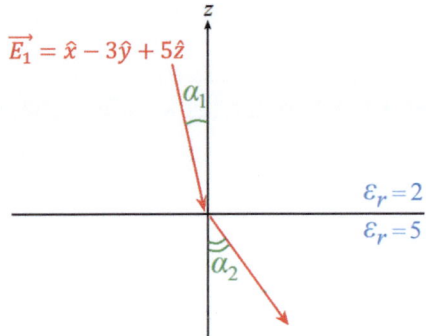

Fig. 11.2 Studying the boundary conditions for the electric field

Partially Solved Exercise

11.3. In Problem 11.2, assume that in the area $z > 0$, $\varepsilon_{r1} = 4$ and $\vec{E_1} = \hat{x} + 2\hat{y} + 4\hat{z}$, and in the area $z \leq 0$, $\varepsilon_{r2} = 2$. Calculate the refracted angle of electric field in the second region.

Solution

As we know, the relation between the incident angle and refracted angle of electric field is as follows.

$$\frac{\tan \alpha_2}{\tan \alpha_1} = \frac{\varepsilon_{r2}}{\varepsilon_{r1}}$$

$$\Rightarrow \frac{\tan \alpha_2}{\dfrac{\sqrt{E_{1x}^2 + E_{1y}^2}}{E_{1z}}} = \frac{\varepsilon_{r2}}{\varepsilon_{r1}}$$

$$\Rightarrow \frac{\tan\alpha_2}{\sqrt{(\quad)^2+(\quad)^2}} = \frac{(\quad)}{(\quad)}$$

$$\Rightarrow \frac{\tan\alpha_2}{\frac{(\quad)}{(\quad)}} = \frac{(\quad)}{(\quad)}$$

$$\Rightarrow \tan\alpha_2 = \frac{(\quad)}{(\quad)} = (\quad)$$

$$\Rightarrow \alpha_2 \approx 15.6°$$

Problem

11.4. In Fig. 11.3, $\vec{E}_1 = 10\hat{x} - 6\hat{y} + 12\hat{z}$ V/m, $\varepsilon_1 = 3\varepsilon_0$, and $\varepsilon_2 = 4.5\varepsilon_0$. Calculate the ratio of electric energy density in the second region to that of the first region. Herein, on the boundary of the environments, there is no free charge.

Difficulty level ○ Easy ○ Normal ● Hard
Calculation amount ○ Small ○ Normal ● Large

1) $\dfrac{39}{28}$
2) $\dfrac{28}{39}$
3) $\dfrac{420}{517}$
4) $\dfrac{517}{420}$

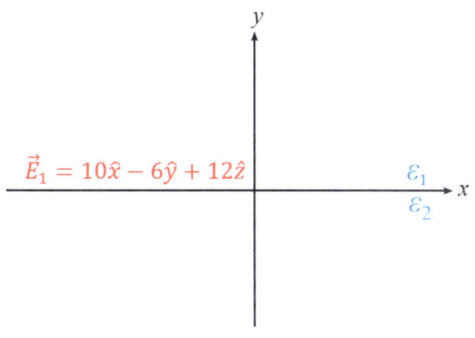

Fig. 11.3 Studying the boundary conditions for the electric field

Problem

11.5. On the boundary of two environments, there is no free charge. Calculate the value of $\dfrac{P_{2n}}{P_{1n}}$ in which P_{1n} and P_{2n} are the normal component of the electric polarization field of the first and second regions. Herein, assume that ε_{1r} and ε_{2r} are the dielectric constants (relative permittivity) of the first and second regions.

Difficulty level ○ Easy ● Normal ○ Hard
Calculation amount ○ Small ● Normal ○ Large

1) $\dfrac{\varepsilon_{1r}}{\varepsilon_{2r}}$

2) $\dfrac{\varepsilon_{2r}}{\varepsilon_{1r}}$

3) $\dfrac{\varepsilon_{2r}-1}{\varepsilon_{1r}-1}$

4) $\dfrac{\varepsilon_{1r}(\varepsilon_{2r}-1)}{\varepsilon_{2r}(\varepsilon_{1r}-1)}$

11.2 Polarization and Electric Field in Dielectrics in Cylindrical Coordinate System

Problem

11.6. In Fig. 11.4, the solid cylinder with length l is a conductor, placed at the origin of the coordinate system, and includes the charge Q. Calculate the magnitude of electric field at an arbitrary point in the first region.

Difficulty level   ○ Easy   ○ Normal   ● Hard
Calculation amount   ○ Small   ● Normal   ○ Large

1) $\vec{E_1} = \dfrac{Q}{\pi l \varepsilon_1 r}\hat{r}$

2) $\vec{E_1} = \dfrac{Q\varepsilon_1}{\pi l(\varepsilon_1+\varepsilon_2)r}\hat{r}$

3) $\vec{E_1} = \dfrac{Q\varepsilon_1}{\pi l \varepsilon_2(\varepsilon_1+\varepsilon_2)r}\hat{r}$

4) $\vec{E_1} = \dfrac{Q}{\pi l(\varepsilon_1+\varepsilon_2)r}\hat{r}$

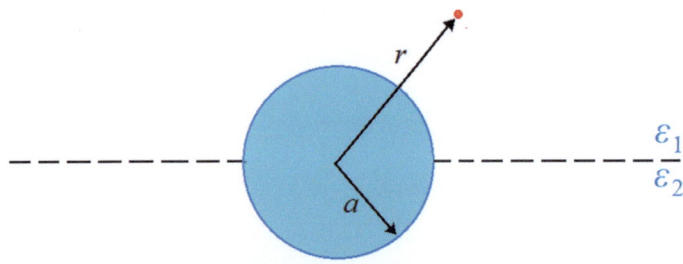

Fig. 11.4 A conducting solid cylinder with length l and charge Q placed at the origin of the coordinate system

Exercise

11.7. In Problem 11.6, determine the electric flux density in both regions.

Final Answer

$\vec{D_1} = \dfrac{Q\varepsilon_1}{\pi l(\varepsilon_1+\varepsilon_2)r}\hat{r},\ \vec{D_2} = \dfrac{Q\varepsilon_2}{\pi l(\varepsilon_1+\varepsilon_2)r}\hat{r}$

11.2 Polarization and Electric Field in Dielectrics in Cylindrical Coordinate System

Problem

11.8. In the coaxial cable shown in Fig. 11.5, the electric field inside the cable is uniform. Which one of the following cases is correct about its dielectric.

Difficulty level   ○ Easy   ○ Normal   ● Hard
Calculation amount   ○ Small   ● Normal   ○ Large

1) $\varepsilon = \varepsilon_0$
2) $\varepsilon \propto \dfrac{1}{r^2}$
3) $\varepsilon \propto \dfrac{1}{r}$
4) $\varepsilon \propto r$

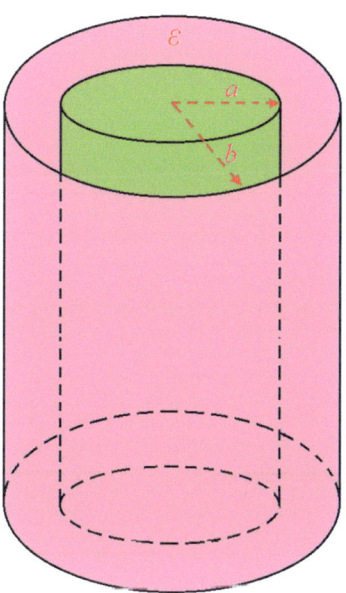

Fig. 11.5 A coaxial cable with a uniform electric field

Problem

11.9. A cylindrical dielectric with radius a and length l that its axis aligned with z-axis is polarized with polarization vector $\vec{P} = \dfrac{1}{r}\hat{r} + z\hat{z}$. Calculate the polarization surface charge density of the cylinder.

Difficulty level   ○ Easy   ○ Normal   ● Hard
Calculation amount   ○ Small   ● Normal   ○ Large

1) $\dfrac{1}{l}$
2) $\dfrac{1}{a}$
3) 0
4) l

Partially Solved Exercise

11.10. A cylindrical dielectric with radius a and length l that its axis aligned with z-axis is polarized with polarization vector $\vec{P} = \frac{1}{r^2}\hat{r}$. Calculate the polarization surface charge density of the cylinder.

Solution

As we know, the polarized surface charge density on the surface of a dielectric can be calculated as follows.

$$\rho_{sb} = \vec{P} \cdot \hat{n}$$

Thus, for the cylindrical surface at $r = a$, we can write:

$$\rho_{sb} = \vec{P} \cdot \hat{r}\Big|_{r=a}$$

$$\Rightarrow \rho_{sb} = (\quad) \cdot (\quad)\Big|_{r=(\quad)}$$

$$\Rightarrow \rho_{sb} = (\quad)\Big|_{r=(\quad)}$$

$$\Rightarrow \rho_{sb} = \frac{1}{a^2}$$

Problem

11.11. A cylindrical dielectric with radius a and length l that its axis aligned with z-axis is polarized with polarization vector $\vec{P} = \frac{1}{r}\hat{r} + z\hat{z}$. Calculate the polarization volume charge density of the cylinder.

Difficulty level ○ Easy ○ Normal ● Hard
Calculation amount ○ Small ● Normal ○ Large
1) 1
2) a
3) $-a$
4) -1

Partially Solved Exercise

11.12. A cylindrical dielectric with radius a and length l that its axis aligned with z-axis is polarized with polarization vector $\vec{P} = \frac{1}{r^2}\hat{r}$. Calculate the polarization volume charge density of the cylinder.

Solution

The relation between polarization volume charge density (ρ_b) and electric polarization field ($\vec{P}$) is as follows.

$$\rho_b = -\nabla \cdot \vec{P}$$

Hence:

11.2 Polarization and Electric Field in Dielectrics in Cylindrical Coordinate System

$$\rho_b = -\left(\frac{1}{r}\frac{\partial}{\partial r}(rP_r) + \frac{1}{r}\frac{\partial}{\partial \varphi}(P_\varphi) + \frac{\partial}{\partial z}(P_z)\right)$$

$$\Rightarrow \rho_b = -\left(\frac{1}{r}\frac{\partial}{\partial r}(r\times(\quad)) + \frac{1}{r}\frac{\partial}{\partial \varphi}(\quad) + \frac{\partial}{\partial z}(\quad)\right)$$

$$\Rightarrow \rho_b = -\left(\frac{1}{r}\frac{\partial}{\partial r}(\quad) + \frac{1}{r}\frac{\partial}{\partial \varphi}(\quad) + \frac{\partial}{\partial z}(\quad)\right)$$

$$\Rightarrow \rho_b = -\left(\frac{1}{r}\times(\quad) + (\quad) + (\quad)\right)$$

$$\Rightarrow \rho_b = \frac{1}{r^3}$$

Problem

11.13. For the coaxial cable illustrated in Fig. 11.6, calculate the value of $\frac{q_b}{\rho_L}$, where q_b is the polarized charge per unit length on the surface of dielectric at $r = b$, and ρ_l is the free charge per unit length on the surface of conductor at $r = a$.

Difficulty level   ○ Easy   ○ Normal   ● Hard
Calculation amount   ○ Small   ○ Normal   ● Large

1) $-\frac{1}{6}$
2) $-\frac{2}{3}$
3) $\frac{1}{6}$
4) $\frac{2}{3}$

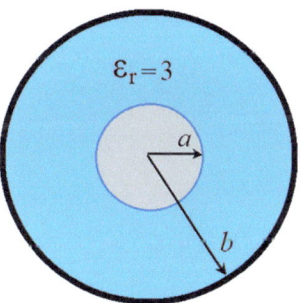

Fig. 11.6 A coaxial cable including free and polarized charges

Problem

11.14. In a cylindrical dielectric, the polarization volume charge density is uniform and equal to ρ_b. Calculate the electric polarization field ($\vec{P}$) inside the cylinder.

Difficulty level   ○ Easy   ○ Normal   ● Hard
Calculation amount   ○ Small   ● Normal   ○ Large

1) $\vec{P} = \dfrac{-\rho_b r^2}{2}\hat{r}$

2) $\vec{P} = \dfrac{-\rho_b r}{3}\hat{r}$

3) $\vec{P} = \dfrac{-\rho_b r}{2}\hat{r}$

4) $\vec{P} = \dfrac{-\rho_b r^2}{3}\hat{r}$

11.3 Polarization and Electric Field in Dielectrics in Spherical Coordinate System

Problem

11.15. In spherical coordinate system, a solid sphere with radius a, nonuniform permittivity $\varepsilon = \varepsilon_0(1 + 2R)$, and nonuniform volume charge density $\rho = \rho_0\left(1 + \dfrac{R}{a}\right)$ is centered at the origin. Calculate the magnitude of electric field at $R = \dfrac{a}{2}$ (Fig. 11.7).

Difficulty level — ○ Easy   ○ Normal   ● Hard
Calculation amount — ○ Small   ○ Normal   ● Large

1) $E = \dfrac{11 a \rho_0}{48 \varepsilon_0 (1 + a)}$

2) $E = \dfrac{22 a \rho_0}{\varepsilon_0 (1 + a)}$

3) $E = \dfrac{11 \rho_0}{24 \varepsilon_0 \left(1 + \frac{1}{a}\right)}$

4) $E = \dfrac{a \rho_0}{\varepsilon_0 (1 + a)}$

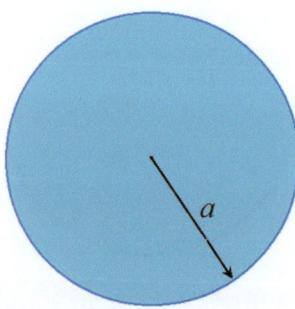

Fig. 11.7 A solid sphere including nonuniform permittivity and nonuniform volume charge density

Problem

11.16. In Fig. 11.8, the solid sphere is a conductor, placed at the origin of the coordinate system, and includes the charge Q. Calculate the magnitude of electric field at an arbitrary point in the first region.

Difficulty level   ○ Easy   ○ Normal   ● Hard
Calculation amount   ○ Small   ● Normal   ○ Large

1) $\vec{E_1} = \dfrac{Q}{2\pi\varepsilon_1 R^2}\hat{R}$

2) $\vec{E_1} = \dfrac{Q}{2\pi(\varepsilon_1+\varepsilon_2)R^2}\hat{R}$

3) $\vec{E_1} = \dfrac{Q\varepsilon_1}{2\pi(\varepsilon_1+\varepsilon_2)R^2}\hat{R}$

4) $\vec{E_1} = \dfrac{Q\varepsilon_1}{2\pi\varepsilon_2(\varepsilon_1+\varepsilon_2)R^2}\hat{R}$

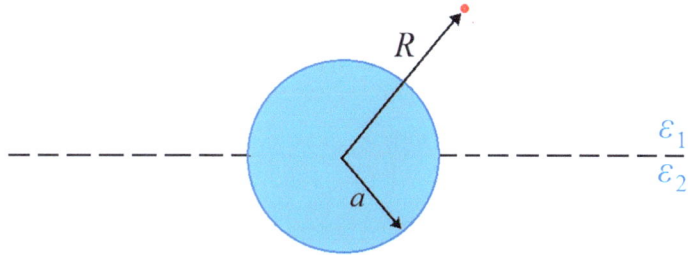

Fig. 11.8 A conducting solid sphere including charge Q placed at the origin of the coordinate system

Exercise

11.17. In Problem 11.6, determine the electric flux density in both regions.

Final Answer

$\vec{D_1} = \dfrac{Q\varepsilon_1}{2\pi(\varepsilon_1+\varepsilon_2)R^2}\hat{R},\ \vec{D_2} = \dfrac{Q\varepsilon_2}{2\pi(\varepsilon_1+\varepsilon_2)R^2}\hat{R}$

Problem

11.18. In Fig. 11.9, the electric field between two concentric spheres is uniform. Which one of the following cases is correct about its dielectric.

Difficulty level   ○ Easy   ○ Normal   ● Hard
Calculation amount   ○ Small   ● Normal   ○ Large

1) $\varepsilon = \varepsilon_0$
2) $\varepsilon \propto \dfrac{1}{R^2}$
3) $\varepsilon \propto \dfrac{1}{R}$
4) $\varepsilon \propto R$

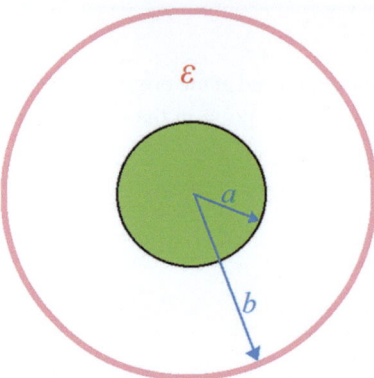

Fig. 11.9 Two concentric spheres with uniform electric field between them

Problem

11.19. A spherical dielectric with radius a is polarized with the polarization vector $\vec{P} = R\hat{R}$. Calculate the polarization surface charge density of the sphere.

Difficulty level   ○ Easy   ● Normal   ○ Hard
Calculation amount   ○ Small   ● Normal   ○ Large
1) a
2) $\dfrac{1}{a}$
3) 0
4) R

Partially Solved Exercise

11.20. A spherical dielectric with radius a is polarized with the polarization vector $\vec{P} = R^2\hat{R}$. Calculate the polarization surface charge density of the sphere.

Solution

As we know, the polarized surface charge density on the surface of a dielectric can be calculated as follows.

$$\rho_{sb} = \vec{P}.\hat{n}$$

Thus, for the spherical surface at $R = a$, we can write:

$$\rho_{sb} = \vec{P}.\hat{R}\bigg|_{R=a}$$

$$\Rightarrow \rho_{sb} = (\quad).(\quad)\bigg|_{R=(\quad)}$$

$$\Rightarrow \rho_{sb} = (\quad)\Big|_{R=(\quad)}$$

$$\Rightarrow \rho_{sb} = a^2$$

Problem

11.21. A spherical dielectric with radius a is polarized with the polarization vector $\vec{P} = \dfrac{1}{R}\hat{R}$. Calculate the polarization volume charge density of the sphere.

Difficulty level   ○ Easy   ○ Normal   ● Hard
Calculation amount   ○ Small   ● Normal   ○ Large

1) $-\dfrac{1}{a^2}$
2) a^2
3) $-\dfrac{1}{R^2}$
4) $-\dfrac{1}{R}$

Partially Solved Exercise

11.22. A spherical dielectric with radius a is polarized with the polarization vector $\vec{P} = \dfrac{1}{R^2}\hat{R}$. Calculate the polarization volume charge density of the sphere.

Solution

The relation between polarization volume charge density (ρ_b) and electric polarization field ($\vec{P}$) is as follows.

$$\rho_b = -\nabla \cdot \vec{P}$$

Hence:

$$\rho_b = -\left(\frac{1}{R^2}\frac{\partial}{\partial R}(R^2 \times (\quad)) + \frac{1}{R\sin\theta}\frac{\partial}{\partial \theta}(\sin\theta \times (\quad)) + \frac{1}{R\sin\theta}\frac{\partial}{\partial \varphi}(\quad)\right)$$

$$\Rightarrow \rho_b = -\left(\frac{1}{R^2}\frac{\partial}{\partial R}(\quad) + (\quad) + (\quad)\right)$$

$$\Rightarrow \rho_b = 0$$

Notes

In this problem, the relations below have been used.

$$\nabla \cdot \vec{F} = \frac{1}{R^2}\frac{\partial}{\partial R}(R^2 F_R) + \frac{1}{R\sin\theta}\frac{\partial}{\partial \theta}(\sin\theta F_\theta) + \frac{1}{R\sin\theta}\frac{\partial}{\partial \varphi}F_\varphi$$

Problem

11.23. In a spherical dielectric, the polarization volume charge density is uniform and equal to ρ_b. Calculate the electric polarization field ($\vec{P}$) inside the sphere.

Difficulty level ○ Easy ○ Normal ● Hard
Calculation amount ○ Small ○ Normal ● Large

1) $\vec{P} = \dfrac{-\rho_b R}{3}\hat{R}$
2) $\vec{P} = \dfrac{-\rho_b}{3R^2}\hat{R}$
3) $\vec{P} = \dfrac{-\rho_b}{3R^3}\hat{R}$
4) $\vec{P} = \dfrac{-\rho_b}{3R}\hat{R}$

References

1. Rahmani-Andebili, M., General Physics II – Practice Problems, Methods, and Solutions, Springer Nature, 2025.
2. Rahmani-Andebili, M., General Physics I – Practice Problems, Methods, and Solutions, Springer Nature, 2025.
3. Rahmani-Andebili, M., Mathematics of Engineering and Science – Practice Problems, Methods, and Solutions, Springer Nature, 2024.
4. Rahmani-Andebili, M., Differential Equations – Practice Problems, Methods, and Solutions, Springer Nature, 2022.
5. Rahmani-Andebili, M., Calculus III – Practice Problems, Methods, and Solutions, Springer Nature, 2023.
6. Rahmani-Andebili, M., Calculus II – Practice Problems, Methods, and Solutions, Springer Nature, 2023.
7. Rahmani-Andebili, M., Calculus I (2nd Ed.) – Practice Problems, Methods, and Solutions, Springer Nature, 2023.
8. Rahmani-Andebili, M., Precalculus (2nd Ed.) – Practice Problems, Methods, and Solutions, Springer Nature, 2024.

12. Polarization and Electric Field in Dielectrics and Boundary Conditions: Part B

Abstract

In this chapter, the problems of the eleventh chapter are fully solved, in detail, step-by-step, and with different methods.

12.1 Polarization and Electric Field in Dielectrics and Boundary Conditions in the Cartesian Coordinate System

12.1. Based on the information given in the problem, we have [1–8]:

$$\vec{D}_1 = 3\hat{x} - 4\hat{y}$$

$$\rho_s = 0$$

As we know, the tangential component of electric field on the boundary of two regions is continuous. In other words:

$$E_{1t} = E_{2t}$$

Moreover, the normal component of electric flux density on the boundary of two regions is discontinuous and the amount of discontinuity is equal to the free surface charge density. Herein, $\hat{n}$ is the unit normal vector of the first region. In other words:

$$\vec{D}_2 \cdot \hat{n} - \vec{D}_1 \cdot \hat{n} = \rho_s$$

Therefore, for this problem, we can write:

$$E_{1y} = E_{2y}$$

$$\Rightarrow \frac{D_{1y}}{\varepsilon_0 \varepsilon_{r1}} = \frac{D_{2y}}{\varepsilon_0 \varepsilon_{r2}}$$

$$\Rightarrow D_{2y} = \frac{\varepsilon_{r2}}{\varepsilon_{r1}} D_{1y} = \frac{3}{1} \times (-4)$$

$$\Rightarrow D_{2y} = -12$$

$$\Rightarrow \overrightarrow{D_{2t}} = -12\widehat{y}$$

In addition:

$$\overrightarrow{D_2}.\widehat{x} - \overrightarrow{D_1}.\widehat{x} = 0$$

$$D_{2x} - D_{1x} = 0$$

$$\Rightarrow D_{2x} = D_{1x}$$

$$\Rightarrow D_{2x} = 3$$

$$\Rightarrow \overrightarrow{D_{2n}} = 3\widehat{x}$$

Hence:

$$\overrightarrow{D_2} = \overrightarrow{D_{2t}} + \overrightarrow{D_{2n}}$$

$$\Rightarrow \overrightarrow{D_2} = 3\widehat{x} - 12\widehat{y}$$

Choice (3) is the answer (Fig. 12.1).

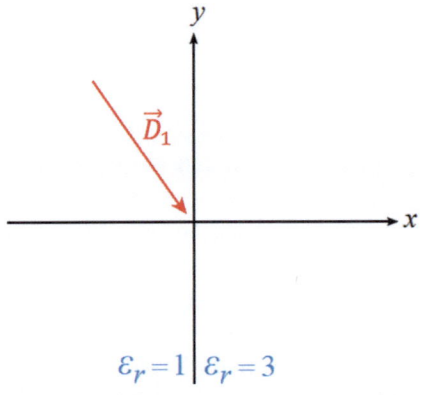

Fig. 12.1 Studying the boundary conditions for the electric flux density

12.2. As we know, the relation between the incident angle and refracted angle of electric field is as follows.

$$\frac{\tan \alpha_2}{\tan \alpha_1} = \frac{\varepsilon_{r2}}{\varepsilon_{r1}}$$

$$\Rightarrow \frac{\tan \alpha_2}{\frac{\sqrt{E_{1x}^2 + E_{1y}^2}}{E_{1z}}} = \frac{\varepsilon_{r2}}{\varepsilon_{r1}}$$

$$\Rightarrow \frac{\tan \alpha_2}{\frac{\sqrt{(1)^2 + (-3)^2}}{5}} = \frac{5}{2}$$

$$\Rightarrow \frac{\tan \alpha_2}{\frac{\sqrt{10}}{5}} = \frac{5}{2}$$

$$\Rightarrow \tan \alpha_2 = \frac{5}{2} \times \frac{\sqrt{10}}{5} = \frac{\sqrt{10}}{2} = 1.581$$

$$\Rightarrow \alpha_2 \approx 57.7°$$

Choice (1) is the answer (Fig. 12.2).

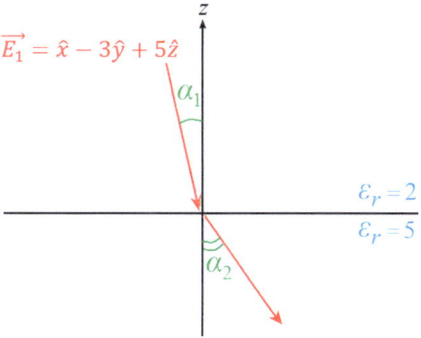

Fig. 12.2 Studying the boundary conditions for the electric field

12.4. Based on the information given in the problem, we have:

$$\varepsilon_1 = 3\varepsilon_0$$

$$\varepsilon_2 = 4.5\varepsilon_0$$

$$\vec{E}_1 = 10\hat{x} - 6\hat{y} + 12\hat{z} \ V/m$$

$$\rho_s = 0$$

As we know, the tangential component of electric field on the boundary of two regions is continuous. In other words:

$$E_{1t} = E_{2t} \qquad (12.1)$$

Moreover, the normal component of electric flux density on the boundary of two regions is discontinuous and the amount of discontinuity is equal to the free surface charge density. Herein, $\hat{n}$ is the unit normal vector of the first region. In other words:

$$\vec{D}_2 \cdot \hat{n} - \vec{D}_1 \cdot \hat{n} = \rho_s \qquad (12.2)$$

From (12.1) we have:

$$\vec{E}_{2t} = 10\hat{x} + 12\hat{z}$$

Also, from (12.2) we have:

$$\vec{D}_2 \cdot (-\hat{y}) - \vec{D}_1 \cdot (-\hat{y}) = 0$$

$$\Rightarrow -D_{2y} + D_{1y} = 0$$

$$\Rightarrow \varepsilon_2 E_{2y} = \varepsilon_1 E_{1y}$$

$$\Rightarrow 4.5\varepsilon_0 E_{2y} = 3\varepsilon_0(-6)$$

$$\Rightarrow E_{2y} = -4$$

$$\Rightarrow \vec{E}_{2n} = -4\hat{y}$$

Therefore:

$$\vec{E}_2 = \vec{E}_{2t} + \vec{E}_{2n}$$

$$\Rightarrow \vec{E}_2 = 10\hat{x} - 4\hat{y} + 12\hat{z}$$

On the other hand, we know that electric energy density in an environment can be calculated as follows.

$$u = \frac{1}{2}\varepsilon E^2$$

Thus:

$$\frac{u_2}{u_1} = \frac{\frac{1}{2}\varepsilon_2 E_2^2}{\frac{1}{2}\varepsilon_1 E_1^2}$$

$$\Rightarrow \frac{u_2}{u_1} = \frac{\frac{1}{2} \times 4.5\varepsilon_0 \left(\sqrt{(10)^2 + (-4)^2 + (12)^2}\right)^2}{\frac{1}{2} \times 3\varepsilon_0 \left(\sqrt{(10)^2 + (-6)^2 + (12)^2}\right)^2}$$

$$\Rightarrow \frac{u_2}{u_1} = \frac{3(100 + 16 + 144)}{2(100 + 36 + 144)}$$

$$\Rightarrow \frac{u_2}{u_1} = \frac{3(260)}{2(280)}$$

$$\Rightarrow \frac{u_2}{u_1} = \frac{39}{28}$$

Choice (1) is the answer (Fig. 12.3).

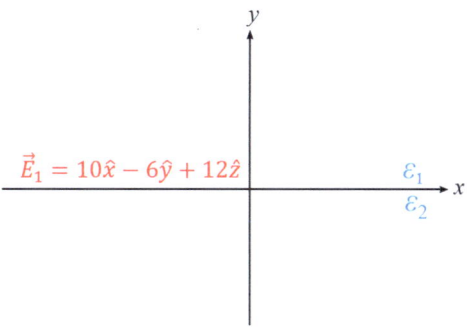

Fig. 12.3 Studying the boundary conditions for the electric field

12.5. As we know, the normal component of electric flux density on the boundary of two regions is discontinuous and the amount of discontinuity is equal to the free surface charge density. Herein, $\hat{n}$ is the unit normal vector of the first region. In other words:

$$\vec{D}_2 \cdot \hat{n} - \vec{D}_1 \cdot \hat{n} = \rho_s$$

Therefore:

$$\varepsilon_0 \varepsilon_{2r} E_{2n} - \varepsilon_0 \varepsilon_{1r} E_{1n} = 0$$

$$\Rightarrow \varepsilon_{2r} E_{2n} = \varepsilon_{1r} E_{1n}$$

$$\Rightarrow E_{1n} = \frac{\varepsilon_{2r}}{\varepsilon_{1r}} E_{2n}$$

In addition, the relation between electric field and electric polarization field is as follows.

$$\vec{P} = \vec{D} - \varepsilon_0 \vec{E}$$

$$\Rightarrow \vec{P} = \varepsilon_0 (\varepsilon_r - 1) \vec{E}$$

Thus:

$$\frac{P_{2n}}{P_{1n}} = \frac{\varepsilon_0 (\varepsilon_{2r} - 1) E_{2n}}{\varepsilon_0 (\varepsilon_{1r} - 1) \frac{\varepsilon_{2r}}{\varepsilon_{1r}} E_{2n}}$$

$$\Rightarrow \frac{P_{2n}}{P_{1n}} = \frac{\varepsilon_{1r}(\varepsilon_{2r} - 1)}{\varepsilon_{2r}(\varepsilon_{1r} - 1)}$$

Choice (4) is the answer.

12.2 Polarization and Electric Field in Dielectrics in Cylindrical Coordinate System

12.6. As we know, the tangential component of electric field on the boundary of two regions is continuous. In other words:

$$E_{1t} = E_{2t}$$

Hence, for this problem, we can write:

$$E_{1r} = E_{2r} \triangleq E_r$$

Now, by applying Gauss's law for the cylindrical Gaussian surface, we have:

$$Q_{in} = \int \vec{D} \cdot \vec{dS}$$

$$\Rightarrow Q = \int \varepsilon \vec{E} \cdot \vec{dS}$$

$$\Rightarrow Q = \int_{z=0}^{l} \int_{\varphi=0}^{\pi} \varepsilon_1 E_r r d\varphi dz + \int_{z=0}^{l} \int_{\varphi=\pi}^{2\pi} \varepsilon_2 E_r r d\varphi dz$$

$$\Rightarrow Q = \varepsilon_1 E_r r \int_{z=0}^{l} \int_{\varphi=0}^{\pi} d\varphi dz + \varepsilon_2 E_r r \int_{z=0}^{l} \int_{\varphi=\pi}^{2\pi} d\varphi dz$$

$$\Rightarrow Q = \varepsilon_1 E_r r \left[\varphi\right]_0^{\pi} \left[z\right]_0^l + \varepsilon_2 E_r r \left[\varphi\right]_{\pi}^{2\pi} \left[z\right]_0^l$$

$$\Rightarrow Q = \varepsilon_1 E_r r (\pi - 0)(l - 0) + \varepsilon_2 E_r r (\pi - 0)(l - 0)$$

$$\Rightarrow Q = \varepsilon_1 E_r r \pi l + \varepsilon_2 E_r r \pi l$$

$$\Rightarrow E_r = \frac{Q}{\pi l (\varepsilon_1 + \varepsilon_2) r}$$

$$\Rightarrow \vec{E_1} = \vec{E_2} = \frac{Q}{\pi l (\varepsilon_1 + \varepsilon_2) r} \hat{r}$$

Choice (4) is the answer (Fig. 12.4).

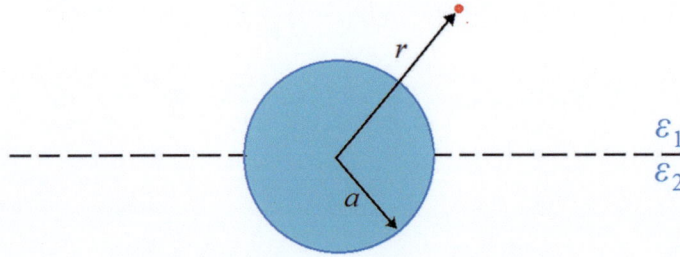

Fig. 12.4 A conducting solid cylinder with length l and charge Q placed at the origin of the coordinate system

12.2 Polarization and Electric Field in Dielectrics in Cylindrical Coordinate System

12.8. First, we need to calculate the electric field inside the coaxial cable. As is shown in Fig. 12.5, a cylindrical Gaussian surface is drawn around the inner cable. Now, by applying Gauss's law we have:

$$Q_{in} = \int \vec{D} \cdot \vec{dS}$$

$$\Rightarrow \int \rho_v dV = \int \varepsilon \vec{E} \cdot \vec{dS}$$

$$\Rightarrow \int_{z=0}^{z=l} \int_{\varphi=0}^{2\pi} \int_{r=0}^{a} \rho_v r\, dr\, d\varphi\, dz = \int_{z=0}^{z=l} \int_{\varphi=0}^{2\pi} \varepsilon E_r r\, d\varphi\, dz$$

$$\Rightarrow \rho_v \left[\frac{r^2}{2}\right]_0^a \int_{z=0}^{z=l} \int_{\varphi=0}^{2\pi} d\varphi\, dz = \varepsilon E_r r \int_{z=0}^{z=l} \int_{\varphi=0}^{2\pi} d\varphi\, dz$$

$$\Rightarrow \rho_v \left(\frac{a^2}{2} - 0\right) = \varepsilon E_r r$$

$$\Rightarrow E_r = \frac{\rho_v a^2}{2\varepsilon r}$$

$$\Rightarrow \vec{E} = \frac{\rho_v a^2}{2\varepsilon r} \hat{r}$$

As can be noticed, to make the magnitude of electric field constant, we must have:

$$\varepsilon r = \text{Constant}$$

$$\Rightarrow \varepsilon \propto \frac{1}{r}$$

In other words, the permittivity of the dielectric must be inversely proportional to the radius of the cylinder. Choice (3) is the answer.

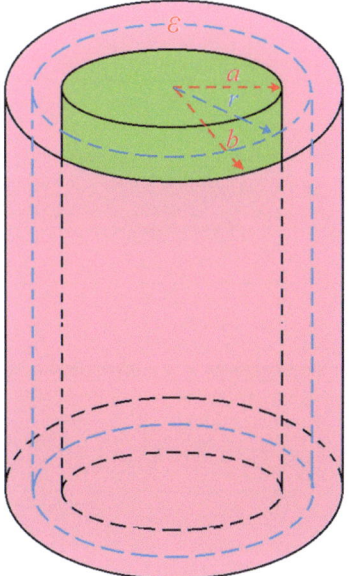

Fig. 12.5 Applying a cylindrical Gaussian surface around the inner cable to calculate the electric field

Notes

In this problem, the relation below has been used.

$$\int x^n dx = \frac{x^{n+1}}{n+1}$$

12.9. Based on the information given in the problem, we have:

$$\vec{P} = \frac{1}{r}\hat{r} + z\hat{z}$$

The polarized surface charge density on the surface of a dielectric can be calculated as follows.

$$\rho_{sb} = \vec{P} \cdot \hat{n}$$

Thus, for the cylindrical surface at $r = a$, we can write:

$$\rho_{sb} = \vec{P} \cdot \hat{r}\bigg|_{r=a}$$

$$\Rightarrow \rho_{sb} = \left(\frac{1}{r}\hat{r} + z\hat{z}\right) \cdot \hat{r}\bigg|_{r=a}$$

$$\Rightarrow \rho_{sb} = \frac{1}{r}\bigg|_{r=a}$$

$$\Rightarrow \rho_{sb} = \frac{1}{a}$$

Choice (2) is the answer.

Notes

In this problem, the relations below have been used.

$$\vec{a} \cdot \vec{b} = \left(a_1\hat{i} + a_2\hat{j} + a_3\hat{k}\right) \cdot \left(b_1\hat{i} + b_2\hat{j} + b_3\hat{k}\right) = a_1b_1 + a_2b_2 + a_3b_3$$

12.2 Polarization and Electric Field in Dielectrics in Cylindrical Coordinate System

12.11. Based on the information given in the problem, we have:

$$\vec{P} = \frac{1}{r}\hat{r} + z\hat{z}$$

The relation between polarization volume charge density (ρ_b) and electric polarization field ($\vec{P}$) is as follows.

$$\rho_b = -\nabla \cdot \vec{P}$$

Hence:

$$\rho_b = -\left(\frac{1}{r}\frac{\partial}{\partial r}(rP_r) + \frac{1}{r}\frac{\partial}{\partial \varphi}(P_\varphi) + \frac{\partial}{\partial z}(P_z)\right)$$

$$\Rightarrow \rho_b = -\left(\frac{1}{r}\frac{\partial}{\partial r}\left(r \times \frac{1}{r}\right) + \frac{1}{r}\frac{\partial}{\partial \varphi}(0) + \frac{\partial}{\partial z}(z)\right)$$

$$\Rightarrow \rho_b = -\left(\frac{1}{r}\frac{\partial}{\partial r}(1) + \frac{1}{r}\frac{\partial}{\partial \varphi}(0) + \frac{\partial}{\partial z}(z)\right)$$

$$\Rightarrow \rho_b = -\left(\frac{1}{r}\times 0 + \frac{1}{r}\times 0 + 1\right)$$

$$\Rightarrow \rho_b = -1$$

Choice (4) is the answer.

> **Notes**
> In this problem, the relations below have been used.
>
> $$\nabla \cdot \vec{F} = \frac{1}{r}\frac{\partial}{\partial r}(rF_r) + \frac{1}{r}\frac{\partial}{\partial \varphi}(F_\varphi) + \frac{\partial}{\partial z}(F_z)$$

12.13. From the previous chapters, we know that the electric field in the region $a < r < b$ is as follows:

$$\vec{E} = \frac{Q}{2\pi \varepsilon r l}\hat{r}$$

Since ρ_l is the free charge per unit length, we can rewrite the electric field equation as follows.

$$\vec{E} = \frac{\rho_l l}{2\pi \varepsilon r l}\hat{r}$$

$$\Rightarrow \vec{E} = \frac{\rho_l}{2\pi \varepsilon r}\hat{r}$$

On the other hand, we know that the relation between electric field and electric polarization field is as follows.

$$\vec{P} = \vec{D} - \varepsilon_0 \vec{E}$$

$$\Rightarrow \vec{P} = \varepsilon_0(\varepsilon_r - 1)\vec{E}$$

Hence:

$$\vec{P} = \varepsilon_0(\varepsilon_r - 1)\frac{\rho_l}{2\pi\varepsilon r}\hat{r}$$

$$\Rightarrow \vec{P} = (\varepsilon_r - 1)\frac{\rho_l}{2\pi\varepsilon_r r}\hat{r}$$

$$\Rightarrow \vec{P} = (3-1)\frac{\rho_l}{2\pi(3)r}\hat{r}$$

$$\Rightarrow \vec{P} = \frac{\rho_l}{3\pi r}\hat{r}$$

The polarized surface density on the surface of a dielectric can be calculated as follows.

$$\rho_{sb} = \vec{P}\cdot\hat{n}$$

Thus, for the cylindrical surface at $r = b$, we have:

$$\rho_{sb} = \vec{P}\cdot\hat{r}\Big|_{r=b}$$

$$\Rightarrow \rho_{sb} = \frac{\rho_l}{3\pi r}\hat{r}\cdot\hat{r}\Big|_{r=b}$$

$$\Rightarrow \rho_{sb} = \frac{\rho_l}{3\pi b}$$

And the polarized charge per unit length on the surface of dielectric at $r = b$ can be calculated as follows.

$$q_b = \rho_{sb}(2\pi b)$$

$$\Rightarrow q_b = \frac{\rho_l}{3\pi b}(2\pi b) = \frac{2\rho_L}{3}$$

$$\Rightarrow \frac{q_b}{\rho_l} = \frac{2}{3}$$

Choice (4) is the answer (Fig. 12.6).

12.2 Polarization and Electric Field in Dielectrics in Cylindrical Coordinate System

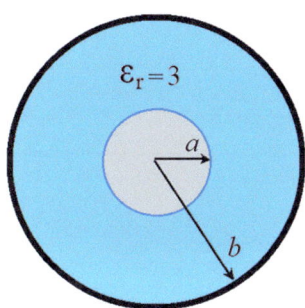

Fig. 12.6 A coaxial cable including free and polarized charges

> **Notes**
> In this problem, the relations below have been used.
> $$\vec{a} \cdot \vec{b} = \left(a_1\hat{i} + a_2\hat{j} + a_3\hat{k}\right) \cdot \left(b_1\hat{i} + b_2\hat{j} + b_3\hat{k}\right) = a_1b_1 + a_2b_2 + a_3b_3$$

12.14. The relation between polarization volume charge density (ρ_b) and electric polarization field ($\vec{P}$) in differential and integral forms is as follows.

$$\rho_b = -\nabla \cdot \vec{P}$$

$$Q_b = -\int \vec{P} \cdot \overrightarrow{dS}$$

Hence, for this problem, we can write:

$$\int_{z=0}^{l} \int_{\varphi=0}^{2\pi} \int_{r=0}^{r} \rho_b r \, dr \, d\varphi \, dz = -\int_{z=0}^{l} \int_{\varphi=0}^{2\pi} P_r r \, d\varphi \, dz$$

$$\Rightarrow \rho_b \int_{z=0}^{l} \int_{\varphi=0}^{2\pi} \int_{r=0}^{r} r \, dr \, d\varphi \, dz = -P_r r \int_{z=0}^{l} \int_{\varphi=0}^{2\pi} d\varphi \, dz$$

$$\Rightarrow \rho_b \left[\frac{r^2}{2}\right]_0^r \int_{z=0}^{l} \int_{\varphi=0}^{2\pi} d\varphi \, dz = -P_r r \int_{z=0}^{l} \int_{\varphi=0}^{2\pi} d\varphi \, dz$$

$$\Rightarrow \rho_b \left(\frac{r^2}{2} - 0\right) = -P_r r$$

$$\Rightarrow P_r = \frac{-\rho_b r}{2}$$

$$\Rightarrow \vec{P} = \frac{-\rho_b r}{2}\hat{r}$$

Choice (3) is the answer.

> **Notes**
> In this problem, the relation below has been used.
> $$\int x^n dx = \frac{x^{n+1}}{n+1}$$

12.3 Polarization and Electric Field in Dielectrics in Spherical Coordinate System

12.15. Based on the information given in the problem, we have:

$$\varepsilon = \varepsilon_0(1+2R)$$

$$\rho = \rho_0\left(1+\frac{R}{a}\right)$$

$$R = \frac{a}{2}$$

The problem can be solved by applying Gauss's law for the Gaussian surface shown in Fig. 12.7.

$$Q_{in} = \int \vec{D}\cdot \vec{dS}$$

$$\Rightarrow \int \rho dV = \int \varepsilon \vec{E}\cdot \vec{dS}$$

$$\Rightarrow \int \rho_0\left(1+\frac{R}{a}\right)dV = \int \varepsilon_0(1+2R)\vec{E}\cdot\vec{dS}$$

$$\Rightarrow \int_{\varphi=0}^{2\pi}\int_{\theta=0}^{\pi}\int_{R=0}^{R}\rho_0\left(1+\frac{R}{a}\right)R^2\sin\theta\, dRd\theta d\varphi = \int_{\varphi=0}^{2\pi}\int_{\theta=0}^{\pi}\varepsilon_0(1+2R)E_R R^2\sin\theta\, d\theta d\varphi$$

$$\Rightarrow \rho_0\int_{\varphi=0}^{2\pi}\int_{\theta=0}^{\pi}\int_{R=0}^{R}\left(R^2+\frac{R^3}{a}\right)\sin\theta\, dRd\theta d\varphi = \varepsilon_0(1+2R)E_R R^2\int_{\varphi=0}^{2\pi}\int_{\theta=0}^{\pi}\sin\theta\, d\theta d\varphi$$

$$\Rightarrow \rho_0\left[\frac{R^3}{3}+\frac{R^4}{4a}\right]_0^R \int_{\varphi=0}^{2\pi}\int_{\theta=0}^{\pi}\sin\theta\, d\theta d\varphi = \varepsilon_0(1+2R)E_R R^2 \int_{\varphi=0}^{2\pi}\int_{\theta=0}^{\pi}\sin\theta\, d\theta d\varphi$$

$$\Rightarrow \rho_0 \left(\frac{R^3}{3} + \frac{R^4}{4a} - 0 \right) = \varepsilon_0 (1 + 2R) E_R R^2$$

$$\Rightarrow E_R = \frac{\rho_0 \left(\frac{R^3}{3} + \frac{R^4}{4a} \right)}{\varepsilon_0 (1 + 2R) R^2}$$

$$\Rightarrow E_R \bigg|_{R = \frac{a}{2}} = \frac{\rho_0 \left(\frac{\left(\frac{a}{2}\right)^3}{3} + \frac{\left(\frac{a}{2}\right)^4}{4a} \right)}{\varepsilon_0 \left(1 + 2\left(\frac{a}{2}\right) \right) \left(\frac{a}{2}\right)^2}$$

$$\Rightarrow E_R \bigg|_{R = \frac{a}{2}} = \frac{\rho_0 \left(\frac{a^3}{24} + \frac{a^3}{64} \right)}{\varepsilon_0 (1 + a) \frac{a^2}{4}}$$

$$\Rightarrow E_R \bigg|_{R = \frac{a}{2}} = \frac{11 \rho_0 a}{48 \varepsilon_0 (1 + a)}$$

Choice (1) is the answer.

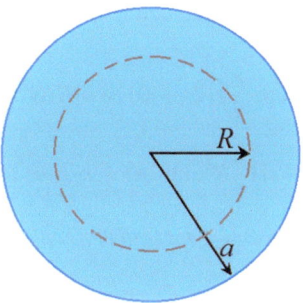

Fig. 12.7 Applying Gauss's law for the Gaussian surface

> **Notes**
> In this problem, the relation below has been used.
> $$\int x^n dx = \frac{x^{n+1}}{n+1}$$

12.16. As we know, the tangential component of electric field on the boundary of two regions is continuous. In other words:

$$E_{1t} = E_{2t}$$

Hence, for this problem, we can write:

$$E_{1R} = E_{2R} \triangleq E_R$$

Now, by applying Gauss's law for the spherical Gaussian surface, we have:

$$Q_{in} = \int \vec{D}.\overrightarrow{dS}$$

$$\Rightarrow Q = \int \varepsilon \vec{E}.\overrightarrow{dS}$$

$$\Rightarrow Q = \int_{\varphi=0}^{2\pi} \int_{\theta=0}^{\frac{\pi}{2}} \varepsilon_1 E_R R^2 \sin\theta\, d\theta d\varphi + \int_{\varphi=0}^{2\pi} \int_{\theta=\frac{\pi}{2}}^{\pi} \varepsilon_2 E_R R^2 \sin\theta\, d\theta d\varphi$$

$$\Rightarrow Q = \varepsilon_1 E_R R^2 \int_{\varphi=0}^{2\pi} \int_{\theta=0}^{\frac{\pi}{2}} \sin\theta\, d\theta d\varphi + \varepsilon_2 E_R R^2 \int_{\varphi=0}^{2\pi} \int_{\theta=\frac{\pi}{2}}^{\pi} \varepsilon_2 E_R R^2 \sin\theta\, d\theta d\varphi$$

$$\Rightarrow Q = \varepsilon_1 E_R R^2 \left[-\cos\theta\right]_0^{\frac{\pi}{2}} \left[\varphi\right]_0^{2\pi} + \varepsilon_2 E_R R^2 \left[-\cos\theta\right]_{\frac{\pi}{2}}^{\pi} \left[\varphi\right]_0^{2\pi}$$

$$\Rightarrow Q = \varepsilon_1 E_R R^2 \left(\cos 0 - \cos\frac{\pi}{2}\right)(2\pi - 0) + \varepsilon_2 E_R R^2 \left(\cos\frac{\pi}{2} - \cos\pi\right)(2\pi - 0)$$

$$\Rightarrow Q = \varepsilon_1 E_R R^2 (1)(2\pi) + \varepsilon_2 E_R R^2 (1)(2\pi)$$

$$\Rightarrow Q = E_R 2\pi R^2 (\varepsilon_1 + \varepsilon_2)$$

$$\Rightarrow E_R = \frac{Q}{2\pi(\varepsilon_1 + \varepsilon_2)R^2}$$

$$\Rightarrow \vec{E_1} = \vec{E_2} = \frac{Q}{2\pi(\varepsilon_1 + \varepsilon_2)R^2}\hat{R}$$

Choice (2) is the answer (Fig. 12.8).

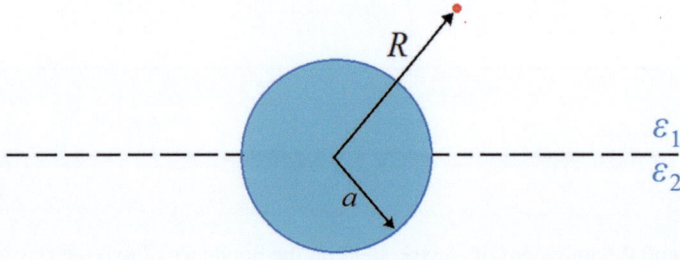

Fig. 12.8 A conducting solid sphere including charge Q placed at the origin of the coordinate system

12.3 Polarization and Electric Field in Dielectrics in Spherical Coordinate System

Notes

In this problem, the relation below has been used.

$$\int \sin x\, dx = -\cos x$$

$$\cos 0 = 1$$

$$\cos \frac{\pi}{2} = 0$$

$$\cos \pi = -1$$

12.18. First, we need to calculate the electric field between the two concentric spheres. As is shown in Fig. 12.9, a spherical Gaussian surface is drawn around the inner sphere. Now, by applying Gauss's law we have:

$$Q_{in} = \int \vec{D}\cdot \vec{dS}$$

$$\Rightarrow \int \rho\, dV = \int \varepsilon \vec{E}\cdot \vec{dS}$$

$$\Rightarrow \int \rho_v\, dV = \int \varepsilon \vec{E}\cdot \vec{dS}$$

$$\Rightarrow \int_{\varphi=0}^{2\pi}\int_{\theta=0}^{\pi}\int_{R=0}^{R} \rho_v R^2 \sin\theta\, dR\, d\theta\, d\varphi = \int_{\varphi=0}^{2\pi}\int_{\theta=0}^{\pi} \varepsilon E_R R^2 \sin\theta\, d\theta\, d\varphi$$

$$\Rightarrow \rho_v \left[\frac{R^3}{3}\right]_0^a \int_{\varphi=0}^{2\pi}\int_{\theta=0}^{\pi} \sin\theta\, d\theta\, d\varphi = \varepsilon E_R R^2 \int_{\varphi=0}^{2\pi}\int_{\theta=0}^{\pi} \sin\theta\, d\theta\, d\varphi$$

$$\Rightarrow \rho_v\left(\frac{a^3}{3} - 0\right) = \varepsilon E_R R^2$$

$$\Rightarrow E_R = \frac{\rho_v a^3}{3\varepsilon R^2}$$

$$\Rightarrow \vec{E} = \frac{\rho_v a^3}{3\varepsilon R^2}\hat{R}$$

As can be seen, to make the magnitude of electric field constant, we must have:

$$\varepsilon R^2 = \text{Constant}$$

$$\Rightarrow \varepsilon \propto \frac{1}{R^2}$$

In other words, the permittivity of the dielectric must be inversely proportional to squared value of radius of concentric spheres. Choice (2) is the answer.

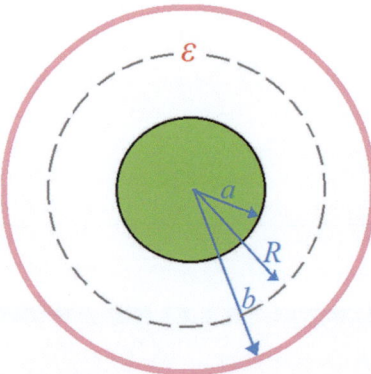

Fig. 12.9 Two concentric spheres with uniform electric field between them

Notes

In this problem, the relation below has been used.

$$\int x^n dx = \frac{x^{n+1}}{n+1}$$

12.19. Based on the information given in the problem, we have:

$$\vec{P} = R\hat{R}$$

The polarized surface charge density on the surface of a dielectric can be calculated as follows.

$$\rho_{sb} = \vec{P} \cdot \hat{n}$$

Thus, for the spherical surface at $R = a$, we can write:

$$\rho_{sb} = \vec{P} \cdot \hat{R} \bigg|_{R=a}$$

$$\Rightarrow \rho_{sb} = \left(R\hat{R} \right) \cdot \hat{R} \bigg|_{R=a}$$

12.3 Polarization and Electric Field in Dielectrics in Spherical Coordinate System

$$\Rightarrow \rho_{sb} = R \Big|_{R=a}$$

$$\Rightarrow \rho_{sb} = a$$

Choice (1) is the answer.

> **Notes**
>
> In this problem, the relations below have been used.
>
> $$\vec{a} \cdot \vec{b} = \left(a_1\hat{i} + a_2\hat{j} + a_3\hat{k}\right) \cdot \left(b_1\hat{i} + b_2\hat{j} + b_3\hat{k}\right) = a_1b_1 + a_2b_2 + a_3b_3$$

12.21. Based on the information given in the problem, we have:

$$\vec{P} = \frac{1}{R}\hat{R}$$

The relation between polarization volume charge density (ρ_b) and electric polarization field ($\vec{P}$) is as follows.

$$\rho_b = -\nabla \cdot \vec{P}$$

Hence:

$$\rho_b = -\left(\frac{1}{R^2}\frac{\partial}{\partial R}\left(R^2 \times \frac{1}{R}\right) + \frac{1}{R\sin\theta}\frac{\partial}{\partial \theta}(\sin\theta \times 0) + \frac{1}{R\sin\theta}\frac{\partial}{\partial \varphi}(0)\right)$$

$$\Rightarrow \rho_b = -\left(\frac{1}{R^2}\frac{\partial}{\partial R}(R) + 0 + 0\right)$$

$$\Rightarrow \rho_b = -\left(\frac{1}{R^2} \times 1\right)$$

$$\Rightarrow \rho_b = -\frac{1}{R^2}$$

Choice (3) is the answer.

> **Notes**
>
> In this problem, the relations below have been used.
>
> $$\nabla \cdot \vec{F} = \frac{1}{R^2}\frac{\partial}{\partial R}(R^2 F_R) + \frac{1}{R\sin\theta}\frac{\partial}{\partial \theta}(\sin\theta F_\theta) + \frac{1}{R\sin\theta}\frac{\partial}{\partial \varphi}F_\varphi$$

12.23. The relation between polarization volume charge density (ρ_b) and electric polarization field ($\vec{P}$) in differential and integral forms is as follows.

$$\rho_b = -\nabla \cdot \vec{P}$$

$$Q_b = -\int \vec{P} \cdot \vec{dS}$$

Hence, for this problem, we can write:

$$\int_{\varphi=0}^{2\pi}\int_{\theta=0}^{\pi}\int_{R=0}^{R}\rho_b R^2 \sin\theta\, dR d\theta d\varphi = -\int_{\varphi=0}^{2\pi}\int_{\theta=0}^{\pi} P_R R^2 \sin\theta\, d\theta d\varphi$$

$$\Rightarrow \rho_b \int_{\varphi=0}^{2\pi}\int_{\theta=0}^{\pi}\int_{R=0}^{R} R^2 \sin\theta\, dR d\theta d\varphi = -P_R R^2 \int_{\varphi=0}^{2\pi}\int_{\theta=0}^{\pi} \sin\theta\, d\theta d\varphi$$

$$\Rightarrow \rho_b \left[\frac{R^3}{3}\right]_0^R \int_{\varphi=0}^{2\pi}\int_{\theta=0}^{\pi} \sin\theta\, d\theta d\varphi = -P_R R^2 \int_{\varphi=0}^{2\pi}\int_{\theta=0}^{\pi} \sin\theta\, d\theta d\varphi$$

$$\Rightarrow \rho_b \left(\frac{R^3}{3} - 0\right) = -P_R R^2$$

$$\Rightarrow P_R = \frac{-\rho_b R}{3}$$

$$\Rightarrow \vec{P} = \frac{-\rho_b R}{3}\hat{R}$$

Choice (1) is the answer.

> **Notes**
>
> In this problem, the relation below has been used.
>
> $$\int x^n dx = \frac{x^{n+1}}{n+1}$$

References

1. Rahmani-Andebili, M., General Physics II – Practice Problems, Methods, and Solutions, Springer Nature, 2025.
2. Rahmani-Andebili, M., General Physics I – Practice Problems, Methods, and Solutions, Springer Nature, 2025.
3. Rahmani-Andebili, M., Mathematics of Engineering and Science – Practice Problems, Methods, and Solutions, Springer Nature, 2024.
4. Rahmani-Andebili, M., Differential Equations – Practice Problems, Methods, and Solutions, Springer Nature, 2022.
5. Rahmani-Andebili, M., Calculus III – Practice Problems, Methods, and Solutions, Springer Nature, 2023.
6. Rahmani-Andebili, M., Calculus II – Practice Problems, Methods, and Solutions, Springer Nature, 2023.
7. Rahmani-Andebili, M., Calculus I (2nd Ed.) – Practice Problems, Methods, and Solutions, Springer Nature, 2023.
8. Rahmani-Andebili, M., Precalculus (2nd Ed.) – Practice Problems, Methods, and Solutions, Springer Nature, 2024.

Flat, Cylindrical, and Spherical Capacitors: Part A 13

Abstract

In this chapter, the basic and advanced problems of flat, cylindrical, and spherical capacitors with uniform and nonuniform dielectrics are studied. Herein, different types of problems and exercises are presented that are categorized as follows.

- **Problems with detailed solution**: They have been designed to teach students the subjects in detail. Moreover, they have been categorized in different levels based on their difficulty levels (easy, normal, and hard) and calculation amounts (small, normal, and large).
- **Partially solved exercises**: They have been designed to encourage students to practice problems while guiding them through the problem-solving procedure and hinting the required formulas.
- **Exercises with final answer**: They have been designed to encourage students to practice more by themselves while hinting them by the final answer as well as to help instructors to give tests or quizzes.

13.1 Flat Capacitors

Problem

13.1. Figure 13.1, the surface area of each plate of a flat capacitor is A. Moreover, the distance between its plates is d where air fills the gap. If a perfect conductor fills 50% of the gap, calculate the capacitance of the capacitor [1–8].

Difficulty level ● Easy   ○ Normal   ○ Hard
Calculation amount ● Small   ○ Normal   ○ Large

1) $\dfrac{\varepsilon_0 A}{d}$
2) $\dfrac{2\varepsilon_0 A}{d}$
3) $\dfrac{\varepsilon_0 A}{2d}$
4) $\dfrac{4\varepsilon_0 A}{d}$

Fig. 13.1 A flat capacitor

Problem

13.2. In Fig. 13.2, calculate the capacitance of the capacitor. Herein, the surface area of each plate of the flat capacitor is A.

Difficulty level ○ Easy ● Normal ○ Hard
Calculation amount ● Small ○ Normal ○ Large

1) $\dfrac{(\varepsilon_1 + \varepsilon_2)A}{2d}$
2) $\dfrac{(\varepsilon_1 + \varepsilon_2)A}{d}$
3) $\dfrac{(\varepsilon_1 + \varepsilon_2)A}{\varepsilon_1 \varepsilon_2 d}$
4) $\dfrac{\varepsilon_1 \varepsilon_2 A}{2(\varepsilon_1 + \varepsilon_2)d}$

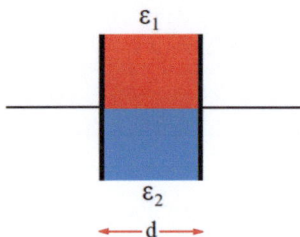

Fig. 13.2 A flat capacitor including two different electrical permittivity

Problem

13.3. In Fig. 13.3, calculate the capacitance of the capacitor. Herein, the surface area of each plate of the flat capacitor is A.

Difficulty level ○ Easy ● Normal ○ Hard
Calculation amount ○ Small ● Normal ○ Large

1) $\dfrac{2(\varepsilon_1 + \varepsilon_2)A}{d}$
2) $\dfrac{(\varepsilon_1 + \varepsilon_2)A}{d}$
3) $\dfrac{\varepsilon_1 \varepsilon_2 A}{(\varepsilon_1 + \varepsilon_2)d}$
4) $\dfrac{2\varepsilon_1 \varepsilon_2 A}{(\varepsilon_1 + \varepsilon_2)d}$

13.1 Flat Capacitors

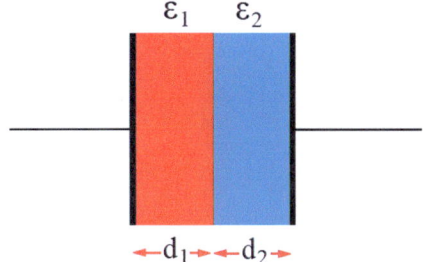

Fig. 13.3 A flat capacitor including two different electrical permittivity

Problem

13.4. Figure 13.4 illustrates two regions with different electrical permittivity. If they are connected to a voltage source, calculate the ratio of electric potential of the first region to that of the second one.

Difficulty level   ○ Easy   ● Normal   ○ Hard
Calculation amount ○ Small   ● Normal   ○ Large

1) $\dfrac{\varepsilon_2 d_1}{\varepsilon_1 d_2}$

2) $\dfrac{\varepsilon_2}{\varepsilon_1}$

3) $\dfrac{\varepsilon_1}{\varepsilon_2}$

4) $\dfrac{\varepsilon_2 d_2}{\varepsilon_1 d_1}$

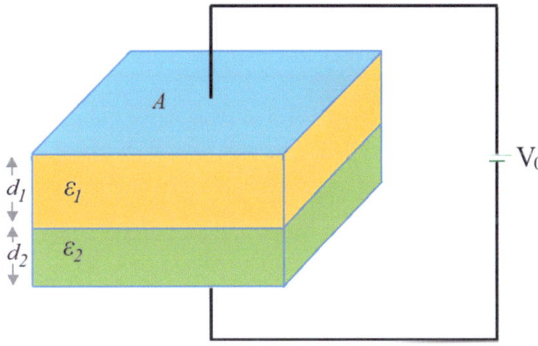

Fig. 13.4 A flat capacitor including two different electrical permittivity connected to a voltage source

Problem

13.5. The flat capacitor, shown in Fig. 13.5, is connected to a voltage source. Calculate the amount of force exerted on the top plate of the capacitor.

Difficulty level   ○ Easy   ○ Normal   ● Hard
Calculation amount ○ Small   ○ Normal   ● Large

1) $-\dfrac{\varepsilon_0 \varepsilon_r^2 A V_0^2}{2(d + \varepsilon_r(D - d))^2}$

2) $-\dfrac{\varepsilon_0 \varepsilon_r^2 A V_0^2}{2(d + \varepsilon_r(D - d))}$

3) $-\dfrac{\varepsilon_0 A V_0^2}{\varepsilon_r D^2}$

4) $\dfrac{\varepsilon_0 A V_0^2}{2[D - d]^2}$

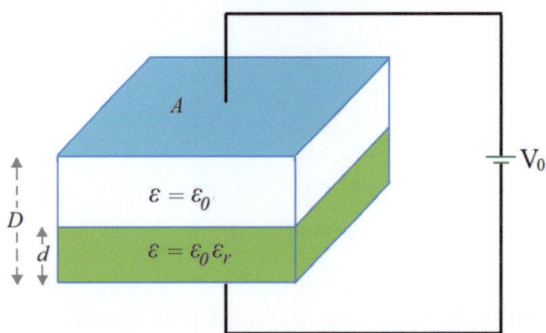

Fig. 13.5 A flat capacitor including two different electrical permittivity connected to a voltage source

Problem

13.6. Calculate the capacitance of a flat capacitor with the surface area A, distance d, and nonuniform relative permittivity $\varepsilon_r = e^{\alpha x}$. The axis of capacitor is aligned with x-axis and its left plate is placed at the origin.

Difficulty level ○ Easy ○ Normal ● Hard
Calculation amount ○ Small ● Normal ○ Large

1) $\varepsilon_0 A \alpha e^{-\alpha d}$
2) $\varepsilon_0 A \alpha (e^{-\alpha d} - 1)$
3) $\dfrac{\varepsilon_0 A \alpha}{1 + e^{-\alpha d}}$
4) $\dfrac{\varepsilon_0 A \alpha}{1 - e^{-\alpha d}}$

Partially Solved Exercise

13.7. Calculate the capacitance of a flat capacitor with the surface area A, distance d, and nonuniform permittivity $\varepsilon = \varepsilon_0 x$. The axis of capacitor is aligned with x-axis and its left plate is placed at $x = d$.

Solution

The permittivity of the capacitor is not uniform. Therefore, the following relation needs to be used to calculate its capacitance.

$$\frac{1}{C} = \int \frac{dx}{\int \varepsilon dS}$$

$$\Rightarrow \frac{1}{C} = \int \frac{dx}{(\quad) \int dS}$$

$$\Rightarrow \frac{1}{C} = \int \frac{dx}{(\quad)}$$

$$\Rightarrow \frac{1}{C} = \frac{(\quad)}{(\quad)} \int_{x=d}^{x=2d} \frac{dx}{(\quad)}$$

$$\Rightarrow \frac{1}{C} = \frac{(\quad)}{(\quad)} [\ln x]_d^{2d}$$

$$\Rightarrow \frac{1}{C} = \frac{(\quad)}{(\quad)} ((\quad) - (\quad))$$

$$\Rightarrow \frac{1}{C} = \frac{(\quad)}{(\quad)}$$

$$\Rightarrow C = \frac{\varepsilon_0 A}{\ln 2}$$

Problem

13.8. The plates of a large flat capacitor are placed at $z = 0$ and $z = d$. The permittivity of the dielectric of the capacitor is $\varepsilon = \varepsilon_0 \left(1 + \dfrac{z^2}{d^2}\right)$ and the surface charge density of the plates are $\pm \rho_s$. Calculate the electric potential difference between the plates of the capacitor.

Difficulty level — ○ Easy ○ Normal ● Hard
Calculation amount — ○ Small ● Normal ○ Large

1) $\dfrac{\rho_s d}{2\varepsilon_0}$
2) $\dfrac{\rho_s}{2\pi\varepsilon_0}$
3) $\dfrac{2\pi \rho_s d}{\varepsilon_0}$
4) $\dfrac{\rho_s \pi d}{4\varepsilon_0}$

13.2 Cylindrical Capacitors

Problem

13.9. Calculate the capacitance of the cylindrical capacitor illustrated in Fig. 13.6. The length of the capacitor is l.

Difficulty level — ○ Easy ● Normal ○ Hard
Calculation amount — ○ Small ● Normal ○ Large

1) $C = \dfrac{2\pi\varepsilon_0 l}{\ln\left(\dfrac{b}{a}\right)}$
2) $C = \dfrac{\pi\varepsilon_0 l}{\ln\left(\dfrac{b}{a}\right)}$
3) $C = \dfrac{2\pi\varepsilon_0 l}{\ln\left(\dfrac{a}{b}\right)}$
4) $C = \dfrac{\pi\varepsilon_0 l}{\ln\left(\dfrac{a}{b}\right)}$

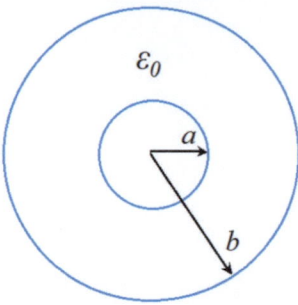

Fig. 13.6 A cylindrical capacitor with free space permittivity

Problem

13.10. Calculate the total capacitance of the cylindrical capacitor illustrated in Fig. 13.7. The length of the capacitor is l.
Difficulty level ○ Easy ● Normal ○ Hard
Calculation amount ○ Small ○ Normal ● Large

1) $C = \dfrac{2\pi\varepsilon_0(1+\varepsilon_r)l}{\ln\left(\dfrac{b}{a}\right)}$

2) $C = \dfrac{4\pi\varepsilon_0(1+\varepsilon_r)l}{\ln\left(\dfrac{b}{a}\right)}$

3) $C = \dfrac{\pi\varepsilon_0(1+\varepsilon_r)l}{\ln\left(\dfrac{b}{a}\right)}$

4) $C = \dfrac{2\pi\varepsilon_0\varepsilon_r l}{\ln\left(\dfrac{b}{a}\right)}$

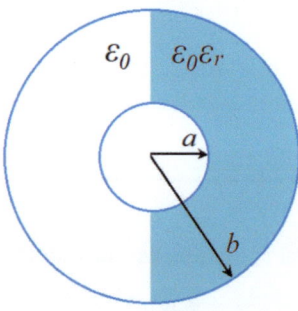

Fig. 13.7 A cylindrical capacitor with different electrical permittivity

Problem

13.11. In Fig. 13.8, $\varepsilon_1 = \varepsilon_0$, $\varepsilon_2 = 2\varepsilon_0$, $r_1 = 2\ cm$, $r_2 = 2.25\ cm$, $r_3 = 2.5\ cm$, and $V_0 = 100\ V$. Calculate the voltage across each region.
Difficulty level ○ Easy ○ Normal ● Hard
Calculation amount ○ Small ○ Normal ● Large

1) $V_1 = 69.09\ V$, $V_2 = 30.96\ V$
2) $V_1 = 30.96\ V$, $V_2 = 69.09\ V$
3) $V_1 = V_2 = 69.09\ V$
4) $V_1 = V_2 = 30.96\ V$

13.2 Cylindrical Capacitors

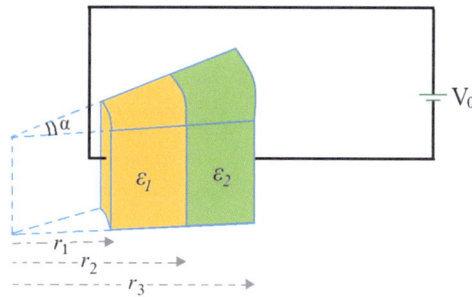

Fig. 13.8 A cylindrical capacitor with different electrical permittivity connected to a voltage source

Problem

13.12. As is shown in Fig. 13.9, a capacitor is made by two cylindrical shells with radiuses a and b. The dielectric of the capacitor is air, and the surface charge density of the smaller cylinder is ρ_s. Find a relation between a and b so that the maximum value of electric energy is stored in the capacitor.

Difficulty level ○ Easy ○ Normal ● Hard
Calculation amount ○ Small ○ Normal ● Large

1) $b = a\sqrt{e}$
2) $b = (e-1)a$
3) $b = ea$
4) $b = e^2 a$

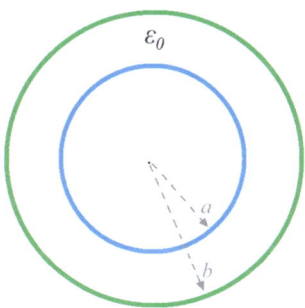

Fig. 13.9 A cylindrical capacitor with free space permittivity

Problem

13.13. Figure 13.10 illustrates a cylindrical capacitor. As can be seen, the capacitor is connected to a voltage source. Calculate the electric field at point A.

Difficulty level ○ Easy ○ Normal ● Hard
Calculation amount ○ Small ○ Normal ● Large

1) $E_A = \dfrac{V_0}{\ln\left(\frac{b}{a}\right)}$

2) $E_A = \dfrac{V_0}{r \ln\left(\frac{b}{a}\right)}$

3) $E_A = \dfrac{V_0}{r^2 \ln\left(\frac{b}{a}\right)}$

4) $E_A = \dfrac{r V_0}{\ln\left(\frac{b}{a}\right)}$

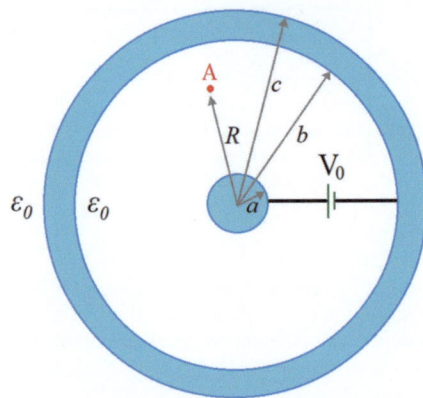

Fig. 13.10 A cylindrical capacitor with free space permittivity connected to a voltage source

13.3 Spherical Capacitors

Problem

13.14. Calculate the capacitance of a solid sphere with radius a placed in free space.
Difficulty level ○ Easy ● Normal ○ Hard
Calculation amount ○ Small ● Normal ○ Large
1) $C = 4\pi\varepsilon_0 a$
2) $C = \dfrac{1}{4\pi\varepsilon_0 a}$
3) ∞
4) 0

Problem

13.15. Calculate the electric potential energy of a solid sphere with radius a and electric potential V_0 placed in free space.
Difficulty level ● Easy ○ Normal ○ Hard
Calculation amount ● Small ○ Normal ○ Large
1) $\dfrac{2\pi\varepsilon_0 a^2 V_0^2}{\sqrt{a^2-2}}$
2) $\tfrac{1}{2}\pi\varepsilon_0 a V_0^2$
3) $\dfrac{2\pi\varepsilon_0 a V_0^2}{\sqrt{a^2-2}}$
4) $2\pi\varepsilon_0 a V_0^2$

Exercise

13.16. The electric potential of a spherical shell with radius a is V_0. Calculate the electric energy stored in it.

Final Answer

$U = 2\pi\varepsilon_0 a V_0^2$

13.3 Spherical Capacitors

Problem

13.17. Calculate the total capacitance of the spherical capacitor illustrated in Fig. 13.11.

Difficulty level ○ Easy ● Normal ○ Hard
Calculation amount ○ Small ○ Normal ● Large

1) $\dfrac{2\pi\varepsilon_0(\varepsilon_r + 1)ab}{b - a}$
2) $\dfrac{4\pi\varepsilon_0(\varepsilon_r + 1)ab}{b - a}$
3) $\dfrac{4\pi\varepsilon_0(\varepsilon_r - 1)ab}{b - a}$
4) $\dfrac{2\pi\varepsilon_0(\varepsilon_r - 1)ab}{b - a}$

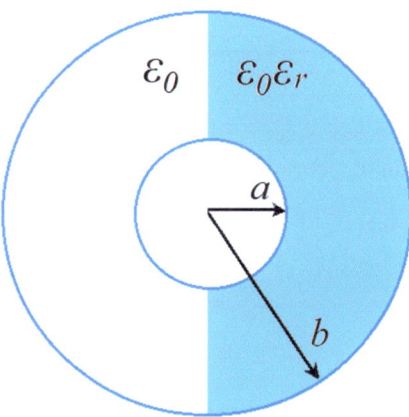

Fig. 13.11 A spherical capacitor with different electrical permittivity

Problem

13.18. As is shown in Fig. 13.12, a capacitor is made by two spherical shells with radius a and b. The dielectric of the capacitor is air, and the surface charge density of the smaller sphere is ρ_s. Find a relation between a and b so that the maximum value of electric energy is stored in the capacitor.

Difficulty level ○ Easy ○ Normal ● Hard
Calculation amount ○ Small ○ Normal ● Large

1) $b = \dfrac{1}{3}a$
2) $b = \dfrac{3}{4}a$
3) $b = \dfrac{4}{3}a$
4) $b = 3a$

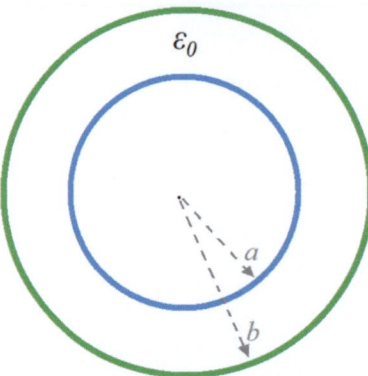

Fig. 13.12 A spherical capacitor with free space permittivity

Problem

13.19. As is shown in Fig. 13.13, in the first test, a spherical conductor with radius a is placed in a very large space with nonuniform dielectric constant $\varepsilon_r = 1 + \dfrac{a}{R}$. In the second experiment, the sphere is placed in free space. Calculate the ratio of capacitances of the sphere in the two tests, that is, $\dfrac{C_2}{C_1}$.

Difficulty level   ○ Easy   ○ Normal   ● Hard
Calculation amount ○ Small   ○ Normal   ● Large
1) $\ln 2$
2) $\dfrac{1}{\ln 2}$
3) $\dfrac{1}{\ln 3}$
4) $\ln 3$

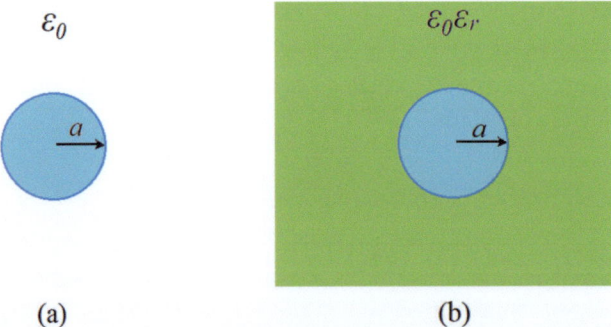

Fig. 13.13 A spherical conductor placed in (a) a nonuniform dielectric and (b) free space

Problem

13.20. As is shown in Fig. 13.14, in the first experiment, a spherical capacitor with radius a is connected to a voltage source with the voltage of V_0 and placed at free space. In the second test, the capacitor is placed in environment with the permittivity $\varepsilon = \varepsilon_0 \left(1 + \dfrac{a^2}{R^2}\right)$. Calculate the change in the electric energy stored in the capacitor.

Difficulty level   ○ Easy   ○ Normal   ● Hard
Calculation amount ○ Small   ○ Normal   ● Large

13.3 Spherical Capacitors

1) $U_2 - U_1 = -\left(\frac{4}{\pi} - 1\right) 4\pi\varepsilon_0 a V_0^2$

2) $U_2 - U_1 = \left(\frac{4}{\pi} - 1\right) 2\pi\varepsilon_0 a V_0^2$

3) $U_2 - U_1 = \left(\frac{4}{\pi} - 1\right) 4\pi\varepsilon_0 a V_0$

4) $U_2 - U_1 = -\left(\frac{4}{\pi} - 1\right) 4\pi\varepsilon_0 a V_0$

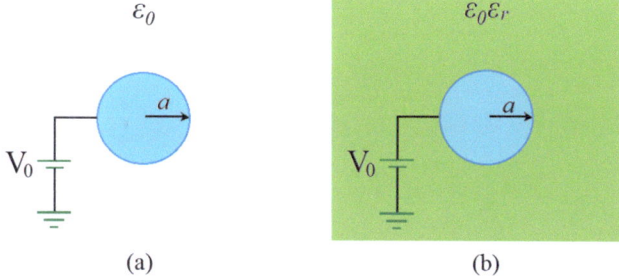

Fig. 13.14 A spherical conductor connected to a voltage source and placed in (a) a nonuniform dielectric and (b) free space

Problem

13.21. Figure 13.15 illustrates a spherical capacitor. As can be seen, the capacitor is connected to a voltage source. Calculate the electric field at point A.

Difficulty level ○ Easy ○ Normal ● Hard
Calculation amount ○ Small ○ Normal ● Large

1) $E_A = \dfrac{(b-a)V_0}{R^2 ab}$

2) $E_A = \dfrac{abV_0}{R^2}$

3) $E_A = \dfrac{V_0}{R^2(b-a)}$

4) $E_A = \dfrac{abV_0}{R^2(b-a)}$

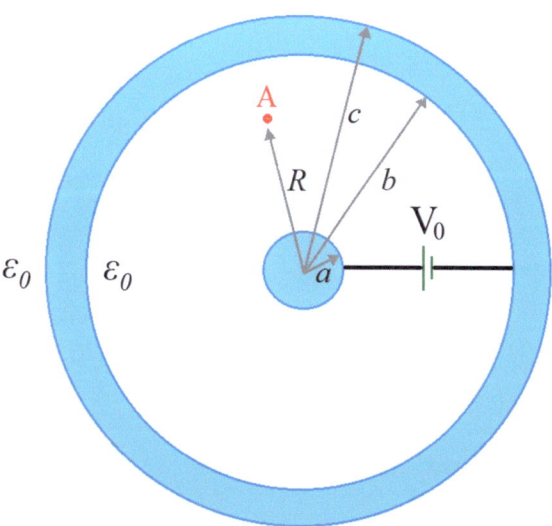

Fig. 13.15 A spherical capacitor with free space permittivity connected to a voltage source

References

1. Rahmani-Andebili, M., General Physics II – Practice Problems, Methods, and Solutions, Springer Nature, 2025.
2. Rahmani-Andebili, M., General Physics I – Practice Problems, Methods, and Solutions, Springer Nature, 2025.
3. Rahmani-Andebili, M., Mathematics of Engineering and Science – Practice Problems, Methods, and Solutions, Springer Nature, 2024.
4. Rahmani-Andebili, M., Differential Equations – Practice Problems, Methods, and Solutions, Springer Nature, 2022.
5. Rahmani-Andebili, M., Calculus III – Practice Problems, Methods, and Solutions, Springer Nature, 2023.
6. Rahmani-Andebili, M., Calculus II – Practice Problems, Methods, and Solutions, Springer Nature, 2023.
7. Rahmani-Andebili, M., Calculus I (2nd Ed.) – Practice Problems, Methods, and Solutions, Springer Nature, 2023.
8. Rahmani-Andebili, M., Precalculus (2nd Ed.) – Practice Problems, Methods, and Solutions, Springer Nature, 2024.

Flat, Cylindrical, and Spherical Capacitors: Part B

Abstract

In this chapter, the problems of the thirteenth chapter are fully solved, in detail, step-by-step, and with different methods.

14.1 Flat Capacitors

14.1. Based on the information given in the problem, the initial capacitance of the flat capacitor is as follows [1–8].

$$C = \frac{\varepsilon_0 A}{d}$$

If part of the dielectric of a capacitor is replaced by a perfect conductor with the surface area A and depth x, the capacitance of capacitor decreases and can be calculated as follows.

$$C' = \frac{\varepsilon_0 A}{d - x}$$

In this problem, $x = 0.5d$, therefore:

$$C' = \frac{\varepsilon_0 A}{d - 0.5d} = \frac{\varepsilon_0 A}{0.5d}$$

$$\Rightarrow C' = \frac{2\varepsilon_0 A}{d}$$

Choice (2) is the answer (Fig. 14.1).

Fig. 14.1 A flat capacitor

> **Notes**
>
> In this problem, the relation below has been used.
>
> The capacitance of a flat capacitor with the surface area A, distance d, and electrical permittivity ε is as follows.
>
> $$C = \frac{\varepsilon A}{d} = \frac{\varepsilon_0 \varepsilon_r A}{d}$$
>
> Herein, ε_0 and ε_r are the free space electrical permittivity and relative electrical permittivity.

14.2. The capacitor, shown in Fig. 14.2, includes two parts with the electrical permittivity ε_1 and ε_2. These two parts are, in fact, two individual capacitors with the surface area $\frac{A}{2}$ that have been connected in parallel. Thus, the total capacitance of the capacitors can be calculated as follows.

$$C = C_1 + C_2$$

$$\Rightarrow C = \frac{\varepsilon_1 \frac{A}{2}}{d} + \frac{\varepsilon_2 \frac{A}{2}}{d}$$

$$\Rightarrow C = \frac{A}{2d}(\varepsilon_1 + \varepsilon_2)$$

Choice (1) is the answer.

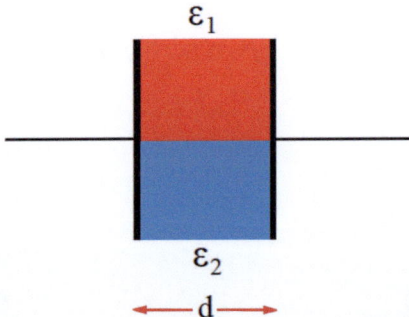

Fig. 14.2 A flat capacitor including two different electrical permittivity

14.1 Flat Capacitors

> **Notes**
>
> In this problem, the relations below have been used.
>
> The capacitance of a flat capacitor with the surface area A, distance d, and electrical permittivity ε is as follows.
>
> $$C = \frac{\varepsilon A}{d} = \frac{\varepsilon_0 \varepsilon_r A}{d}$$
>
> Herein, ε_0 and ε_r are the free space electrical permittivity and relative electrical permittivity.
>
> The equivalent capacitance of n parallel capacitors can be calculated as follows.
>
> $$C_{eq} = C_1 + \ldots + C_n$$

14.3. The capacitor, shown in Fig. 14.3, includes two parts with the electrical permittivity ε_1 and ε_2. These two parts are, in fact, two individual capacitors with the distance $\frac{d}{2}$ that have been connected in series. Hence, the total capacitance of the capacitors can be calculated as follows.

$$C = \frac{C_1 C_2}{C_1 + C_2}$$

$$\Rightarrow C = \frac{\left(\frac{\varepsilon_1 A}{\frac{d}{2}}\right)\left(\frac{\varepsilon_2 A}{\frac{d}{2}}\right)}{\left(\frac{\varepsilon_1 A}{\frac{d}{2}}\right) + \left(\frac{\varepsilon_2 A}{\frac{d}{2}}\right)} = \frac{\frac{2\varepsilon_1 A}{d} \frac{2\varepsilon_2 A}{d}}{\frac{2A}{d}(\varepsilon_1 + \varepsilon_2)}$$

$$\Rightarrow C = \frac{2A}{d} \frac{\varepsilon_1 \varepsilon_2}{(\varepsilon_1 + \varepsilon_2)}$$

Choice (4) is the answer.

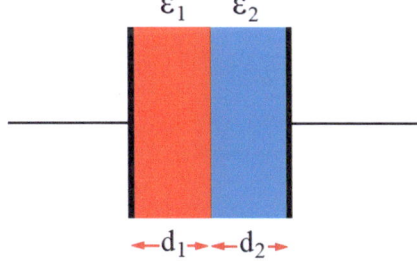

Fig. 14.3. A flat capacitor including two different electrical permittivity

> **Notes**
>
> In this problem, the relations below have been used.
>
> The capacitance of a flat capacitor with the surface area A, distance d, and electrical permittivity ε is as follows.
>
> $$C = \frac{\varepsilon A}{d} = \frac{\varepsilon_0 \varepsilon_r A}{d}$$
>
> Herein, ε_0 and ε_r are the free space electrical permittivity and relative electrical permittivity.
>
> The equivalent capacitance of n series capacitors can be calculated as follows.
>
> $$\frac{1}{C_{eq}} = \frac{1}{C_1} + \ldots + \frac{1}{C_n}$$

14.4. From General Physics II, we know that the electric potential of a capacitor is inversely proportional to its capacitance. Thus:

$$\frac{V_1}{V_2} = \frac{C_2}{C_1}$$

The capacitor, shown in Fig. 14.4, includes two series capacitors as follows.

$$C_1 = \frac{\varepsilon_1 A}{d_1}$$

$$C_2 = \frac{\varepsilon_2 A}{d_2}$$

Therefore:

$$\frac{V_1}{V_2} = \frac{\frac{\varepsilon_2 A}{d_2}}{\frac{\varepsilon_1 A}{d_1}}$$

$$\Rightarrow \frac{V_1}{V_2} = \frac{\varepsilon_2 d_1}{\varepsilon_1 d_2}$$

Choice (1) is the answer.

14.1 Flat Capacitors

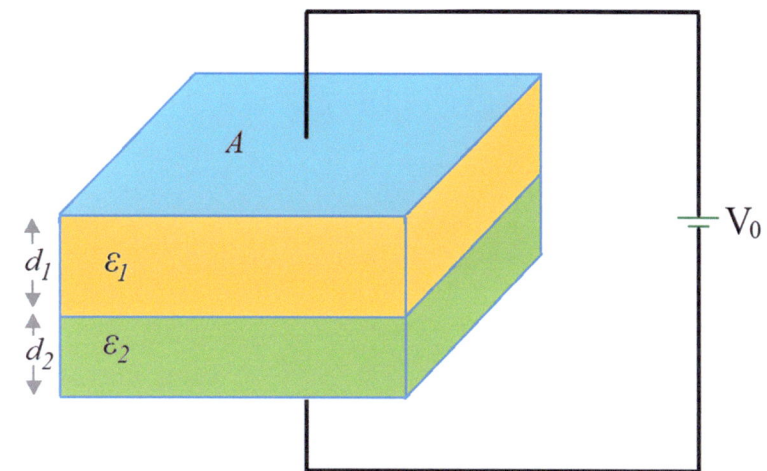

Fig. 14.4 A flat capacitor including two different electrical permittivity connected to a voltage source

> **Notes**
> In this problem, the relation below has been used.
>
> The capacitance of a flat capacitor with the surface area A, distance d, and electrical permittivity ε is as follows.
>
> $$C = \frac{\varepsilon A}{d} = \frac{\varepsilon_0 \varepsilon_r A}{d}$$
>
> Herein, ε_0 and ε_r are the free space electrical permittivity and relative electrical permittivity.

14.5. First, we need to calculate the total electric energy stored in the system. Herein, the relation below must be applied since the capacitor is connected to a voltage source.

$$U = \frac{1}{2} C V_0^2$$

Then, the amount of force can be calculated as follows.

$$F = \frac{dU}{dx}$$

The system shown in Fig. 14.5 is the series connection of two capacitors. The equivalent capacitance of the system can be calculated as follows.

$$\begin{cases} C_1 = \dfrac{\varepsilon_0 A}{D - d} \\ C_2 = \dfrac{\varepsilon_0 \varepsilon_r A}{d} \end{cases}$$

$$\Rightarrow C = \frac{C_1 C_2}{C_1 + C_2} = \frac{\left(\frac{\varepsilon_0 A}{D-d}\right)\left(\frac{\varepsilon_0 \varepsilon_r A}{d}\right)}{\left(\frac{\varepsilon_0 A}{D-d}\right) + \left(\frac{\varepsilon_0 \varepsilon_r A}{d}\right)}$$

$$\Rightarrow C = \frac{\left(\frac{\varepsilon_0 A}{D-d}\right)\left(\frac{\varepsilon_0 \varepsilon_r A}{d}\right)}{\frac{\varepsilon_0 A d + \varepsilon_0 \varepsilon_r A(D-d)}{d(D-d)}} = \frac{\left(\frac{\varepsilon_0 A}{D-d}\right)\left(\frac{\varepsilon_0 \varepsilon_r A}{d}\right)}{\frac{\varepsilon_0 A(d + \varepsilon_r(D-d))}{d(D-d)}}$$

$$\Rightarrow C = \frac{\varepsilon_0 \varepsilon_r A}{d + \varepsilon_r(D-d)}$$

The total electric energy stored in the system can be calculated as follows.

$$U = \frac{1}{2} C V_0^2$$

$$\Rightarrow U = \frac{1}{2} \frac{\varepsilon_0 \varepsilon_r A}{d + \varepsilon_r(D-d)} V_0^2$$

Finally, the amount of force exerted on the top plate of the capacitor can be calculated as follows.

$$F = \frac{d}{dD}\left(\frac{1}{2} \frac{\varepsilon_0 \varepsilon_r A}{d + \varepsilon_r(D-d)} V_0^2\right)$$

$$\Rightarrow F = \frac{1}{2} \varepsilon_0 \varepsilon_r A V_0^2 \frac{d}{dD}\left(\frac{1}{d + \varepsilon_r(D-d)}\right)$$

$$\Rightarrow F = \frac{1}{2} \varepsilon_0 \varepsilon_r A V_0^2 \frac{0 - \varepsilon_r}{(d + \varepsilon_r(D-d))^2}$$

$$\Rightarrow F = -\frac{\varepsilon_0 \varepsilon_r^2 A V_0^2}{2(d + \varepsilon_r(D-d))^2}$$

Choice (1) is the answer.

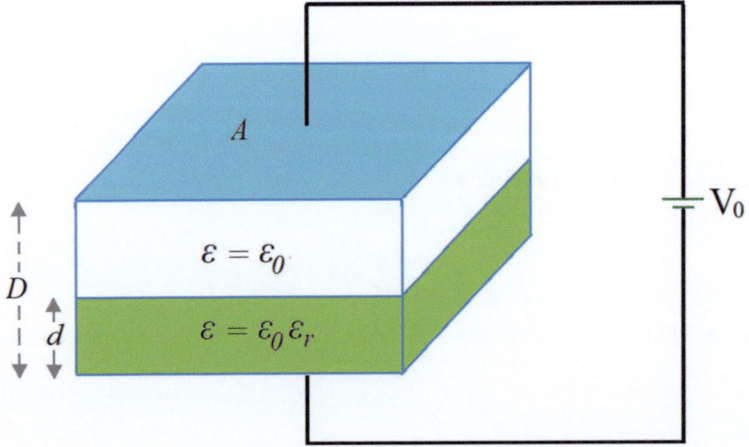

Fig. 14.5 A flat capacitor including two different electrical permittivity connected to a voltage source

14.1 Flat Capacitors

14.6. In this problem, the permittivity of the capacitor is not uniform. Thus, we need to use the following relation to calculate its capacitance.

$$\frac{1}{C} = \int \frac{dx}{\int \varepsilon dS}$$

$$\Rightarrow \frac{1}{C} = \int \frac{dx}{\varepsilon_0 e^{\alpha x} \int dS}$$

$$\Rightarrow \frac{1}{C} = \int \frac{dx}{\varepsilon_0 e^{\alpha x} A}$$

$$\Rightarrow \frac{1}{C} = \frac{1}{\varepsilon_0 A} \int_{x=0}^{x=d} e^{-\alpha x} dx$$

$$\Rightarrow \frac{1}{C} = \frac{1}{\varepsilon_0 A} \left[\frac{1}{-\alpha} e^{-\alpha x} \right]_0^d$$

$$\Rightarrow \frac{1}{C} = \frac{1}{\varepsilon_0 A} \left(\frac{e^{-\alpha d} - e^0}{-\alpha} \right)$$

$$\Rightarrow \frac{1}{C} = \frac{1 - e^{-\alpha d}}{\varepsilon_0 A \alpha}$$

$$\Rightarrow C = \frac{\varepsilon_0 A \alpha}{1 - e^{-\alpha d}}$$

Choice (4) is the answer.

> **Notes**
> In this problem, the relation below has been used.
>
> $$\int e^{u(x)} du = e^{u(x)}$$

14.8. As we know, the magnitude of electric field around a large flat plate can be calculated as follows.

$$E = \frac{\rho_s}{2\varepsilon}$$

Hence, for this problem, the electric field between two plates of the capacitor is as follows.

$$\vec{E} = 2 \times \frac{\rho_s}{2\varepsilon_0 \left(1 + \frac{z^2}{d^2}\right)} \hat{z} = \frac{\rho_s}{\varepsilon_0 \left(1 + \frac{z^2}{d^2}\right)} \hat{z}$$

Moreover, the electric potential difference between the plates of the capacitor can be calculated as follows.

$$\Delta V = \int_0^d \vec{E} \cdot d\vec{l}$$

$$\Rightarrow \Delta V = \int_0^d \frac{\rho_s}{\varepsilon_0 \left(1 + \frac{z^2}{d^2}\right)} dz$$

$$\Rightarrow \Delta V = \frac{\rho_s}{\varepsilon_0} \int_0^d \frac{1}{\left(1 + \frac{z^2}{d^2}\right)} dz$$

$$\Rightarrow \Delta V = \frac{\rho_s}{\varepsilon_0} \left[d\left(\arctan \frac{z}{d}\right)\right]_0^d$$

$$\Rightarrow \Delta V = \frac{\rho_s d}{\varepsilon_0} (\arctan 1 - \arctan 0)$$

$$\Rightarrow \Delta V = \frac{\rho_s d}{\varepsilon_0} \left(\frac{\pi}{4} - 0\right)$$

$$\Rightarrow \Delta V = \frac{\rho_s \pi d}{4\varepsilon_0}$$

Choice (4) is the answer.

> **Notes**
>
> In this problem, the relations below have been used.
>
> $$\int \frac{dx}{a^2 + x^2} = \frac{1}{a} \arctan \frac{x}{a}$$
>
> $$\arctan 1 = \frac{\pi}{4}$$
>
> $$\arctan 0 = 0$$

14.2 Cylindrical Capacitors

14.9. The capacitance of the capacitor can be calculated as follows.

$$\frac{1}{C} = \int \frac{dr}{\int \varepsilon dS}$$

14.2 Cylindrical Capacitors

$$\Rightarrow \frac{1}{C} = \int \frac{dr}{\varepsilon_0 r \int_{z=0}^{z=l} \int_{\varphi=0}^{\varphi=2\pi} d\varphi dz}$$

$$\Rightarrow \frac{1}{C} = \int \frac{dr}{\varepsilon_0 r [\varphi]_0^{2\pi} [z]_0^l}$$

$$\Rightarrow \frac{1}{C} = \int \frac{dr}{\varepsilon_0 r (2\pi) l}$$

$$\Rightarrow \frac{1}{C} = \frac{1}{2\pi \varepsilon_0 l} \int_{r=a}^{r=b} \frac{dr}{r}$$

$$\Rightarrow \frac{1}{C} = \frac{1}{2\pi \varepsilon_0 l} [\ln r]_{r_1}^{r_2}$$

$$\Rightarrow \frac{1}{C} = \frac{1}{2\pi \varepsilon_0 l} (\ln b - \ln a)$$

$$\Rightarrow C = \frac{2\pi \varepsilon_0 l}{\ln \left(\frac{b}{a}\right)}$$

Choice (1) is the answer (Fig. 14.6).

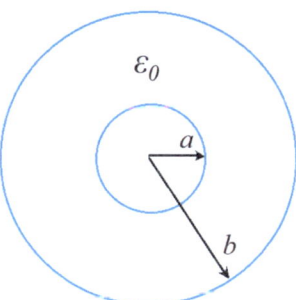

Fig. 14.6 A cylindrical capacitor with free space permittivity

> **Notes**
>
> In this problem, the relations below have been used.
>
> $$\int \frac{dx}{x} = \ln x + c$$
>
> $$\ln a - \ln b = \ln \left(\frac{a}{b}\right)$$

14.10. The cylindrical capacitor illustrated in Fig. 14.7 can be considered a system of two parallel capacitors.

The capacitance of the left-side capacitor can be calculated as follows.

$$\frac{1}{C_1} = \int \frac{dr}{\int \varepsilon dS}$$

$$\Rightarrow \frac{1}{C_1} = \int \frac{dr}{\varepsilon_0 r \int_{z=0}^{z=l} \int_{\varphi=0}^{\varphi=\pi} d\varphi dz}$$

$$\Rightarrow \frac{1}{C_1} = \int \frac{dr}{\varepsilon_0 r [\varphi]_0^\pi [z]_0^l}$$

$$\Rightarrow \frac{1}{C_1} = \int \frac{dr}{\varepsilon_0 r \pi l}$$

$$\Rightarrow \frac{1}{C_1} = \frac{1}{\pi \varepsilon_0 l} \int_{r=a}^{r=b} \frac{dr}{r}$$

$$\Rightarrow \frac{1}{C_1} = \frac{1}{\pi \varepsilon_0 l} [\ln r]_{r_1}^{r_2}$$

$$\Rightarrow \frac{1}{C_1} = \frac{1}{\pi \varepsilon_0 l} (\ln b - \ln a)$$

$$\Rightarrow C_1 = \frac{\pi \varepsilon_0 l}{\ln\left(\frac{b}{a}\right)}$$

Likewise, the capacitance of the right-side capacitor is as follows.

$$\Rightarrow C_2 = \frac{\pi \varepsilon_0 \varepsilon_r l}{\ln\left(\frac{b}{a}\right)}$$

Thus, the equivalent capacitance of the system can be calculated as follows.

$$C = C_1 + C_2$$

$$\Rightarrow C = \frac{\pi \varepsilon_0 l}{\ln\left(\frac{b}{a}\right)} + \frac{\pi \varepsilon_0 \varepsilon_r l}{\ln\left(\frac{b}{a}\right)}$$

$$\Rightarrow C = \frac{\pi \varepsilon_0 (1 + \varepsilon_r) l}{\ln\left(\frac{b}{a}\right)}$$

Choice (3) is the answer.

14.2 Cylindrical Capacitors

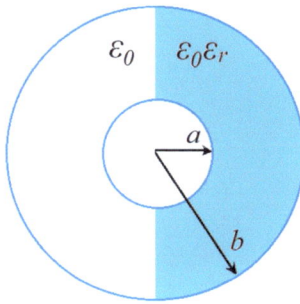

Fig. 14.7 A cylindrical capacitor with different electrical permittivity

> **Notes**
> In this problem, the relations below have been used.
> $$\int \frac{dx}{x} = \ln x + c$$
> $$\ln a - \ln b = \ln\left(\frac{a}{b}\right)$$

14.11. Based on the information given in the problem, we have:

$$\varepsilon_1 = \varepsilon_0$$

$$\varepsilon_2 = 2\varepsilon_0$$

$$r_1 = 0.02 \; m$$

$$r_2 = 0.0225 \; m$$

$$r_3 = 0.025 \; m$$

$$V_0 = 100 \; V$$

As can be noticed, the circuit includes two series capacitors that each of them is part of a cylindrical capacitor. Hence, we need to use the following relation to calculate the capacitance of each part.

$$\frac{1}{C} = \int \frac{dr}{\int \varepsilon dS}$$

For the first part, we have:

$$\frac{1}{C_1} = \int \frac{dr}{\int \varepsilon_1 dS}$$

$$\Rightarrow \frac{1}{C_1} = \int \frac{dr}{\varepsilon_1 r \int_{z=0}^{z=l} \int_{\varphi=0}^{\varphi=\alpha} d\varphi dz}$$

$$\Rightarrow \frac{1}{C_1} = \int \frac{dr}{\varepsilon_1 r [\varphi]_0^\alpha [z]_0^l}$$

$$\Rightarrow \frac{1}{C_1} = \int \frac{dr}{\varepsilon_1 r \alpha l}$$

$$\Rightarrow \frac{1}{C_1} = \frac{1}{\alpha \varepsilon_1 l} \int_{r=r_1}^{r=r_2} \frac{dr}{r}$$

$$\Rightarrow \frac{1}{C_1} = \frac{1}{\alpha \varepsilon_1 l} [\ln r]_{r_1}^{r_2}$$

$$\Rightarrow \frac{1}{C_1} = \frac{1}{\alpha \varepsilon_1 l} (\ln r_2 - \ln r_1)$$

$$\Rightarrow C_1 = \frac{\alpha \varepsilon_1 l}{\ln \left(\frac{r_2}{r_1} \right)}$$

Likewise, for the first part, we have:

$$C_2 = \frac{\alpha \varepsilon_2 l}{\ln \left(\frac{r_3}{r_2} \right)}$$

Now, by putting the quantities in their parameters, we have:

$$C_1 = \frac{\alpha \varepsilon_0 l}{\ln \left(\frac{2.25}{2} \right)} = 8.49 \alpha \varepsilon_0 l$$

$$C_2 = \frac{\alpha 2 \varepsilon_0 l}{\ln \left(\frac{2.5}{2.25} \right)} = 18.98 \alpha \varepsilon_0 l$$

From General Physics II, we know that the electric potential of a capacitor in a circuit with two series capacitors can be calculated as follows.

$$V_1 = \frac{C_2}{C_1 + C_2} V_0$$

$$V_2 = \frac{C_1}{C_1 + C_2} V_0$$

Therefore:

$$V_1 = \frac{18.98 \alpha \varepsilon_0 l}{8.49 \alpha \varepsilon_0 l + 18.98 \alpha \varepsilon_0 l} \times 100 \Rightarrow V_1 = 69.09 \, V$$

$$V_2 = \frac{8.49\alpha\varepsilon_0 l}{8.49\alpha\varepsilon_0 l + 18.98\alpha\varepsilon_0 l} \times 100 \Rightarrow V_2 = 30.96 \text{ V}$$

Choice (1) is the answer (Fig. 14.8).

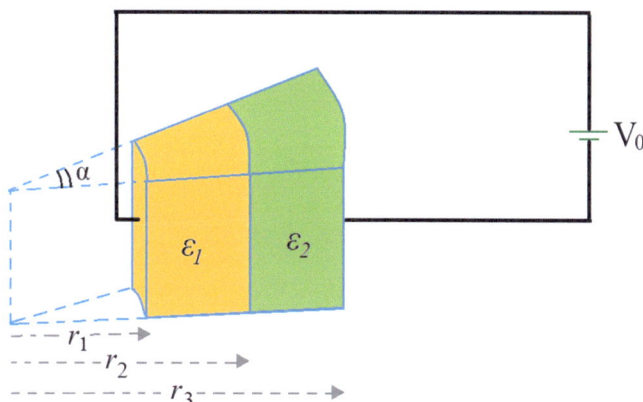

Fig. 14.8 A cylindrical capacitor with different electrical permittivity connected to a voltage source

---

Notes

In this problem, the relations below have been used.

$$\int \frac{dx}{x} = \ln x + c$$

$$\ln a - \ln b = \ln\left(\frac{a}{b}\right)$$

14.12. The following relation can be used to calculate the capacitance of the cylindrical capacitor.

$$\frac{1}{C} = \int \frac{dr}{\int \varepsilon dS}$$

$$\Rightarrow \frac{1}{C} = \int \frac{dr}{\varepsilon_0 r \int_{z=0}^{z=l} \int_{\varphi=0}^{\varphi=2\pi} d\varphi dz}$$

$$\Rightarrow \frac{1}{C} = \int \frac{dr}{\varepsilon_0 r [\varphi]_0^{2\pi} [z]_0^l}$$

$$\Rightarrow \frac{1}{C} = \int \frac{dr}{\varepsilon_0 r (2\pi) l}$$

$$\Rightarrow \frac{1}{C} = \frac{1}{2\pi\varepsilon_0 l} \int_{r=a}^{r=b} \frac{dr}{r}$$

$$\Rightarrow \frac{1}{C} = \frac{1}{2\pi\varepsilon_0 l} [\ln r]_{r_1}^{r_2}$$

$$\Rightarrow \frac{1}{C} = \frac{1}{2\pi\varepsilon_0 l} (\ln b - \ln a)$$

$$\Rightarrow C = \frac{2\pi\varepsilon_0 l}{\ln\left(\frac{b}{a}\right)}$$

Now, we can calculate the total electric energy stored in the capacitor. Herein, the relation below must be applied since the capacitor is disconnected from a voltage source.

$$U = \frac{1}{2}\frac{Q^2}{C}$$

$$\Rightarrow U = \frac{1}{2}\frac{(2\pi a l \rho_s)^2}{\frac{2\pi\varepsilon_0 l}{\ln\left(\frac{b}{a}\right)}}$$

$$\Rightarrow U = \frac{(2\pi a l \rho_s)^2 \ln\left(\frac{b}{a}\right)}{4\pi\varepsilon_0 l}$$

$$\Rightarrow U = \frac{(2\pi l \rho_s)^2}{4\pi\varepsilon_0 l} a^2 \ln\left(\frac{b}{a}\right)$$

To find a relation between a and b for the maximum value of electric energy, we need to solve the following equation.

$$\frac{dU}{da} = 0$$

Hence:

$$\frac{d}{da}\left(\frac{(2\pi l \rho_s)^2}{4\pi\varepsilon_0 l} a^2 \ln\left(\frac{b}{a}\right)\right) = 0$$

$$\Rightarrow \frac{(2\pi l \rho_s)^2}{4\pi\varepsilon_0 l} \frac{d}{da}\left(a^2(\ln b - \ln a)\right) = 0$$

$$\Rightarrow 2a(\ln b - \ln a) + a^2\left(-\frac{1}{a}\right) = 0$$

$$\Rightarrow 2a \ln\left(\frac{b}{a}\right) - a = 0$$

$$\Rightarrow 2 \ln\left(\frac{b}{a}\right) - 1 = 0$$

14.2 Cylindrical Capacitors

$$\Rightarrow \ln\left(\frac{b}{a}\right) = \frac{1}{2}$$

$$\Rightarrow \frac{b}{a} = \sqrt{e}$$

$$\Rightarrow b = a\sqrt{e}$$

Choice (1) is the answer (Fig. 14.9).

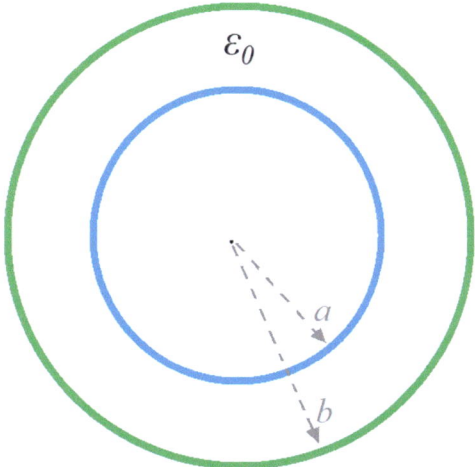

Fig. 14.9 A cylindrical capacitor with free space permittivity

Notes

In this problem, the relations below have been used.

$$\int \frac{dx}{x} = \ln x + c$$

$$\ln a - \ln b = \ln\left(\frac{a}{b}\right)$$

$$\frac{d}{dx}x^n = nx^{n-1}$$

$$\frac{d}{dx}\ln x = \frac{1}{x}$$

14.13. We know that the capacitance of the cylindrical capacitor is as follows.

$$C = \frac{2\pi\varepsilon_0 l}{\ln\left(\frac{b}{a}\right)}$$

Moreover, we know that the electric charge of a capacitor can be calculated as follows.

$$Q = VC$$

$$\Rightarrow Q = V_0 \times \frac{2\pi\varepsilon_0 l}{\ln\left(\frac{b}{a}\right)}$$

$$\Rightarrow Q = \frac{2\pi\varepsilon_0 l V_0}{\ln\left(\frac{b}{a}\right)} \tag{14.1}$$

On the other hand, we can calculate the electric field at point A by applying Gauss's law as is shown in Fig. 14.10.

$$Q = \int \vec{D} \cdot \vec{dS}$$

$$\Rightarrow Q = \int \varepsilon_0 \vec{E} \cdot \vec{dS}$$

$$\Rightarrow Q = \int_{z=0}^{l} \int_{\varphi=0}^{2\pi} \varepsilon_0 E_r r \, d\varphi \, dz$$

$$\Rightarrow Q = \varepsilon_0 E_r r (2\pi l)$$

$$\Rightarrow E_r = \frac{Q}{2\pi\varepsilon_0 r l} \tag{14.2}$$

Solving (14.1) and (14.2):

$$E_A = \frac{\dfrac{2\pi\varepsilon_0 l V_0}{\ln\left(\frac{b}{a}\right)}}{2\pi\varepsilon_0 r l}$$

$$\Rightarrow E_A = \frac{V_0}{r \ln\left(\frac{b}{a}\right)}$$

Choice (2) is the answer.

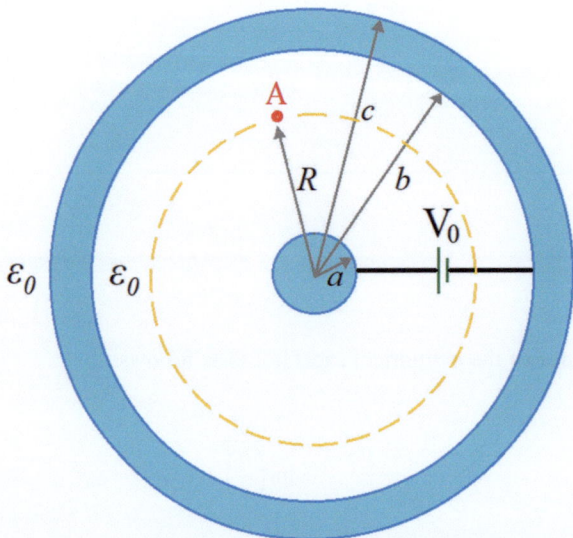

Fig. 14.10 A cylindrical capacitor with free space permittivity connected to a voltage source

Notes

In this problem, the relation below has been used.

$$\int_{z=0}^{l}\int_{\varphi=0}^{2\pi} d\varphi dz = \left[\varphi\right]_0^{2\pi}\left[z\right]_0^l = (2\pi - 0)(l - 0) = 2\pi l$$

14.3 Spherical Capacitors

14.14. A solid sphere with radius a can be considered a capacitor. The following relation can be used to calculate the capacitance of a spherical capacitor.

$$\frac{1}{C} = \int \frac{dR}{\int \varepsilon dS}$$

$$\Rightarrow \frac{1}{C} = \int \frac{dR}{\int_{\varphi=0}^{2\pi}\int_{\theta=0}^{\pi} \varepsilon_0 R^2 \sin\theta d\theta d\varphi}$$

$$\Rightarrow \frac{1}{C} = \int \frac{dR}{\varepsilon_0 R^2 \int_{\varphi=0}^{2\pi}\int_{\theta=0}^{\pi} \sin\theta d\theta d\varphi}$$

$$\Rightarrow \frac{1}{C} = \int \frac{dR}{\varepsilon_0 R^2 [-\cos\theta]_0^{\pi}[\varphi]_0^{2\pi}}$$

$$\Rightarrow \frac{1}{C} = \int \frac{dR}{\varepsilon_0 R^2 (\cos 0 - \cos\pi)(2\pi - 0)}$$

$$\Rightarrow \frac{1}{C} = \frac{1}{4\pi\varepsilon_0}\int_a^{\infty} \frac{dR}{R^2}$$

$$\Rightarrow \frac{1}{C} = \frac{1}{4\pi\varepsilon_0}\left[-\frac{1}{R}\right]_a^{\infty}$$

$$\Rightarrow \frac{1}{C} = \frac{1}{4\pi\varepsilon_0}\left(\frac{1}{a} - \frac{1}{\infty}\right)$$

$$\Rightarrow \frac{1}{C} = \frac{1}{4\pi\varepsilon_0 a}$$

$$\Rightarrow C = 4\pi\varepsilon_0 a$$

Choice (1) is the answer.

> **Notes**
>
> In this problem, the relations below have been used.
>
> $$\int \sin x \, dx = -\cos x$$
>
> $$\cos \pi = -1$$
>
> $$\cos 0 = 1$$
>
> $$\int x^n \, dx = \frac{x^{n+1}}{n+1}$$

14.15. From the solution of the previous problem, we know that the capacitance of a solid sphere with radius a placed in free space is as follows.

$$C = 4\pi\varepsilon_0 a$$

Since the capacitor is connected to a voltage source, the relation below must be used to calculate the electric potential energy of the capacitor.

$$U = \frac{1}{2}CV_0^2$$

Therefore:

$$U = 2\pi\varepsilon_0 a V_0^2$$

Choice (4) is the answer.

14.17. The spherical capacitor illustrated in Fig. 14.11 can be considered a system of two parallel capacitors.

The capacitance of the left-side capacitor can be calculated as follows.

$$\frac{1}{C_1} = \int \frac{dR}{\int \varepsilon dS}$$

14.3 Spherical Capacitors

$$\Rightarrow \frac{1}{C_1} = \int \frac{dR}{\int_{\varphi=0}^{\pi} \int_{\theta=0}^{\pi} \varepsilon_0 R^2 \sin\theta d\theta d\varphi}$$

$$\Rightarrow \frac{1}{C_1} = \int \frac{dR}{\varepsilon_0 R^2 \int_{\varphi=0}^{\pi} \int_{\theta=0}^{\pi} \sin\theta d\theta d\varphi}$$

$$\Rightarrow \frac{1}{C_1} = \int \frac{dR}{\varepsilon_0 R^2 [-\cos\theta]_0^{\pi} [\varphi]_0^{\pi}}$$

$$\Rightarrow \frac{1}{C_1} = \int \frac{dR}{\varepsilon_0 R^2 (\cos 0 - \cos \pi)(\pi - 0)}$$

$$\Rightarrow \frac{1}{C_1} = \frac{1}{2\pi\varepsilon_0} \int_a^b \frac{dR}{R^2}$$

$$\Rightarrow \frac{1}{C_1} = \frac{1}{2\pi\varepsilon_0} \left[-\frac{1}{R}\right]_a^b$$

$$\Rightarrow \frac{1}{C_1} = \frac{1}{2\pi\varepsilon_0} \left(\frac{1}{a} - \frac{1}{b}\right)$$

$$\Rightarrow \frac{1}{C_1} = \frac{b-a}{2\pi\varepsilon_0 ab}$$

$$\Rightarrow C_1 = \frac{2\pi\varepsilon_0 ab}{b-a}$$

Likewise, the capacitance of the right-side capacitor is as follows.

$$\Rightarrow C_2 = \frac{2\pi\varepsilon_0 \varepsilon_r ab}{b-a}$$

Thus, the equivalent capacitance of the system can be calculated as follows.

$$C = C_1 + C_2$$

$$\Rightarrow C = \frac{2\pi\varepsilon_0 ab}{b-a} + \frac{2\pi\varepsilon_0 \varepsilon_r ab}{b-a}$$

$$\Rightarrow C = \frac{2\pi\varepsilon_0 (\varepsilon_r + 1) ab}{b-a}$$

Choice (1) is the answer.

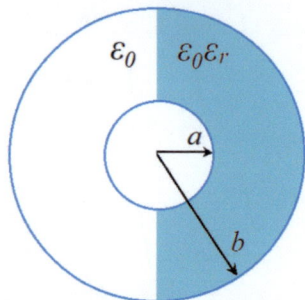

Fig. 14.11 A spherical capacitor with different electrical permittivity

> **Notes**
> In this problem, the relations below have been used.
>
> $$\int \sin x\, dx = -\cos x$$
>
> $$\cos \pi = -1$$
>
> $$\cos 0 = 1$$
>
> $$\int x^n\, dx = \frac{x^{n+1}}{n+1}$$

14.18. The following relation can be used to calculate the capacitance of a spherical capacitor.

$$\frac{1}{C} = \int \frac{dR}{\int \varepsilon\, dS}$$

$$\Rightarrow \frac{1}{C} = \int \frac{dR}{\int_{\varphi=0}^{2\pi} \int_{\theta=0}^{\pi} \varepsilon_0 R^2 \sin\theta\, d\theta\, d\varphi}$$

$$\Rightarrow \frac{1}{C} = \int \frac{dR}{\varepsilon_0 R^2 \int_{\varphi=0}^{2\pi} \int_{\theta=0}^{\pi} \sin\theta\, d\theta\, d\varphi}$$

$$\Rightarrow \frac{1}{C} = \int \frac{dR}{\varepsilon_0 R^2 [-\cos\theta]_0^{\pi} [\varphi]_0^{2\pi}}$$

14.3 Spherical Capacitors

$$\Rightarrow \frac{1}{C} = \int \frac{dR}{\varepsilon_0 R^2 (\cos 0 - \cos \pi)(2\pi - 0)}$$

$$\Rightarrow \frac{1}{C} = \frac{1}{4\pi\varepsilon_0} \int_a^b \frac{dR}{R^2}$$

$$\Rightarrow \frac{1}{C} = \frac{1}{4\pi\varepsilon_0} \left[-\frac{1}{R}\right]_a^b$$

$$\Rightarrow \frac{1}{C} = \frac{1}{4\pi\varepsilon_0} \left(\frac{1}{a} - \frac{1}{b}\right)$$

$$\Rightarrow C = \frac{4\pi\varepsilon_0}{\frac{1}{a} - \frac{1}{b}}$$

Now, we can calculate the total electric energy stored in the capacitor. Herein, the relation below must be applied since the capacitor is disconnected from a voltage source.

$$U = \frac{1}{2} \frac{Q^2}{C}$$

$$\Rightarrow U = \frac{1}{2} \left(4\pi a^2 \rho_s\right)^2 \frac{1}{4\pi\varepsilon_0} \left(\frac{1}{a} - \frac{1}{b}\right)$$

$$\Rightarrow U = \frac{2\pi \rho_s^2}{\varepsilon_0} \left(a^3 - \frac{a^4}{b}\right)$$

To find a relation between a and b for the maximum value of electric energy, we need to solve the following equation.

$$\frac{dU}{da} = 0$$

Hence:

$$\frac{d}{da}\left(\frac{2\pi \rho_s^2}{\varepsilon_0}\left(a^3 - \frac{a^4}{b}\right)\right) = 0$$

$$\Rightarrow \frac{2\pi \rho_s^2}{\varepsilon_0} \frac{d}{da}\left(a^3 - \frac{a^4}{b}\right) = 0$$

$$\Rightarrow 3a^2 - \frac{4a^3}{b} = 0$$

$$\Rightarrow a^2\left(3 - \frac{4a}{b}\right) = 0$$

$$\Rightarrow 3 - \frac{4a}{b} = 0$$

$$\Rightarrow b = \frac{4}{3}a$$

Choice (3) is the answer (Fig. 14.12).

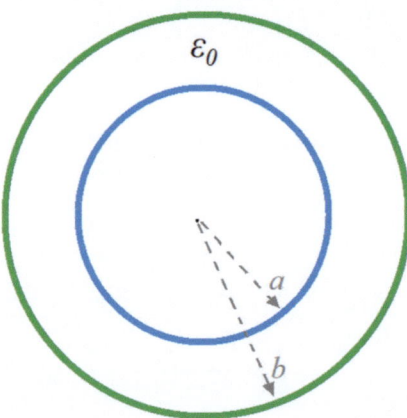

Fig. 14.12 A spherical capacitor with free space permittivity

> **Notes**
> In this problem, the relations below have been used.
>
> $$\int \sin x \, dx = -\cos x$$
>
> $$\cos \pi = -1$$
>
> $$\cos 0 = 1$$
>
> $$\int x^n \, dx = \frac{x^{n+1}}{n+1}$$
>
> $$\frac{d}{dx} x^n = n x^{n-1}$$

14.19. A solid sphere with radius a can be considered a capacitor. The following relation can be used to calculate the capacitance of the spherical capacitor in the first test.

$$\frac{1}{C_1} = \int \frac{dR}{\int \varepsilon \, dS}$$

$$\Rightarrow \frac{1}{C_1} = \int \frac{dR}{\int_{\varphi=0}^{2\pi} \int_{\theta=0}^{\pi} \varepsilon_0 R^2 \sin\theta \, d\theta \, d\varphi}$$

14.3 Spherical Capacitors

$$\Rightarrow \frac{1}{C_1} = \int \frac{dR}{\varepsilon_0 R^2 \int\limits_{\varphi=0}^{2\pi} \int\limits_{\theta=0}^{\pi} \sin\theta \, d\theta \, d\varphi}$$

$$\Rightarrow \frac{1}{C_1} = \int \frac{dR}{\varepsilon_0 R^2 [-\cos\theta]_0^\pi [\varphi]_0^{2\pi}}$$

$$\Rightarrow \frac{1}{C_1} = \int \frac{dR}{\varepsilon_0 R^2 (\cos 0 - \cos \pi)(2\pi - 0)}$$

$$\Rightarrow \frac{1}{C_1} = \frac{1}{4\pi\varepsilon_0} \int\limits_a^\infty \frac{dR}{R^2}$$

$$\Rightarrow \frac{1}{C_1} = \frac{1}{4\pi\varepsilon_0} \left[-\frac{1}{R}\right]_a^\infty$$

$$\Rightarrow \frac{1}{C_1} = \frac{1}{4\pi\varepsilon_0} \left(\frac{1}{a} - \frac{1}{\infty}\right)$$

$$\Rightarrow \frac{1}{C_1} = \frac{1}{4\pi\varepsilon_0 a}$$

$$\Rightarrow C_1 = 4\pi\varepsilon_0 a$$

For the second test, we have:

$$\frac{1}{C_2} = \int \frac{dR}{\int \varepsilon \, dS}$$

$$\Rightarrow \frac{1}{C_2} = \int \frac{dR}{\int\limits_{\psi=0}^{2\pi} \int\limits_{\theta=0}^{\pi} \varepsilon_0 \left(1+\frac{a}{R}\right) R^2 \sin\theta \, d\theta \, d\varphi}$$

$$\Rightarrow \frac{1}{C_2} = \int \frac{dR}{\varepsilon_0 \left(1+\frac{a}{R}\right) R^2 \int\limits_{\varphi=0}^{2\pi} \int\limits_{\theta=0}^{\pi} \sin\theta \, d\theta \, d\varphi}$$

$$\Rightarrow \frac{1}{C_2} = \int \frac{dR}{\varepsilon_0 \left(1+\frac{a}{R}\right) R^2 [-\cos\theta]_0^\pi [\varphi]_0^{2\pi}}$$

$$\Rightarrow \frac{1}{C_2} = \int \frac{dR}{\varepsilon_0\left(1+\frac{a}{R}\right)R^2(\cos 0 - \cos \pi)(2\pi - 0)}$$

$$\Rightarrow \frac{1}{C_2} = \frac{1}{4\pi\varepsilon_0} \int_a^\infty \frac{dR}{R(R+a)}$$

$$\Rightarrow \frac{1}{C_2} = \frac{1}{4\pi\varepsilon_0 a} \int_a^\infty \left(\frac{dR}{R} - \frac{dR}{R+a}\right)$$

$$\Rightarrow \frac{1}{C_2} = \frac{1}{4\pi\varepsilon_0 a} [\ln R - \ln(R+a)]_a^\infty$$

$$\Rightarrow \frac{1}{C_2} = \frac{1}{4\pi\varepsilon_0 a} \left[\ln\left(\frac{R}{R+a}\right)\right]_a^\infty$$

$$\Rightarrow \frac{1}{C_2} = \frac{1}{4\pi\varepsilon_0 a} \left(\ln 1 - \ln\left(\frac{a}{2a}\right)\right)$$

$$\Rightarrow \frac{1}{C_2} = \frac{1}{4\pi\varepsilon_0 a} \ln 2$$

$$\Rightarrow C_2 = \frac{4\pi\varepsilon_0 a}{\ln 2}$$

Finally, the ratio of capacitances of the sphere in the two tests is as follows.

$$\frac{C_2}{C_1} = \frac{1}{\ln 2}$$

Choice (2) is the answer (Fig. 14.13).

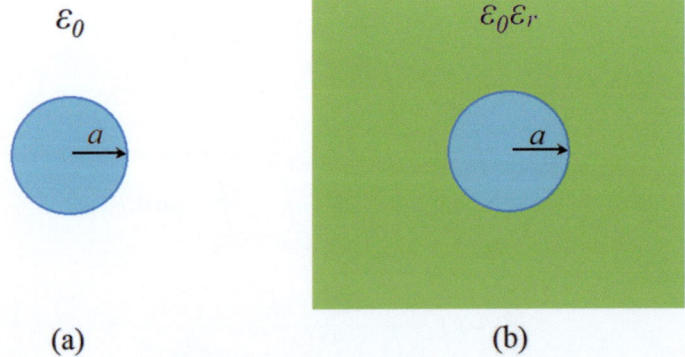

Fig. 14.13 A spherical conductor placed in (**a**) a nonuniform dielectric and (**b**) free space

14.3 Spherical Capacitors

> **Notes**
>
> In this problem, the relations below have been used.
>
> $$\int \sin x \, dx = -\cos x$$
>
> $$\cos \pi = -1$$
>
> $$\cos 0 = 1$$
>
> $$\int x^n dx = \frac{x^{n+1}}{n+1}$$
>
> $$\int \frac{du}{u(x)} = \ln u(x)$$
>
> $$\ln 1 = 0$$
>
> $$\ln\left(\frac{1}{a}\right) = -\ln a$$

14.20. As we know, the capacitance of a solid sphere with radius a is as follows.

$$C_1 = 4\pi\varepsilon_0 a$$

However, the capacitance of the capacitor in the second test, can be calculated as follows.

$$\frac{1}{C_2} = \int \frac{dR}{\int \varepsilon dS}$$

$$\Rightarrow \frac{1}{C_2} = \int \frac{dR}{\int_{\varphi=0}^{2\pi} \int_{\theta=0}^{\pi} \varepsilon_0 \left(1 + \frac{a^2}{R^2}\right) R^2 \sin\theta \, d\theta \, d\varphi}$$

$$\Rightarrow \frac{1}{C_2} = \int \frac{dR}{\varepsilon_0 \left(1 + \frac{a^2}{R^2}\right) R^2 \int_{\varphi=0}^{2\pi} \int_{\theta=0}^{\pi} \sin\theta \, d\theta \, d\varphi}$$

$$\Rightarrow \frac{1}{C_2} = \int \frac{dR}{\varepsilon_0 \left(1 + \frac{a^2}{R^2}\right) R^2 [-\cos\theta]_0^\pi [\varphi]_0^{2\pi}}$$

$$\Rightarrow \frac{1}{C_2} = \int \frac{dR}{\varepsilon_0\left(1+\frac{a^2}{R^2}\right)R^2(\cos 0 - \cos \pi)(2\pi - 0)}$$

$$\Rightarrow \frac{1}{C_2} = \frac{1}{4\pi\varepsilon_0 a^2} \int_a^\infty \frac{dR}{1+\left(\frac{R}{a}\right)^2}$$

$$\Rightarrow \frac{1}{C_2} = \frac{1}{4\pi\varepsilon_0 a} \left[\arctan \frac{R}{a}\right]_a^\infty$$

$$\Rightarrow \frac{1}{C_2} = \frac{1}{4\pi\varepsilon_0 a} (\arctan \infty - \arctan 1)$$

$$\Rightarrow \frac{1}{C_2} = \frac{1}{4\pi\varepsilon_0 a} \left(\frac{\pi}{2} - \frac{\pi}{4}\right)$$

$$\Rightarrow \frac{1}{C_2} = \frac{1}{16\varepsilon_0 a}$$

$$\Rightarrow C_2 = 16\varepsilon_0 a$$

Herein, the relation below must be applied to calculate the electric energy stored in a capacitor since the capacitor is connected to a voltage source.

$$U = \frac{1}{2}CV_0^2$$

Hence:

$$U_2 - U_1 = \frac{1}{2}(16\varepsilon_0 a - 4\pi\varepsilon_0 a)V_0^2$$

$$\Rightarrow U_2 - U_1 = (8\varepsilon_0 a - 2\pi\varepsilon_0 a)V_0^2$$

$$\Rightarrow U_2 - U_1 = \left(\frac{4}{\pi} - 1\right)2\pi\varepsilon_0 a V_0^2$$

Choice (2) is the answer (Fig. 14.14).

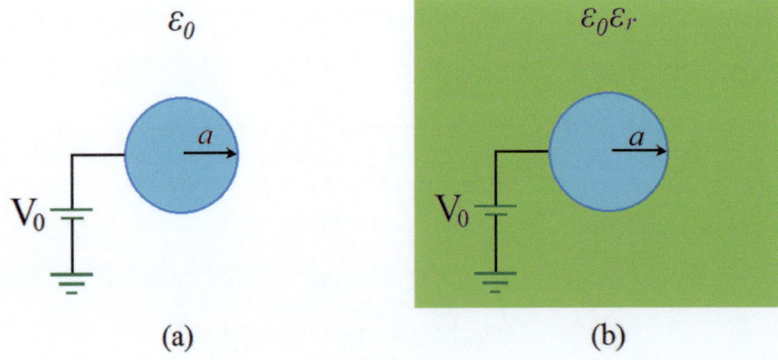

Fig. 14.14 A spherical conductor connected to a voltage source and placed in (**a**) a nonuniform dielectric and (**b**) free space

14.3 Spherical Capacitors

> **Notes**
>
> In this problem, the relations below have been used.
>
> $$\int \sin x\, dx = -\cos x$$
>
> $$\cos \pi = -1$$
>
> $$\cos 0 = 1$$
>
> $$\int \frac{dx}{a^2 + x^2} = \frac{1}{a} \arctan \frac{x}{a}$$
>
> $$\arctan \infty = \frac{\pi}{2}$$
>
> $$\arctan 1 = \frac{\pi}{4}$$

14.21. The capacitance of the capacitor can be calculated as follows.

$$\frac{1}{C} = \int \frac{dR}{\int \varepsilon\, dS}$$

$$\Rightarrow \frac{1}{C} = \int \frac{dR}{\int_{\varphi=0}^{2\pi} \int_{\theta=0}^{\pi} \varepsilon_0 R^2 \sin\theta\, d\theta\, d\varphi}$$

$$\Rightarrow \frac{1}{C} = \int \frac{dR}{\varepsilon_0 R^2 \int_{\varphi=0}^{2\pi} \int_{\theta=0}^{\pi} \sin\theta\, d\theta\, d\varphi}$$

$$\Rightarrow \frac{1}{C} = \int \frac{dR}{\varepsilon_0 R^2 [-\cos\theta]_0^{\pi} [\varphi]_0^{2\pi}}$$

$$\Rightarrow \frac{1}{C} = \int \frac{dR}{\varepsilon_0 R^2 (\cos 0 - \cos \pi)(2\pi - 0)}$$

$$\Rightarrow \frac{1}{C} = \frac{1}{4\pi\varepsilon_0} \int_a^b \frac{dR}{R^2}$$

$$\Rightarrow \frac{1}{C_1} = \frac{1}{4\pi\varepsilon_0}\left[-\frac{1}{R}\right]_a^b$$

$$\Rightarrow \frac{1}{C} = \frac{1}{4\pi\varepsilon_0}\left(\frac{1}{a} - \frac{1}{b}\right)$$

$$\Rightarrow \frac{1}{C} = \frac{b-a}{4\pi\varepsilon_0 ab}$$

$$\Rightarrow C = \frac{4\pi\varepsilon_0 ab}{b-a}$$

Moreover, we know that the electric charge of a capacitor can be calculated as follows.

$$Q = VC$$

$$\Rightarrow Q = V_0 \times \frac{4\pi\varepsilon_0 ab}{b-a}$$

$$\Rightarrow Q = \frac{4\pi\varepsilon_0 ab V_0}{b-a} \tag{14.3}$$

On the other hand, we can calculate the electric field at point A by applying Gauss's law as is shown in Fig. 14.15.

$$Q = \int \vec{D} \cdot \vec{dS}$$

$$\Rightarrow Q = \int \varepsilon_0 \vec{E} \cdot \vec{dS}$$

$$\Rightarrow Q = \int_{\varphi=0}^{2\pi} \int_{\theta=0}^{\pi} \varepsilon_0 E_R R^2 \sin\theta \, d\theta \, d\varphi$$

$$\Rightarrow Q = \varepsilon_0 E_R R^2 \int_{\varphi=0}^{2\pi} \int_{\theta=0}^{\pi} \sin\theta \, d\theta \, d\varphi$$

$$\Rightarrow Q = 4\pi\varepsilon_0 E_R R^2$$

$$\Rightarrow E_R = \frac{Q}{4\pi\varepsilon_0 R^2} \tag{14.4}$$

Solving (14.3) and (14.4):

$$E_A = \frac{\frac{4\pi\varepsilon_0 ab V_0}{b-a}}{4\pi\varepsilon_0 R^2}$$

$$\Rightarrow E_A = \frac{ab V_0}{R^2(b-a)}$$

Choice (4) is the answer.

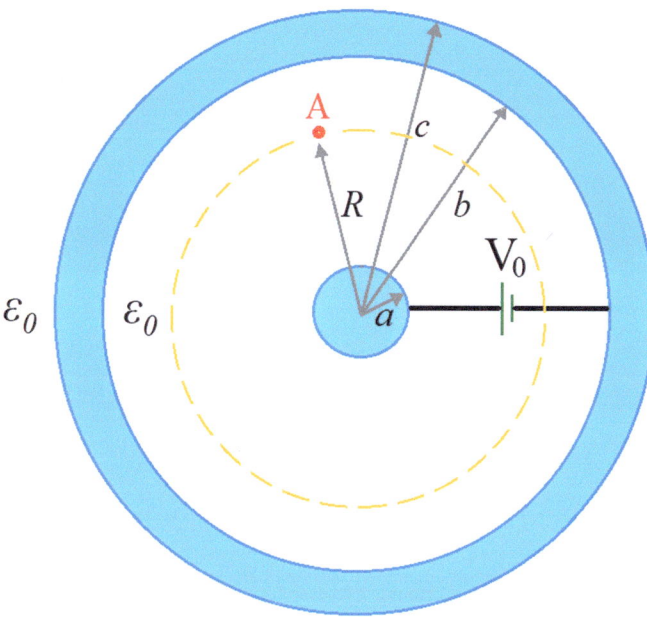

Fig. 14.15 A spherical capacitor with free space permittivity connected to a voltage source

> **Notes**
> In this problem, the relations below have been used.
>
> $$\int \sin x \, dx = -\cos x$$
>
> $$\cos \pi = -1$$
>
> $$\cos 0 = 1$$
>
> $$\int x^n \, dx = \frac{x^{n+1}}{n+1}$$
>
> $$\int_{\varphi=0}^{2\pi} \int_{\theta=0}^{\pi} \sin\theta \, d\theta \, d\varphi = 4\pi$$

References

1. Rahmani-Andebili, M., General Physics II – Practice Problems, Methods, and Solutions, Springer Nature, 2025.
2. Rahmani-Andebili, M., General Physics I – Practice Problems, Methods, and Solutions, Springer Nature, 2025.
3. Rahmani-Andebili, M., Mathematics of Engineering and Science – Practice Problems, Methods, and Solutions, Springer Nature, 2024.
4. Rahmani-Andebili, M., Differential Equations – Practice Problems, Methods, and Solutions, Springer Nature, 2022.
5. Rahmani-Andebili, M., Calculus III – Practice Problems, Methods, and Solutions, Springer Nature, 2023.
6. Rahmani-Andebili, M., Calculus II – Practice Problems, Methods, and Solutions, Springer Nature, 2023.
7. Rahmani-Andebili, M., Calculus I (2nd Ed.) – Practice Problems, Methods, and Solutions, Springer Nature, 2023.
8. Rahmani-Andebili, M., Precalculus (2nd Ed.) – Practice Problems, Methods, and Solutions, Springer Nature, 2024.

Method of Image Charge in Electrostatics: Part A 15

Abstract

In this chapter, the basic and advanced problems of method of image charge in electrostatics are studied. The subjects include the method of image charge for grounded and isolated conductors. Herein, different types of problems and exercises are presented that are categorized as follows.

- **Problems with detailed solution**: They have been designed to teach students the subjects in detail. Moreover, they have been categorized in different levels based on their difficulty levels (easy, normal, and hard) and calculation amounts (small, normal, and large).
- **Partially solved exercises**: They have been designed to encourage students to practice problems while guiding them through the problem-solving procedure and hinting the required formulas.
- **Exercises with final answer**: They have been designed to encourage students to practice more by themselves while hinting them by the final answer as well as to help instructors to give tests or quizzes.

15.1 Method of Image Charge for Grounded Conductors

Problem

15.1. As is shown in Fig. 15.1, the distance of the point charge q from the center of a grounded solid spherical conductor with radius a is $d = 2a$. Calculate the force that the charge and the sphere exert on each other [1–8].

Difficulty level   ○ Easy   ● Normal   ○ Hard
Calculation amount   ○ Small   ● Normal   ○ Large

1) $\dfrac{-q^2}{9\pi\varepsilon_0 a^2}$

2) $\dfrac{-q^2}{36\pi\varepsilon_0 a^2}$

3) $\dfrac{-2q^2}{9\pi\varepsilon_0 a^2}$

4) $\dfrac{-q^2}{18\pi\varepsilon_0 a^2}$

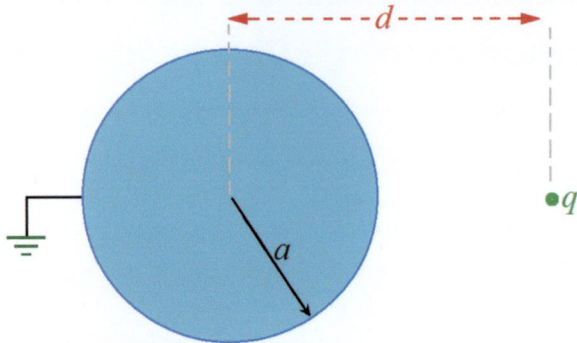

Fig. 15.1 A point charge placed in front of a grounded solid spherical conductor

Partially Solved Exercise

15.2. Solve Problem 15.1 by assuming that $d = 5a$.

Solution

By using the method of image charge in electrostatics, a grounded solid spherical conductor can be replaced by the image of the charge q' at a distance d' from the center of the sphere. The parameters q' and d' can be calculated as follows.

$$q' = -\frac{\text{Radius of sphere}}{\text{Distance of the charge from the center of sphere}} q$$

$$d' = \frac{(\text{Radius of sphere})^2}{\text{Distance of the charge from the center of sphere}}$$

Hence:

$$q' = -\frac{(\quad)}{(\quad)}(\quad) = -\frac{(\quad)}{(\quad)}$$

$$d' = \frac{(\quad)}{(\quad)} = \frac{(\quad)}{(\quad)}$$

Based on the previous chapters, the electric force between two charges can be calculated as follows.

$$F = \frac{qq'}{4\pi\varepsilon_0 x^2}$$

$$\Rightarrow F = \frac{qq'}{4\pi\varepsilon_0 (d-d')^2}$$

$$\Rightarrow F = \frac{q(\quad)}{4\pi\varepsilon_0 ((\quad)-(\quad))^2}$$

$$\Rightarrow F = \frac{-(\quad)q^2}{4\pi\varepsilon_0 (\quad)^2}$$

$$\Rightarrow F = \frac{-5q^2}{324\pi\varepsilon_0 a^2}$$

15.1 Method of Image Charge for Grounded Conductors

Problem

15.3. The distance of a point charge q from the center of a grounded solid spherical conductor with radius a is x, as can be seen in Fig. 15.2. Calculate the amount of work needed to move the charge to infinity.

Difficulty level — ○ Easy ○ Normal ● Hard
Calculation amount — ○ Small ○ Normal ● Large

1) $\dfrac{q^2 a}{4\pi\varepsilon_0 (R^2 - a^2)}$

2) $\dfrac{-q^2 a}{8\pi\varepsilon_0 (R^2 - a^2)}$

3) $\dfrac{-q^2 a}{4\pi\varepsilon_0 (R^2 - a^2)}$

4) $\dfrac{q^2 a}{8\pi\varepsilon_0 (R^2 - a^2)}$

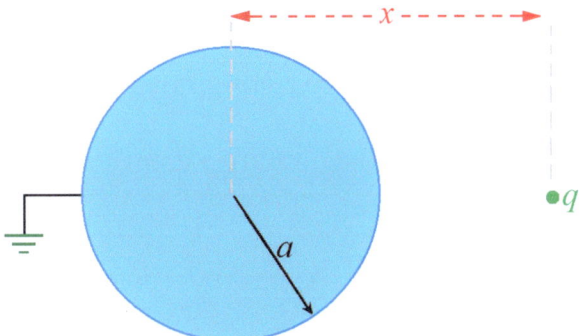

Fig. 15.2 A point charge placed in front of a grounded solid spherical conductor

Problem

15.4. Figure 15.3 shows the point charge $q = 2 \, nC$ at the vicinity of a grounded solid spherical conductor. The switch is opened and then the point charge is gradually moved away from the sphere. Calculate the electric potential of the sphere in this condition. Herein, assume that $a = 2 \, cm$, $d = 4 \, cm$, and $\varepsilon_0 = \dfrac{1}{36\pi} \times 10^{-9} \, F/m$.

Difficulty level — ○ Easy ○ Normal ● Hard
Calculation amount — ○ Small ● Normal ○ Large

1) $-225 \, V$
2) $-450 \, V$
3) $-900 \, V$
4) $-22.5 \, kV$

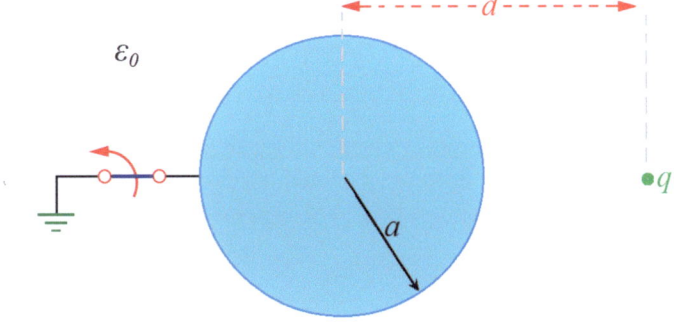

Fig. 15.3 A charge placed in front of a grounded solid spherical conductor

15.2 Method of Image Charge for Isolated Conductors

Problem

15.5. The point charge q is exerted to a solid spherical conductor with radius a as is illustrated in Fig. 15.4. Then, another charge q is placed at the distance $d = 2a$ from the center of the sphere. Calculate the force that the sphere exerts on the point charge.

Difficulty level ○ Easy ○ Normal ● Hard
Calculation amount ○ Small ● Normal ○ Large

1) $\dfrac{q^2}{144\pi\varepsilon_0 a^2}$

2) $\dfrac{17q^2}{144\pi\varepsilon_0 a^2}$

3) $\dfrac{11q^2}{288\pi\varepsilon_0 a^2}$

4) $\dfrac{43q^2}{288\pi\varepsilon_0 a^2}$

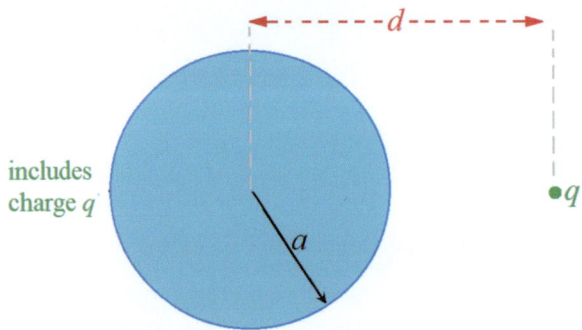

Fig. 15.4 A point charge placed in front of an isolated solid spherical conductor

Partially Solved Exercise

15.6. Solve Problem 15.5 by assuming that $d = 3a$.

Solution

By using the method of image charge in electrostatics, an isolated solid spherical conductor can be replaced by the image of the charge q' at a distance d' from the center of the sphere and the charge q'' placed at the center of the sphere. The parameters q', d', and q'' can be calculated as follows.

$$q' = -\frac{\text{Radius of sphere}}{\text{Distance of the charge from the center of sphere}} q$$

$$d' = \frac{(\text{Radius of sphere})^2}{\text{Distance of the charge from the center of sphere}}$$

$$q'' = (\text{Primary charge of sphere}) - q'$$

Herein, q'' is used because the total charge of the sphere must remain constant since the sphere is isolated. Hence:

15.2 Method of Image Charge for Isolated Conductors

$$q' = -\frac{(\quad)}{(\quad)}(\quad) = -(\quad)$$

$$d' = \frac{(\quad)}{(\quad)} = (\quad)$$

$$q'' = q - (\quad) = (\quad)$$

Based on the previous chapters, the electric force between two charges can be calculated as follows.

$$F = \frac{qq'}{4\pi\varepsilon_0 x^2}$$

The total force that the sphere exerts on the charge is the vector sum of the forces that the charges q' and q'' exerts on the charge q. Hence:

$$F = \frac{qq'}{4\pi\varepsilon_0(d-d')^2} + \frac{qq''}{4\pi\varepsilon_0 d^2}$$

$$\Rightarrow F = \frac{q(\quad)}{4\pi\varepsilon_0((\quad)-(\quad))^2} + \frac{q(\quad)}{4\pi\varepsilon_0(\quad)^2}$$

$$\Rightarrow F = \frac{-(\quad)q^2}{4\pi\varepsilon_0(\quad)^2} + \frac{(\quad)q^2}{4\pi\varepsilon_0(\quad)^2}$$

$$\Rightarrow F = -\frac{(\quad)q^2}{(\quad)\pi\varepsilon_0 a^2} + \frac{(\quad)q^2}{(\quad)\pi\varepsilon_0 a^2}$$

$$\Rightarrow F = \frac{4q^2}{75\pi\varepsilon_0 a^2}$$

Problem

15.7. Consider an isolated spherical conductor with radius a and electric potential V_0 in free space. If a point charge q is placed in front of the sphere at $d = 3a$ distance from its center, the force on the point charge is zero. Calculate the value of V_0. Herein, assume that $\varepsilon_0 = \frac{1}{36\pi} \times 10^{-9} \, F/m$ (Fig. 15.5).

Difficulty level   ○ Easy   ○ Normal   ● Hard
Calculation amount   ○ Small   ● Normal   ○ Large

1) $V_0 = \dfrac{7q}{108\pi\varepsilon_0 a}$

2) $V_0 = \dfrac{17q}{768\pi\varepsilon_0 a}$

3) $V_0 = \dfrac{7q}{768\pi\varepsilon_0 a}$

4) $V_0 = \dfrac{17q}{108\pi\varepsilon_0 a}$

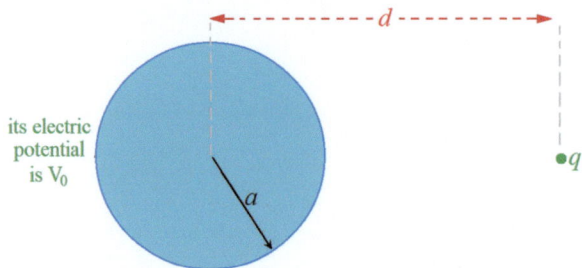

Fig. 15.5 A point charge placed in front of an isolated solid spherical conductor

Problem

15.8. A point charge q is placed in front of a charge-free spherical conductor with radius a. Calculate the total energy of the system if the distance of the charge from the center of sphere is $d = 2a$ (Fig. 15.6).

Difficulty level   ○ Easy   ○ Normal   ● Hard
Calculation amount ○ Small   ○ Normal   ● Large

1) $-\dfrac{7q^2}{24\pi\varepsilon_0 a}$

2) $-\dfrac{7q^2}{96\pi\varepsilon_0 a}$

3) $-\dfrac{7q^2}{12\pi\varepsilon_0 a}$

4) $-\dfrac{7q^2}{48\pi\varepsilon_0 a}$

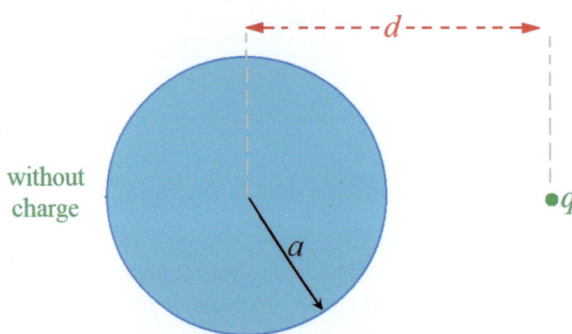

Fig. 15.6 A point charge placed in front of an isolated solid spherical conductor

References

1. Rahmani-Andebili, M., General Physics II – Practice Problems, Methods, and Solutions, Springer Nature, 2025.
2. Rahmani-Andebili, M., General Physics I – Practice Problems, Methods, and Solutions, Springer Nature, 2025.
3. Rahmani-Andebili, M., Mathematics of Engineering and Science – Practice Problems, Methods, and Solutions, Springer Nature, 2024.
4. Rahmani-Andebili, M., Differential Equations – Practice Problems, Methods, and Solutions, Springer Nature, 2022.
5. Rahmani-Andebili, M., Calculus III – Practice Problems, Methods, and Solutions, Springer Nature, 2023.
6. Rahmani-Andebili, M., Calculus II – Practice Problems, Methods, and Solutions, Springer Nature, 2023.
7. Rahmani-Andebili, M., Calculus I (2nd Ed.) – Practice Problems, Methods, and Solutions, Springer Nature, 2023.
8. Rahmani-Andebili, M., Precalculus (2nd Ed.) – Practice Problems, Methods, and Solutions, Springer Nature, 2024.

Method of Image Charge in Electrostatics: Part B 16

Abstract

In this chapter, the problems of the fifteenth chapter are fully solved, in detail, step-by-step, and with different methods.

16.1 Method of Image Charge for Grounded Conductors

16.1. By using the method of image charge in electrostatics, a grounded solid spherical conductor can be replaced by the image of the charge q' at a distance d' from the center of the sphere. The parameters q' and d' can be calculated as follows [1–8].

$$q' = -\frac{\text{Radius of sphere}}{\text{Distance of the charge from the center of sphere}} q$$

$$d' = \frac{(\text{Radius of sphere})^2}{\text{Distance of the charge from the center of sphere}}$$

Hence:

$$q' = -\frac{a}{2a} q = -\frac{q}{2}$$

$$d' = \frac{a^2}{2a} - \frac{a}{2}$$

Based on the previous chapters, the electric force between two charges can be calculated as follows.

$$F = \frac{qq'}{4\pi\varepsilon_0 x^2}$$

$$\Rightarrow F = \frac{qq'}{4\pi\varepsilon_0 (d-d')^2}$$

$$\Rightarrow F = \frac{q\left(-\frac{q}{2}\right)}{4\pi\varepsilon_0 \left(2a - \frac{a}{2}\right)^2}$$

$$\Rightarrow F = \frac{-\frac{q^2}{2}}{4\pi\varepsilon_0 \left(\frac{3a}{2}\right)^2}$$

$$\Rightarrow F = \frac{-q^2}{18\pi\varepsilon_0 a^2}$$

Choice (4) is the answer (Fig. 16.1).

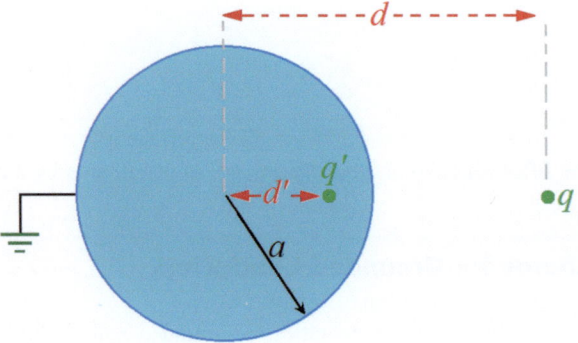

Fig. 16.1 A point charge placed in front of a grounded solid spherical conductor.

16.3. By using the method of image charge in electrostatics, a grounded solid spherical conductor can be replaced by the image of the charge q' at a distance d' from the center of the sphere. The parameters q' and d' can be calculated as follows.

$$q' = -\frac{\text{Radius of sphere}}{\text{Distance of the charge from the center of sphere}} q$$

$$d' = \frac{(\text{Radius of sphere})^2}{\text{Distance of the charge from the center of sphere}}$$

Thus:

$$q' = -\frac{a}{x} q$$

$$x' = \frac{a^2}{x}$$

Based on the previous chapters, the electric force between two charges can be calculated as follows.

$$F = \frac{qq'}{4\pi\varepsilon_0 x^2}$$

$$\Rightarrow F = \frac{qq'}{4\pi\varepsilon_0 (x-x')^2}$$

16.1 Method of Image Charge for Grounded Conductors

$$\Rightarrow F = \frac{q\left(-\frac{a}{x}q\right)}{4\pi\varepsilon_0\left(x - \frac{a^2}{x}\right)^2}$$

$$\Rightarrow F = \frac{-axq^2}{4\pi\varepsilon_0(x^2 - a^2)^2}$$

As can be noticed, the force between the charge and its image is an attractive force. Thus, an opposite force must be applied to move the charge to infinity. In other words:

$$F = \frac{axq^2}{4\pi\varepsilon_0(x^2 - a^2)^2}$$

Moreover, the amount of work to move the charge to infinity can be calculated as follows.

$$W = \int_x^\infty F dx$$

$$\Rightarrow W = \int_x^\infty \frac{axq^2}{4\pi\varepsilon_0(x^2 - a^2)^2} dx$$

$$\Rightarrow W = \frac{aq^2}{4\pi\varepsilon_0} \int_x^\infty \frac{x}{(x^2 - a^2)^2} dx$$

$$\Rightarrow W = \frac{aq^2}{4\pi\varepsilon_0} \left[-\frac{\frac{1}{2}}{x^2 - a^2} \right]_x^\infty$$

$$\Rightarrow W = \frac{aq^2}{4\pi\varepsilon_0} \left(\frac{\frac{1}{2}}{x^2 - a^2} - 0 \right)$$

$$\Rightarrow W = \frac{q^2 a}{8\pi\varepsilon_0(x^2 - a^2)}$$

Choice (4) is the answer (Fig. 16.2).

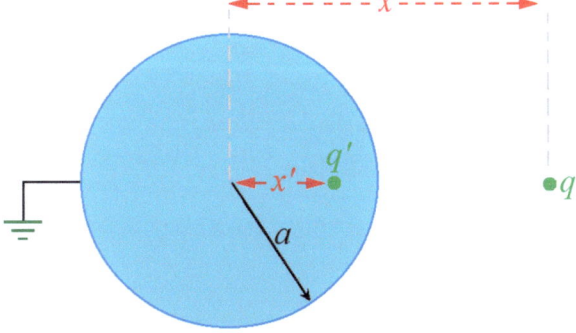

Fig. 16.2 A point charge placed in front of a grounded solid spherical conductor

> **Notes**
>
> In this problem, the relation below has been used.
>
> $$\int u^n(x)\,du = \frac{u^{n+1}(x)}{n+1}$$

16.4. Based on the information given in the problem, we have:

$$q = 2\ nC$$

$$a = 2\ cm$$

$$d = 4\ cm$$

$$\varepsilon_0 = \frac{1}{36\pi} \times 10^{-9}\ F/m$$

By using the method of image charge in electrostatics, a charge q' is induced in the grounded solid spherical conductor that can be calculated as follows.

$$q' = -\frac{\text{Radius of sphere}}{\text{Distance of the charge from the center of sphere}}\,q$$

Therefore:

$$q' = -\frac{a}{d}q = -\frac{2}{4} \times 2\ nC = -1\ nC$$

Since the switch is opened and then the point charge is gradually moved away from the sphere, the charge of sphere remains constant.

Based on the previous chapters, the electric potential of the sphere can be calculated as follows.

$$V = \frac{q'}{4\pi\varepsilon_0 a}$$

$$\Rightarrow V = \frac{-10^{-9}}{4\pi\left(\frac{1}{36\pi} \times 10^{-9}\right)(0.02)}$$

$$\Rightarrow V = -450\ V$$

Choice (2) is the answer (Fig. 16.3).

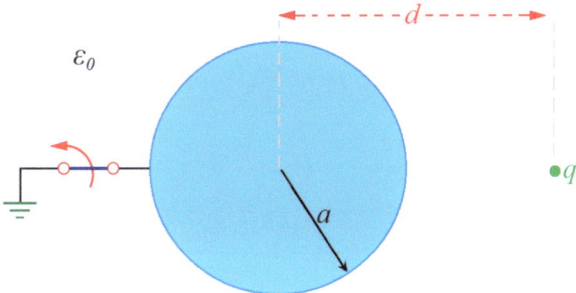

Fig. 16.3 A charge placed in front of a grounded solid spherical conductor

16.2 Method of Image Charge for Isolated Conductors

16.5. By using the method of image charge in electrostatics, an isolated solid spherical conductor can be replaced by the image of the charge q' at a distance d' from the center of the sphere and the charge q' placed at the center of the sphere. The parameters q', d', and q' can be calculated as follows.

$$q' = -\frac{\text{Radius of sphere}}{\text{Distance of the charge from the center of sphere}} q$$

$$d' = \frac{(\text{Radius of sphere})^2}{\text{Distance of the charge from the center of sphere}}$$

$$q'' = (\text{Primary charge of sphere}) - q'$$

Herein, q'' is used because the total charge of the sphere must remain constant since the sphere is isolated.

Hence:

$$q' = -\frac{a}{2a}q = -\frac{q}{2}$$

$$d' = \frac{a^2}{2a} = \frac{a}{2}$$

$$q'' = q - \left(-\frac{q}{2}\right) = \frac{3q}{2}$$

Based on the previous chapters, the electric force between two charges can be calculated as follows.

$$F = \frac{qq'}{4\pi\varepsilon_0 x^2}$$

The total force that the sphere exerts on the point charge is the vector sum of the forces that the charges q' and q'' exert on the charge q. Hence:

$$F = \frac{qq'}{4\pi\varepsilon_0(d-d')^2} + \frac{qq''}{4\pi\varepsilon_0 d^2}$$

$$\Rightarrow F = \frac{q\left(-\frac{q}{2}\right)}{4\pi\varepsilon_0\left(2a - \frac{a}{2}\right)^2} + \frac{q\left(\frac{3q}{2}\right)}{4\pi\varepsilon_0(2a)^2}$$

$$\Rightarrow F = \frac{-\frac{q^2}{2}}{4\pi\varepsilon_0\left(\frac{3a}{2}\right)^2} + \frac{\frac{3q^2}{2}}{4\pi\varepsilon_0(2a)^2}$$

$$\Rightarrow F = -\frac{q^2}{18\pi\varepsilon_0 a^2} + \frac{3q^2}{32\pi\varepsilon_0 a^2}$$

$$\Rightarrow F = \frac{11q^2}{288\pi\varepsilon_0 a^2}$$

Choice (3) is the answer (Fig. 16.4).

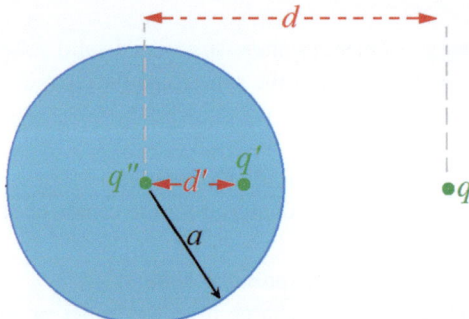

Fig. 16.4 A point charge placed in front of an isolated solid spherical conductor

16.7. Based on the information given in the problem, we have:

$$V = V_0$$

$$d = 3a$$

$$F = 0$$

By using the method of image charge in electrostatics, an isolated solid spherical conductor can be replaced by the image of the charge q' at a distance d' from the center of the sphere and the charge q'' placed at the center of the sphere. The parameters q', d', and q' can be calculated as follows.

$$q' = -\frac{\text{Radius of sphere}}{\text{Distance of the charge from the center of sphere}} q$$

$$d' = \frac{(\text{Radius of sphere})^2}{\text{Distance of the charge from the center of sphere}}$$

$$q'' = (\text{Primary charge of sphere}) - q'$$

16.2 Method of Image Charge for Isolated Conductors

Herein, q'' is used because the total charge of the sphere must remain constant since the sphere is isolated.

First, the primary charge of the sphere needs to be determined. Based on the previous chapters, the relation between the electric potential and total charge of a sphere is as follows.

$$V = \frac{q}{4\pi\varepsilon_0 a}$$

$$\Rightarrow V_0 = \frac{q}{4\pi\varepsilon_0 a}$$

$$\Rightarrow q = 4\pi\varepsilon_0 a V_0$$

The value of q', d', and q'' can be calculated as follows.

$$q' = -\frac{a}{3a}q = -\frac{q}{3}$$

$$d' = \frac{a^2}{3a} = \frac{a}{3}$$

$$q'' = 4\pi\varepsilon_0 a V_0 - \left(-\frac{q}{3}\right) = 4\pi\varepsilon_0 a V_0 + \frac{q}{3}$$

Based on the previous chapters, the electric force between two charges can be calculated as follows.

$$F = \frac{qq'}{4\pi\varepsilon_0 x^2}$$

The total force that the sphere exerts on the charge is the vector sum of the forces that the charges q' and q' exerts on the charge q. Therefore:

$$0 = \frac{qq'}{4\pi\varepsilon_0 (d-d')^2} + \frac{qq''}{4\pi\varepsilon_0 d^2}$$

$$\Rightarrow 0 = \frac{q\left(-\frac{q}{3}\right)}{4\pi\varepsilon_0 \left(3a - \frac{a}{3}\right)^2} + \frac{q\left(4\pi\varepsilon_0 a V_0 + \frac{q}{3}\right)}{4\pi\varepsilon_0 (3a)^2}$$

$$\Rightarrow \frac{\frac{q}{3}}{\left(\frac{8}{3}\right)^2} = \frac{4\pi\varepsilon_0 a V_0 + \frac{q}{3}}{(3)^2}$$

$$\Rightarrow \frac{3q}{64} = \frac{4\pi\varepsilon_0 a V_0}{9} + \frac{q}{27}$$

$$\Rightarrow \frac{17q}{1728} = \frac{4\pi\varepsilon_0 a V_0}{9}$$

$$\Rightarrow V_0 = \frac{17q}{768\pi\varepsilon_0 a}$$

Choice (2) is the answer (Fig. 16.5).

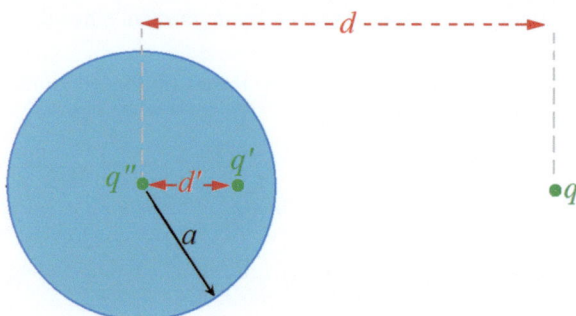

Fig. 16.5 A point charge placed in front of an isolated solid spherical conductor

---

16.8. By using the method of image charge in electrostatics, an isolated solid spherical conductor can be replaced by the image of the charge q' at a distance d' from the center of the sphere and the charge q'' placed at the center of the sphere. The parameters q', d', and q'' can be calculated as follows.

$$q' = -\frac{\text{Radius of sphere}}{\text{Distance of the charge from the center of sphere}} q$$

$$d' = \frac{(\text{Radius of sphere})^2}{\text{Distance of the charge from the center of sphere}}$$

$$q'' = (\text{Primary charge of sphere}) - q'$$

Herein, q'' is used because the total charge of the sphere must remain constant since the sphere is isolated.

Hence:

$$q' = -\frac{a}{2a} q = -\frac{q}{2}$$

$$d' = \frac{a^2}{2a} = \frac{a}{2}$$

$$q'' = 0 - \left(-\frac{q}{2}\right) = \frac{q}{2}$$

Based on the previous chapters, the electric potential energy of a system of N charges can be calculated as follows.

$$U = \frac{1}{2} \sum_{i=1}^{N} q_i V_i$$

Herein, V_i is the total electric potential of the other charges at the place of the charge q_i.

For this problem, we can write:

$$U = \frac{1}{2} \sum_{i=1}^{3} q_i V_i$$

$$\Rightarrow U = \frac{1}{2}\left[q_1\left(\frac{q_2}{4\pi\varepsilon_0 d_{12}} + \frac{q_3}{4\pi\varepsilon_0 d_{13}}\right) + q_2\left(\frac{q_3}{4\pi\varepsilon_0 d_{23}} + \frac{q_1}{4\pi\varepsilon_0 d_{21}}\right) + q_3\left(\frac{q_1}{4\pi\varepsilon_0 d_{31}} + \frac{q_2}{4\pi\varepsilon_0 d_{32}}\right)\right]$$

$$\Rightarrow U = \frac{1}{2}\left[q\left(\frac{-\frac{q}{2}}{4\pi\varepsilon_0\left(2a-\frac{a}{2}\right)} + \frac{\frac{q}{2}}{4\pi\varepsilon_0(2a)}\right) + \left(-\frac{q}{2}\right)\left(\frac{\frac{q}{2}}{4\pi\varepsilon_0\left(\frac{a}{2}\right)} + \frac{q}{4\pi\varepsilon_0\left(2a-\frac{a}{2}\right)}\right) + \frac{q}{2}\left(\frac{q}{4\pi\varepsilon_0(2a)} + \frac{-\frac{q}{2}}{4\pi\varepsilon_0\left(\frac{a}{2}\right)}\right)\right]$$

$$\Rightarrow U = \frac{q^2}{8\pi\varepsilon_0 a}\left[\left(\frac{-\frac{1}{2}}{\frac{3}{2}} + \frac{\frac{1}{2}}{2}\right) + \left(-\frac{1}{2}\right)\left(\frac{\frac{1}{2}}{\frac{1}{2}} + \frac{1}{\frac{3}{2}}\right) + \frac{1}{2}\left(\frac{1}{2} + \frac{-\frac{1}{2}}{\frac{1}{2}}\right)\right]$$

$$\Rightarrow U = \frac{q^2}{8\pi\varepsilon_0 a}\left[\left(-\frac{1}{3} + \frac{1}{4}\right) + \left(-\frac{1}{2}\right)\left(1 + \frac{2}{3}\right) + \frac{1}{2}\left(\frac{1}{2} - 1\right)\right]$$

$$\Rightarrow U = \frac{q^2}{8\pi\varepsilon_0 a}\left[-\frac{1}{12} - \frac{5}{6} - \frac{1}{4}\right]$$

$$\Rightarrow U = \frac{q^2}{8\pi\varepsilon_0 a}\left[-\frac{1+10+3}{12}\right]$$

$$\Rightarrow U = \frac{q^2}{8\pi\varepsilon_0 a}\left[-\frac{14}{12}\right]$$

$$\Rightarrow U = -\frac{7q^2}{48\pi\varepsilon_0 a}$$

Choice (4) is the answer (Fig. 16.6).

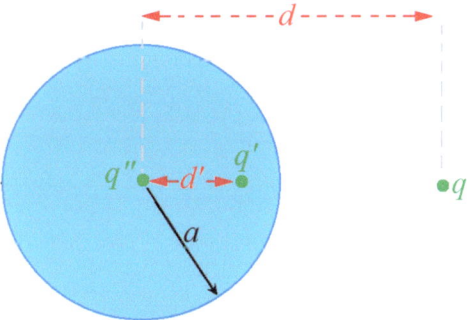

Fig. 16.6 A point charge placed in front of an isolated solid spherical conductor

References

1. Rahmani-Andebili, M., General Physics II – Practice Problems, Methods, and Solutions, Springer Nature, 2025.
2. Rahmani-Andebili, M., General Physics I – Practice Problems, Methods, and Solutions, Springer Nature, 2025.
3. Rahmani-Andebili, M., Mathematics of Engineering and Science – Practice Problems, Methods, and Solutions, Springer Nature, 2024.
4. Rahmani-Andebili, M., Differential Equations – Practice Problems, Methods, and Solutions, Springer Nature, 2022.
5. Rahmani-Andebili, M., Calculus III – Practice Problems, Methods, and Solutions, Springer Nature, 2023.
6. Rahmani-Andebili, M., Calculus II – Practice Problems, Methods, and Solutions, Springer Nature, 2023.
7. Rahmani-Andebili, M., Calculus I (2nd Ed.) – Practice Problems, Methods, and Solutions, Springer Nature, 2023.
8. Rahmani-Andebili, M., Precalculus (2nd Ed.) – Practice Problems, Methods, and Solutions, Springer Nature, 2024.

17. Flat, Cylindrical, and Spherical Resistors and Boundary Conditions for Electric Current: Part A

Abstract

In this chapter, the basic and advanced problems of boundary conditions for electric current as well as flat, cylindrical, and spherical resistors with uniform and nonuniform conductivities are studied. Herein, different types of problems and exercises are presented that are categorized as follows.

- **Problems with detailed solution**: They have been designed to teach students the subjects in detail. Moreover, they have been categorized in different levels based on their difficulty levels (easy, normal, and hard) and calculation amounts (small, normal, and large).
- **Partially solved exercises**: They have been designed to encourage students to practice problems while guiding them through the problem-solving procedure and hinting the required formulas.
- **Exercises with final answer**: They have been designed to encourage students to practice more by themselves while hinting them by the final answer as well as to help instructors to give tests or quizzes.

17.1 Boundary Conditions for Electric Current

Problem

17.1. Figure 17.1 shows two environments with different electrical conductivities and permittivity. Calculate the surface charge density on the boundary of the environments if $\alpha_1 = 60°$ [1–8].

Difficulty level   ○ Easy   ● Normal   ○ Hard
Calculation amount   ○ Small   ● Normal   ○ Large

1) $\rho_s = \frac{\sqrt{3}}{2}\left(\frac{\varepsilon_2}{\sigma_2} - \frac{\varepsilon_1}{\sigma_1}\right)J$

2) $\rho_s = \frac{1}{2}\left(\frac{\varepsilon_1}{\sigma_1} - \frac{\varepsilon_2}{\sigma_2}\right)J$

3) $\rho_s = \frac{\sqrt{3}}{2}\left(\frac{\varepsilon_1}{\sigma_1} - \frac{\varepsilon_2}{\sigma_2}\right)J$

4) $\rho_s = \frac{1}{2}\left(\frac{\varepsilon_2}{\sigma_2} - \frac{\varepsilon_1}{\sigma_1}\right)J$

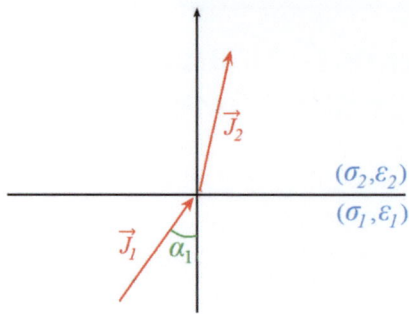

Fig. 17.1 Studying the boundary conditions for the electric current density

Problem

17.2. In Fig. 17.2, in the first region, the volume current density is $\vec{J}_1 = J_0(\hat{x} + 2\hat{y} + 6\hat{z})$. Calculate the surface charge density in the second region if $\sigma_2 = 3\sigma_1$ and $\varepsilon_2 = 2\varepsilon_1$.

Difficulty level ○ Easy ● Normal ○ Hard
Calculation amount ○ Small ● Normal ○ Large

1) $\dfrac{2\varepsilon_1 J_0}{\sigma_1}$
2) $\dfrac{2\varepsilon_2 J_0}{\sigma_2}$
3) $2\left(\dfrac{\varepsilon_1}{\sigma_1} - \dfrac{\varepsilon_2}{\sigma_2}\right) J_0$
4) $\left(\dfrac{\varepsilon_1}{\sigma_1} - \dfrac{\varepsilon_2}{\sigma_2}\right) J_0$

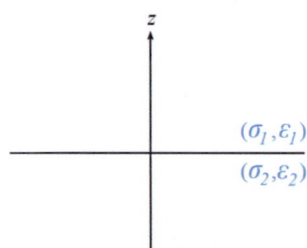

Fig. 17.2 Studying the boundary conditions for the electric current density

Partially Solved Exercise

17.3. Solve Problem 17.2 by assuming that $\vec{J}_1 = \hat{x} + 2\hat{y} + 3\hat{z}$, $\sigma_2 = 0.5\sigma_1$, and $\varepsilon_2 = 2\varepsilon_1$.

Solution

As we know, the normal component of electric flux density on the boundary of two regions is discontinuous and the amount of discontinuity is equal to the free surface charge density. Herein, $\hat{n}$ is the unit normal vector of the first region. In other words:

$$\vec{D}_2 \cdot \hat{n} - \vec{D}_1 \cdot \hat{n} = \rho_s$$

In addition, the normal component of electric current density on the boundary of two regions is continuous. In other words:

17.2 Flat Resistors

$$J_{1n} = J_{2n} \triangleq J_n$$

On the other hand, we know that:

$$D = \varepsilon E$$

$$E = \frac{J}{\sigma}$$

Therefore:

$$-\frac{\varepsilon_2 J_z}{\sigma_2} - \left(-\frac{\varepsilon_1 J_z}{\sigma_1}\right) = \rho_s$$

$$\Rightarrow \rho_s = \left(\frac{\varepsilon_1}{\sigma_1} - \frac{\varepsilon_2}{\sigma_2}\right) J_z$$

By putting the quantities in the parameters, we have:

$$\rho_s = \frac{\varepsilon_1}{\sigma_1}\left(1 - \frac{(\quad)}{(\quad)}\right) \times (\quad)$$

$$\Rightarrow \rho_s = -\frac{9\varepsilon_1}{\sigma_1}$$

17.2 Flat Resistors

Problem

17.4. For the cuboid resistor shown in Fig. 17.3, $I = 2\ mA$, $\sigma = 1\ m/s$, and $a = 1\ mm$. Calculate the voltage across it.
Difficulty level ● Easy ○ Normal ○ Hard
Calculation amount ○ Small ● Normal ○ Large
1) $2\ V$
2) $2\ kV$
3) $2\ mV$
4) $1\ mV$

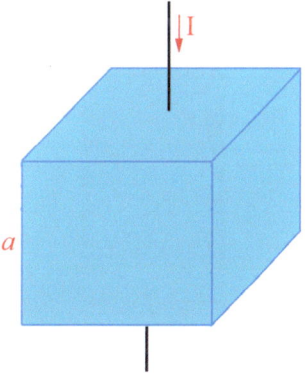

Fig. 17.3 A flat resistor

Problem

17.5. In Fig. 17.4, calculate the resistance of the cuboid resistor with the surface area, length, and conductivities of A, l, σ_1, and σ_2, respectively.

Difficulty level   ○ Easy   ● Normal   ○ Hard
Calculation amount   ● Small   ○ Normal   ○ Large

1) $R = \dfrac{l(\sigma_1 + \sigma_2)}{2A\sigma_1\sigma_2}$
2) $R = \dfrac{l\sigma_1\sigma_2}{2A(\sigma_1 + \sigma_2)}$
3) $R = \dfrac{l(\sigma_1 + \sigma_2)}{A}$
4) $R = \dfrac{l}{A\sigma_1\sigma_2}$

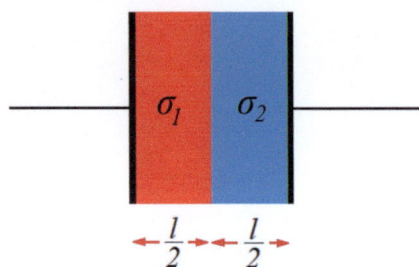

Fig. 17.4 A flat resistor including two different electrical conductivities

Problem

17.6. In Fig. 17.5, calculate the resistance of the cuboid resistor with the surface area, length, and conductivities of A, l, σ_1, and σ_2, respectively.

Difficulty level   ○ Easy   ● Normal   ○ Hard
Calculation amount   ○ Small   ● Normal   ○ Large

1) $R = \dfrac{2l}{\sigma_1\sigma_2 A}$
2) $R = \dfrac{2l}{(\sigma_1 + \sigma_2)A}$
3) $R = \dfrac{2l(\sigma_1 + \sigma_2)}{A}$
4) $R = \dfrac{l}{(\sigma_1 + \sigma_2)A}$

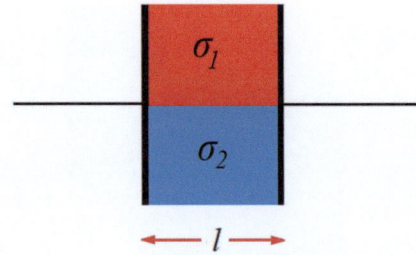

Fig. 17.5 A flat resistor including two different electrical conductivities

17.2 Flat Resistors

Problem

17.7. Figure 17.6 illustrates two regions with different conductivities. Calculate the ratio of electric potential of the first region to that of the second one.

Difficulty level   ○ Easy   ● Normal   ○ Hard
Calculation amount   ○ Small   ● Normal   ○ Large

1) $\dfrac{V_1}{V_2} = \dfrac{l_2 \sigma_1}{l_1 \sigma_2}$
2) $\dfrac{V_1}{V_2} = \dfrac{l_2 \sigma_2}{l_1 \sigma_1}$
3) $\dfrac{V_1}{V_2} = \dfrac{l_1 \sigma_1}{l_2 \sigma_2}$
4) $\dfrac{V_1}{V_2} = \dfrac{l_1 \sigma_2}{l_2 \sigma_1}$

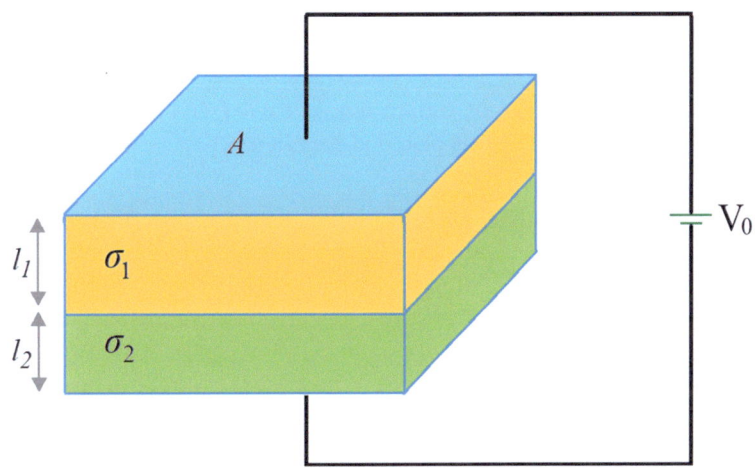

Fig. 17.6 A flat resistor connected to a voltage source including two different electrical conductivities

Problem

17.8. Calculate the surface current density on the boundary of two environments of the resistor shown in Fig. 17.7.

Difficulty level   ○ Easy   ● Normal   ○ Hard
Calculation amount   ○ Small   ● Normal   ○ Large

1) $\dfrac{\sigma_1 \sigma_2 V}{\sigma_1 d_1 + \sigma_2 d_2}$
2) $\dfrac{\sigma_1 \sigma_2 \varepsilon_1}{\sigma_1 d_1 + \sigma_2 d_2}$
3) $\dfrac{\sigma_1 \sigma_2 V}{\sigma_2 d_1 + \sigma_1 d_2}$
4) $\left(\dfrac{\sigma_1}{\sigma_2 d_1} + \dfrac{\sigma_2}{\sigma_1 d_2} \right) V$

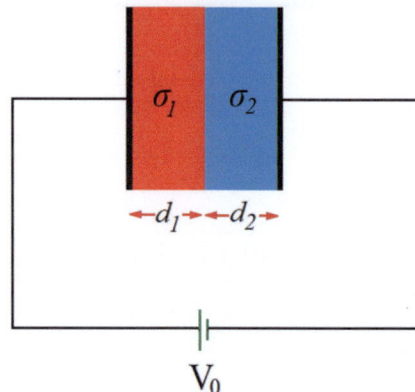

Fig. 17.7 A flat resistor connected to a voltage source including two different electrical conductivities

Problem

17.9. Calculate the resistance of a flat resistor with the surface area A, length l, and nonuniform conductivity $\sigma = e^{\alpha x}$. The axis of resistor is aligned with x-axis and its left side is placed at the origin.

Difficulty level ○ Easy ● Normal ○ Hard
Calculation amount ○ Small ● Normal ○ Large

1) $R = \sigma A \alpha (1 - e^{-\alpha d})$
2) $R = \sigma A \alpha (1 - e^{\alpha d})$
3) $R = \dfrac{1 - e^{-\alpha d}}{\sigma A \alpha}$
4) $R = \dfrac{1 + e^{-\alpha d}}{\sigma A \alpha}$

Partially Solved Exercise

17.10. Calculate the resistance of a flat resistor with the surface area A, length l, and nonuniform conductivity $\sigma = \sigma_0 x$. The axis of resistor is aligned with x-axis and its left side is placed at $x = d$.

Solution

The conductivity of the resistor is not uniform. Therefore, the following relation needs to be used to calculate its capacitance.

$$R = \int \dfrac{dx}{\int \sigma dS}$$

$$\Rightarrow R = \int \dfrac{dx}{\sigma_0 x \int dS}$$

$$\Rightarrow R = \int \dfrac{dx}{\sigma_0 x A}$$

$$\Rightarrow R = \dfrac{1}{\sigma_0 A} \int_{x=d}^{x=2d} \dfrac{dx}{()}$$

17.3 Cylindrical Resistors

$$\Rightarrow R = \frac{1}{\sigma_0 A}[(\quad)]_d^{2d}$$

$$\Rightarrow R = \frac{1}{\sigma_0 A}((\quad) - (\quad))$$

$$\Rightarrow R = \frac{\ln 2}{\sigma_0 A}$$

Problem

17.11. The conductivity of the resistor, shown in Fig. 17.8, is nonuniform and as follows.

$$\sigma = \sigma_0 \left(1 + \frac{z}{h}\right)$$

Calculate the resistance between the top and bottom disks.

Difficulty level   ○ Easy   ● Normal   ○ Hard
Calculation amount   ○ Small   ● Normal   ○ Large

1) $R = \dfrac{l \ln 2}{\sigma_0 \pi a^2}$
2) $R = \dfrac{l \ln 3}{\sigma_0 \pi a}$
3) $R = \dfrac{l \ln 2}{\sigma_0 \pi a}$
4) $R = \dfrac{l \ln 3}{\sigma_0 \pi a^2}$

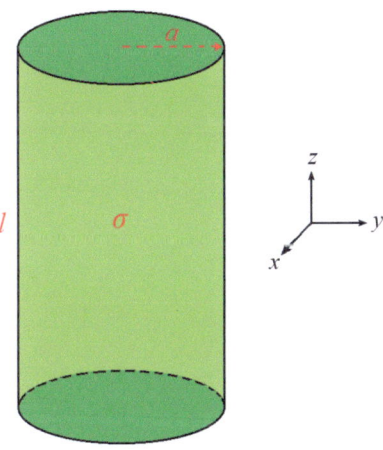

Fig. 17.8 A resistor including nonuniform conductivity

17.3 Cylindrical Resistors

Problem

17.12. Calculate the resistance of the hollow cylinder between its internal and external surfaces shown in Fig. 17.9. The length of the cylinder is l.

Difficulty level   ○ Easy   ● Normal   ○ Hard
Calculation amount   ○ Small   ● Normal   ○ Large

1) $\dfrac{l}{\pi\sigma_0(b^2-a^2)}$

2) $\dfrac{l}{2\pi\sigma_0 ab}$

3) $\dfrac{1}{2\pi l\sigma_0}\left(\dfrac{b}{a}\right)^2$

4) $\dfrac{\ln\left(\dfrac{b}{a}\right)}{2\pi l\sigma_0}$

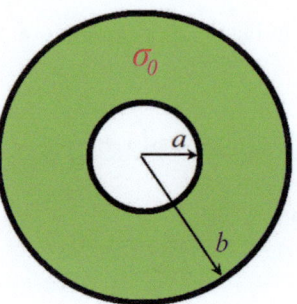

Fig. 17.9 A cylindrical resistor

Problem

17.13. Calculate the total resistance of the cylindrical resistor illustrated in Fig. 17.10. The length of the resistor is l.

Difficulty level — ○ Easy ○ Normal ● Hard
Calculation amount — ○ Small ○ Normal ● Large

1) $R = \dfrac{\ln\left(\dfrac{b}{a}\right)}{\pi l(\sigma_1+\sigma_2)}$

2) $R = \dfrac{\pi l(\sigma_1+\sigma_2)}{\ln\left(\dfrac{b}{a}\right)}$

3) $R = \dfrac{\pi l(\sigma_1+\sigma_2)}{\ln\left(\dfrac{a}{b}\right)}$

4) $R = \dfrac{\ln\left(\dfrac{b}{a}\right)(\sigma_1+\sigma_2)}{\pi l}$

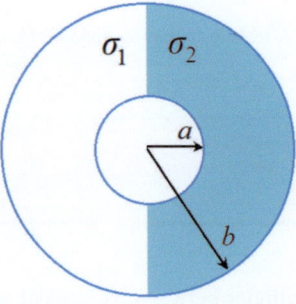

Fig. 17.10 A cylindrical resistor including two different electrical conductivities

17.3 Cylindrical Resistors

Problem

17.14. The electrical conductivity of a coaxial cable is nonuniform and as follows.

$$\sigma = \frac{\sigma_0}{1 + \frac{a}{r}}$$

Calculate the per unit length electrical resistance between the internal and external surface of the cable with the inner and outer radiuses a and b, respectively (Fig. 17.11).

Difficulty level   ○ Easy   ○ Normal   ● Hard
Calculation amount   ○ Small ○ Normal   ● Large

1) $\dfrac{1}{2\pi\sigma_0}\left(\ln\left(\dfrac{2b}{a}\right) + \dfrac{b-a}{b}\right)$

2) $-\dfrac{1}{2\pi\sigma_0}\left(ab + \dfrac{1}{2}\left(a^2 + b^2\right)\right)$

3) $\dfrac{1}{2\pi\sigma_0}\left(\ln\left(\dfrac{b}{a}\right) + \dfrac{b-a}{b}\right)$

4) $-\dfrac{1}{2\pi\sigma_0}\left(a^2 b - \dfrac{1}{2}\left(a^2 + b\right)\right)$

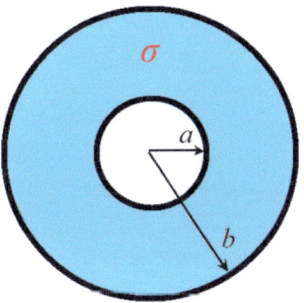

Fig. 17.11 A cylindrical resistor with nonuniform electrical conductivity

Partially Solved Exercise

17.15. In Fig. 17.12, the conductivity of cylinder is nonuniform and as follows.

$$\sigma = \frac{\sigma_0}{r^2}$$

Calculate the resistance of the cylinder between its inner and outer surfaces.

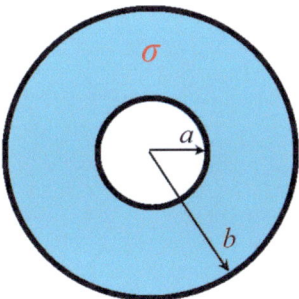

Fig. 17.12 A cylindrical resistor with nonuniform electrical conductivity

Solution

The electrical resistance between the internal and external surface of the cylinder with the nonuniform conductivity can be calculated as follows.

$$R = \int \frac{dr}{\int \sigma dS}$$

$$\Rightarrow R = \int_a^b \frac{dr}{\int_{z=0}^{z=l} \int_{\varphi=0}^{\varphi=2\pi} (\quad\quad)}$$

$$\Rightarrow R = \int_a^b (\quad) \frac{dr}{\int_{z=0}^{z=l} \int_{\varphi=0}^{\varphi=2\pi} d\varphi dz}$$

$$\Rightarrow R = \int_a^b (\quad) \frac{dr}{(\quad)(\quad)}$$

$$\Rightarrow R = \frac{1}{2\pi l \sigma_0} \int_a^b (\quad) dr$$

$$\Rightarrow R = \frac{1}{2\pi l \sigma_0} [(\quad)]_a^b$$

$$\Rightarrow R = \frac{b^2 - a^2}{4\pi l \sigma_0}$$

Problem

17.16. In Fig. 17.13, the conductivity of cylinder is nonuniform and as follows.

$$\sigma = \frac{\sigma_0}{r}$$

Calculate the current of the cylinder per unit length.

Difficulty level   ○ Easy   ○ Normal   ● Hard
Calculation amount   ○ Small   ● Normal   ○ Large

1) $\dfrac{2\pi\sigma_0 V_0}{b-a}$

2) $\dfrac{2\pi\sigma_0 V_0}{\ln\left(\dfrac{b}{a}\right)}$

3) $\dfrac{2\pi\sigma_0 ab V_0}{b^3 - a^3}$

4) $\dfrac{2\pi\sigma_0 V_0}{b^3 - a^3}$

17.4 Spherical Resistors

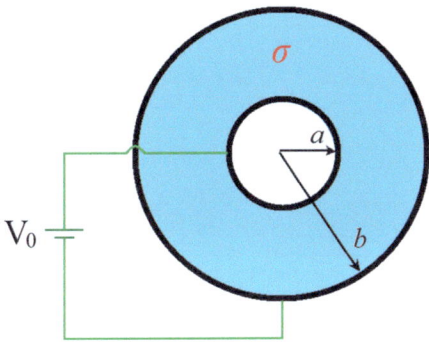

Fig. 17.13 A cylindrical resistor with nonuniform electrical conductivity connected to a voltage source

Problem

17.17. In Fig. 17.14, $\sigma_2 = 2\sigma_1$, $r_1 = 2$ cm, $r_2 = 2.25$ cm, $r_3 = 2.5$ cm, and $V_0 = 100$ V. Calculate the voltage across each region.
Difficulty level ○ Easy ○ Normal ● Hard
Calculation amount ○ Small ○ Normal ● Large
1) $V_1 = 30.77$ V, $V_2 = 69.23$ V
2) $V_1 = 69.23$ V, $V_2 = 30.77$ V
3) $V_1 = V_2 = 69.23$ V
4) $V_1 = V_2 = 30.77$ V

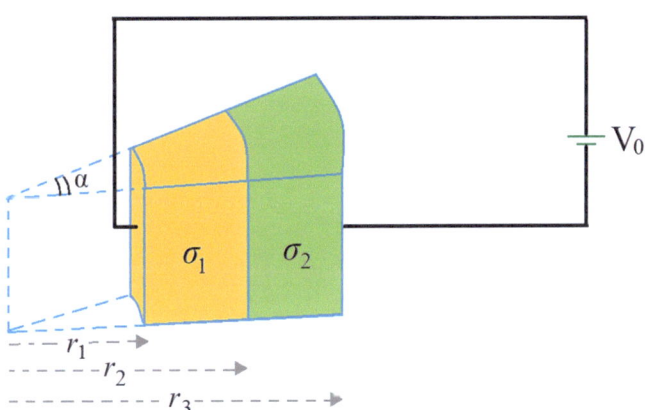

Fig. 17.14 A cylindrical resistor with different electrical conductivities connected to a voltage source

17.4 Spherical Resistors

Problem

17.18. Calculate the resistance of the spherical resistor illustrated in Fig. 17.15. The resistor is a perfect conductor.
Difficulty level ○ Easy ● Normal ○ Hard
Calculation amount ○ Small ● Normal ○ Large
1) $R = \dfrac{1}{2\pi\sigma_0 a}$
2) $R = \dfrac{1}{\pi\sigma_0 a}$
3) $R = \dfrac{a}{2\pi\sigma_0}$
4) $R = \dfrac{1}{4\pi\sigma_0 a}$

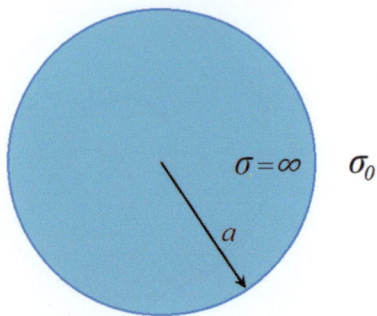

Fig. 17.15 A spherical resistor

Problem

17.19. Calculate the resistance of the hollow sphere between its internal and external surfaces shown in Fig. 17.16.
Difficulty level   ○ Easy   ● Normal   ○ Hard
Calculation amount   ○ Small   ● Normal   ○ Large

1) $R = \dfrac{ab}{4\pi\sigma_0(b-a)}$
2) $R = \dfrac{b-a}{4\pi\sigma_0 ab}$
3) $R = \dfrac{ab}{2\pi\sigma_0(b-a)}$
4) $R = \dfrac{b-a}{2\pi\sigma_0 ab}$

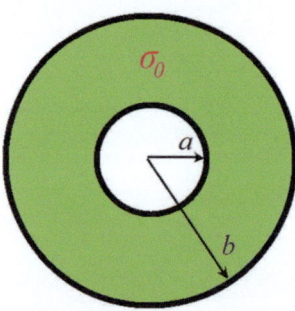

Fig. 17.16 A spherical resistor

Problem

17.20. Calculate the total resistance of the spherical resistor illustrated in Fig. 17.17.
Difficulty level   ○ Easy   ○ Normal   ● Hard
Calculation amount   ○ Small   ○ Normal   ● Large

1) $R = \dfrac{(b-a)(\sigma_1+\sigma_2)}{2\pi ab}$
2) $R = \dfrac{2\pi ab}{(b-a)(\sigma_1+\sigma_2)}$
3) $R = \dfrac{b-a}{2\pi(\sigma_1+\sigma_2)}$
4) $R = \dfrac{b-a}{2\pi ab(\sigma_1+\sigma_2)}$

17.4 Spherical Resistors

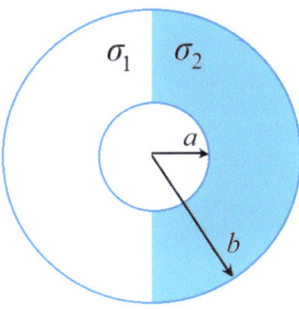

Fig. 17.17 A spherical resistor including different conductivities

Problem

17.21. As is illustrated in Fig. 17.18, the region $a \leq R \leq b$ has been filled with a material with the nonuniform electrical conductivity $\sigma(R) = \dfrac{\sigma_0}{R^2}$. Calculate the electrical resistance between these two surfaces.

Difficulty level   ○ Easy   ○ Normal   ● Hard
Calculation amount   ○ Small   ● Normal   ○ Large

1) $R = \dfrac{b-a}{2\pi\sigma_0}$
2) $R = \dfrac{b-a}{4\pi\sigma_0}$
3) $R = \dfrac{1}{2\pi\sigma_0(b-a)}$
4) $R = \dfrac{\ln\left(\dfrac{b}{a}\right)}{2\pi\sigma_0}$

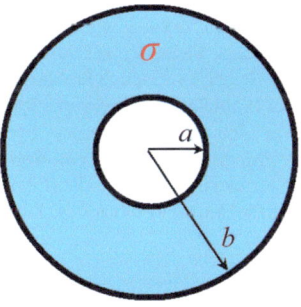

Fig. 17.18 A spherical resistor including nonuniform conductivity

Problem

17.22. As is illustrated in Fig. 17.19, the region $a \leq R \leq b$ has been filled with a material with the nonuniform electrical conductivity $\sigma(R) = \dfrac{\sigma_0}{R}$. The surfaces are connected to the electric potential V_0. Calculate the volume current density between these two surfaces.

Difficulty level   ○ Easy   ○ Normal   ● Hard
Calculation amount   ○ Small   ○ Normal   ● Large

1) $J = \dfrac{\sigma_0 V_0}{R^2 \ln\left(\dfrac{b}{a}\right)}$

2) $J = \dfrac{\sigma_0 V_0 R^2}{\ln\left(\dfrac{b}{a}\right)}$

3) $J = \dfrac{\sigma_0 V_0 \ln\left(\dfrac{b}{a}\right)}{R}$

4) $J = \dfrac{\sigma_0 V_0}{R \ln\left(\dfrac{b}{a}\right)}$

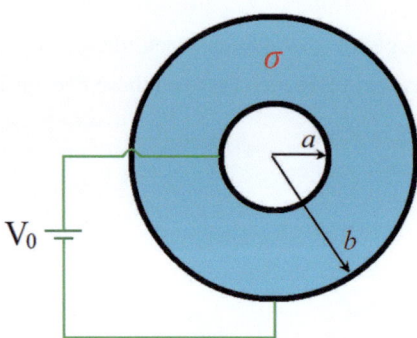

Fig. 17.19 A spherical resistor including nonuniform conductivity connected to a voltage source

References

1. Rahmani-Andebili, M., General Physics II – Practice Problems, Methods, and Solutions, Springer Nature, 2025.
2. Rahmani-Andebili, M., General Physics I – Practice Problems, Methods, and Solutions, Springer Nature, 2025.
3. Rahmani-Andebili, M., Mathematics of Engineering and Science – Practice Problems, Methods, and Solutions, Springer Nature, 2024.
4. Rahmani-Andebili, M., Differential Equations – Practice Problems, Methods, and Solutions, Springer Nature, 2022.
5. Rahmani-Andebili, M., Calculus III – Practice Problems, Methods, and Solutions, Springer Nature, 2023.
6. Rahmani-Andebili, M., Calculus II – Practice Problems, Methods, and Solutions, Springer Nature, 2023.
7. Rahmani-Andebili, M., Calculus I (2nd Ed.) – Practice Problems, Methods, and Solutions, Springer Nature, 2023.
8. Rahmani-Andebili, M., Precalculus (2nd Ed.) – Practice Problems, Methods, and Solutions, Springer Nature, 2024.

18 Flat, Cylindrical, and Spherical Resistors and Boundary Conditions for Electric Current: Part B

Abstract

In this chapter, the problems of the seventeenth chapter are fully solved, in detail, step-by-step, and with different methods.

18.1 Boundary Conditions for Electric Current

18.1. Based on the information given in the problem, we have [1–8]:

$$\alpha_1 = 60°$$

As we know, the normal component of electric flux density on the boundary of two regions is discontinuous and the amount of discontinuity is equal to the free surface charge density. Herein, $\hat{n}$ is the unit normal vector of the first region. In other words:

$$\vec{D}_2 \cdot \hat{n} - \vec{D}_1 \cdot \hat{n} = \rho_s$$

In addition, the normal component of electric current density on the boundary of two regions is continuous. In other words:

$$J_{1n} = J_{2n} \triangleq J_n$$

On the other hand, we know that:

$$D = \varepsilon E$$

$$E = \frac{J}{\sigma}$$

Therefore:

$$\Rightarrow \frac{\varepsilon_2 J_n}{\sigma_2} - \frac{\varepsilon_1 J_n}{\sigma_1} = \rho_s$$

$$\Rightarrow \rho_s = \left(\frac{\varepsilon_2}{\sigma_2} - \frac{\varepsilon_1}{\sigma_1}\right) J_n$$

From Fig. 18.1, it is noticed that:

$$J_n = J\cos 60° = \frac{1}{2}J$$

Hence:

$$\Rightarrow \rho_s = \frac{1}{2}\left(\frac{\varepsilon_2}{\sigma_2} - \frac{\varepsilon_1}{\sigma_1}\right)J$$

Choice (4) is the answer.

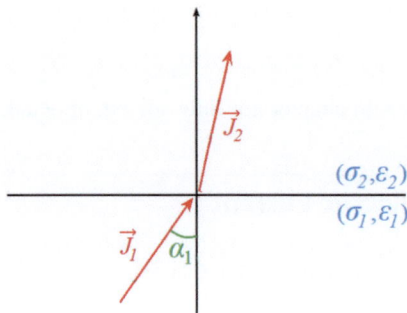

Fig. 18.1 Studying the boundary conditions for the electric current density

18.2. Based on the information given in the problem, we have:

$$\vec{J}_1 = J_0(\hat{x} + 2\hat{y} + 6\hat{z})$$

$$\sigma_2 = 3\sigma_1$$

$$\varepsilon_2 = 2\varepsilon_1$$

As we know, the normal component of electric flux density on the boundary of two regions is discontinuous and the amount of discontinuity is equal to the free surface charge density. Herein, $\hat{n}$ is the unit normal vector of the first region. In other words:

$$\vec{D}_2 \cdot \hat{n} - \vec{D}_1 \cdot \hat{n} = \rho_s$$

In addition, the normal component of electric current density on the boundary of two regions is continuous. In other words:

$$J_{1n} = J_{2n} \triangleq J_n$$

On the other hand, we know that:

$$D = \varepsilon E$$

$$E = \frac{J}{\sigma}$$

Therefore:

$$-\frac{\varepsilon_2 J_z}{\sigma_2} - \left(-\frac{\varepsilon_1 J_z}{\sigma_1}\right) = \rho_s$$

$$\Rightarrow \rho_s = \left(\frac{\varepsilon_1}{\sigma_1} - \frac{\varepsilon_2}{\sigma_2}\right) J_z$$

By putting the quantities in the parameters, we have:

$$\rho_s = \frac{\varepsilon_1}{\sigma_1}\left(1 - \frac{2}{3}\right) 6J_0$$

$$\Rightarrow \rho_s = \frac{2\varepsilon_1 J_0}{\sigma_1}$$

Choice (1) is the answer (Fig. 18.2).

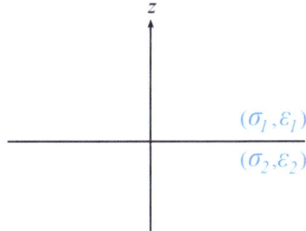

Fig. 18.2 Studying the boundary conditions for the electric current density

18.2 Flat Resistors

18.4. Based on the information given in the problem, we have:

$$I = 0.002 \, A$$

$$\sigma = 1 \, m/s$$

$$a = 1 \, mm$$

The resistance of the resistor can be calculated as follows.

$$R = \frac{a}{\sigma a^2} = \frac{1}{\sigma a}$$

Moreover, the voltage of the resistor can be calculated by using Ohm's law (study General Physics II) as follows.

$$V = RI$$

$$V = \frac{I}{\sigma a}$$

$$\Rightarrow V = \frac{0.002}{1 \times 10^{-3}}$$

$$\Rightarrow V = 2 \, V$$

Choice (1) is the answer (Fig. 18.3).

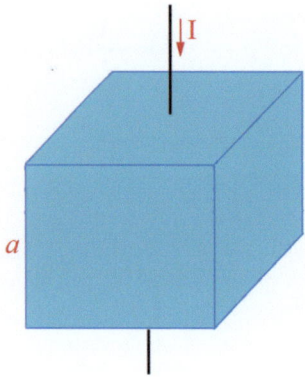

Fig. 18.3 A flat resistor

18.5. The cuboid resistor, shown in Fig. 18.4, includes two parts with the electrical conductivities σ_1 and σ_2. These two parts are, in fact, two individual resistors with the length $\frac{l}{2}$ that have been connected in series. Thus, the total resistance of the resistors can be calculated as follows.

$$R = R_1 + R_2$$

$$\Rightarrow R = \frac{\frac{l}{2}}{\sigma_1 A} + \frac{\frac{l}{2}}{\sigma_2 A}$$

$$\Rightarrow R = \frac{l}{2A} \left(\frac{1}{\sigma_1} + \frac{1}{\sigma_2} \right)$$

$$\Rightarrow R = \frac{l(\sigma_1 + \sigma_2)}{2A \sigma_1 \sigma_2}$$

Choice (1) is the answer.

18.2 Flat Resistors

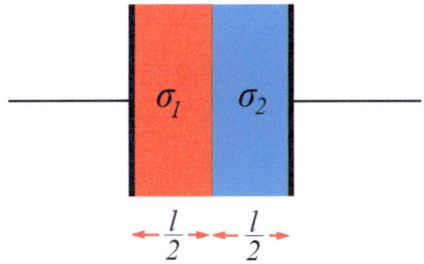

Fig. 18.4 A flat resistor including two different electrical conductivities

> **Notes**
>
> In this problem, the relations below have been used.
>
> The resistance of a flat resistor with the surface area A, length l, and electrical conductivity σ is as follows.
>
> $$R = \frac{l}{\sigma A}$$
>
> The equivalent resistance of n series resistors can be calculated as follows.
>
> $$R_{eq} = R_1 + \dots + R_n$$

18.6. The cuboid resistor, shown in Fig. 18.5, includes two parts with the electrical conductivities σ_1 and σ_2. These two parts are, in fact, two individual resistors with the surface area $\frac{A}{2}$ that have been connected in parallel. Thus, the total resistance of the resistors can be calculated as follows.

$$R = \frac{R_1 R_2}{R_1 + R_2}$$

$$\Rightarrow R = \frac{\left(\dfrac{l}{\sigma_1 \frac{A}{2}}\right)\left(\dfrac{l}{\sigma_2 \frac{A}{2}}\right)}{\left(\dfrac{l}{\sigma_1 \frac{A}{2}}\right) + \left(\dfrac{l}{\sigma_2 \frac{A}{2}}\right)}$$

$$\Rightarrow R = \frac{\left(\dfrac{l}{\frac{A}{2}}\right)\left(\dfrac{l}{\sigma_1 \sigma_2 \frac{A}{2}}\right)}{\left(\dfrac{l}{\frac{A}{2}}\right)\left(\dfrac{\sigma_1 + \sigma_2}{\sigma_1 \sigma_2}\right)}$$

$$\Rightarrow R = \frac{\frac{2l}{A}}{\frac{\sigma_1 + \sigma_2}{1}}$$

$$\Rightarrow R = \frac{2l}{(\sigma_1 + \sigma_2)A}$$

Choice (2) is the answer.

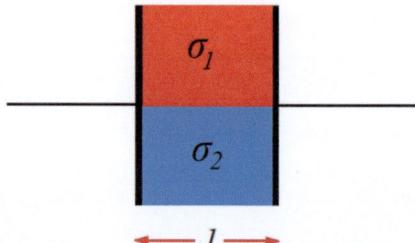

Fig. 18.5 A flat resistor including two different electrical conductivities

> **Notes**
>
> In this problem, the relations below have been used.
>
> The resistance of a flat resistor with the surface area A, length l, and electrical conductivity σ is as follows.
>
> $$R = \frac{l}{\sigma A}$$
>
> The equivalent resistance of n parallel resistors can be calculated as follows.
>
> $$\frac{1}{R_{eq}} = \frac{1}{R_1} + \ldots + \frac{1}{R_n}$$

18.7. From General Physics II, we know that the electric potential of a resistor is proportional to its resistance. Hence:

$$\frac{V_1}{V_2} = \frac{R_1}{R_2}$$

The resistor, shown in Fig. 18.6, includes two series resistors as follows.

$$R_1 = \frac{l_1}{\sigma_1 A}$$

$$R_2 = \frac{l_2}{\sigma_2 A}$$

Therefore:

$$\frac{V_1}{V_2} = \frac{\frac{l_1}{\sigma_1 A}}{\frac{l_2}{\sigma_2 A}}$$

$$\Rightarrow \frac{V_1}{V_2} = \frac{l_1 \sigma_2}{l_2 \sigma_1}$$

Choice (4) is the answer.

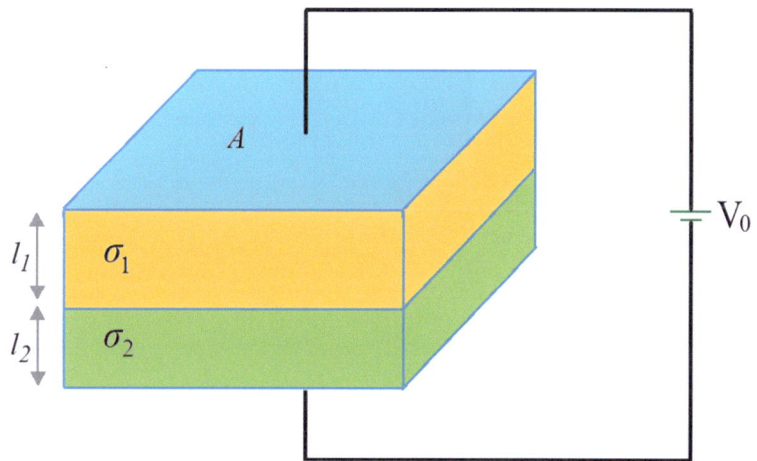

Fig. 18.6 A flat resistor connected to a voltage source including two different electrical conductivities

18.8. The cuboid resistor, shown in Fig. 18.7, includes two parts with the electrical conductivities σ_1 and σ_2. These two parts are, in fact, two individual resistors with different lengths that have been connected in series. Thus, the total resistance of the resistors can be calculated as follows.

$$R = R_1 + R_2$$

$$\Rightarrow R = \frac{d_1}{\sigma_1 A} + \frac{d_2}{\sigma_2 A}$$

$$\Rightarrow R = \frac{d_1 \sigma_2 + d_2 \sigma_1}{\sigma_1 \sigma_2 A}$$

According to Ohm's law (study General Physics II), the current in the resistor can be calculated as follows.

$$I = \frac{V_0}{R}$$

$$\Rightarrow I = \frac{\sigma_1 \sigma_2 A V_0}{d_1 \sigma_2 + d_2 \sigma_1}$$

On the other hand, the surface current density on the boundary of the two environments can be calculated as follows.

$$J = \frac{I}{A}$$

$$\Rightarrow I = \frac{\sigma_1 \sigma_2 V_0}{d_1 \sigma_2 + d_2 \sigma_1}$$

Choice (3) is the answer.

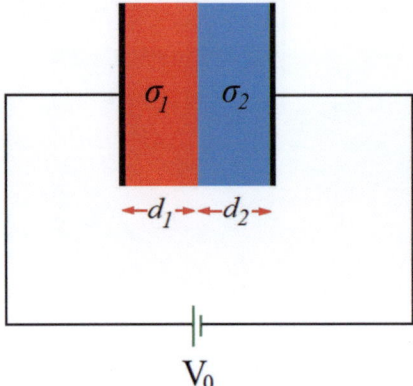

Fig. 18.7 A flat resistor connected to a voltage source including two different electrical conductivities

18.9. In this problem, the conductivity of the resistor is not uniform. Therefore, we need to use the following relation to calculate its resistance.

$$R = \int \frac{dx}{\int \sigma dS}$$

$$\Rightarrow R = \int \frac{dx}{\sigma e^{\alpha x} \int dS}$$

$$\Rightarrow R = \int \frac{dx}{\sigma e^{\alpha x} A}$$

$$\Rightarrow R = \frac{1}{\sigma A} \int_{x=0}^{x=d} e^{-\alpha x} dx$$

18.2 Flat Resistors

$$\Rightarrow R = \frac{1}{\sigma A}\left[\frac{1}{-\alpha}e^{-\alpha x}\right]_0^d$$

$$\Rightarrow R = \frac{1}{\sigma A}\left(\frac{e^{-\alpha d}-e^0}{-\alpha}\right)$$

$$\Rightarrow R = \frac{1-e^{-\alpha d}}{\sigma A \alpha}$$

Choice (3) is the answer.

> **Notes**
> In this problem, the relation below has been used.
> $$\int e^{u(x)} du = e^{u(x)}$$

18.11. Despite the appearance of the resistor, it is not a cylindrical resistor but a flat resistor.

Since the conductivity of the resistor is not uniform, the following relation must be used to calculate its resistance.

$$R = \int \frac{dz}{\int \sigma dS}$$

$$\Rightarrow R = \int \frac{dz}{\sigma_0\left(1+\frac{z}{l}\right)\int dS}$$

$$\Rightarrow R = \int \frac{dz}{\sigma_0\left(1+\frac{z}{l}\right)\pi a^2}$$

$$\Rightarrow R = \frac{1}{\sigma_0 \pi a^2}\int_{z=0}^{z=l} \frac{dz}{1+\frac{z}{l}}$$

$$\Rightarrow R = \frac{1}{\sigma_0 \pi a^2}\left[l \ln\left(1+\frac{z}{l}\right)\right]_0^l$$

$$\Rightarrow R = \frac{l}{\sigma_0 \pi a^2}(\ln 2 - \ln 1)$$

$$\Rightarrow R = \frac{l \ln 2}{\sigma_0 \pi a^2}$$

Choice (1) is the answer (Fig. 18.8).

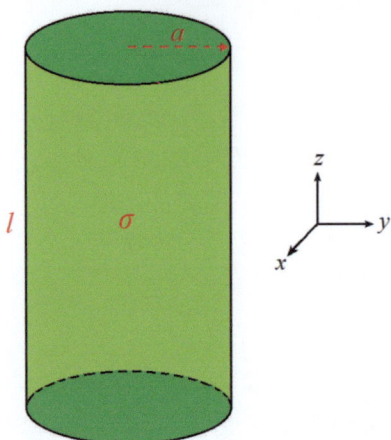

Fig. 18.8 A resistor including nonuniform conductivity

> **Notes**
> In this problem, the relations below have been used.
> $$\int \frac{du}{u(x)} = \ln u(x)$$
> $$\ln 1 = 0$$

18.3 Cylindrical Resistors

18.12. *Method 1*: The electrical resistance between the internal and external surface of the hollow cylinder with the uniform conductivity can be calculated as follows.

$$R = \int \frac{dr}{\int \sigma dS}$$

$$\Rightarrow R = \int_a^b \frac{dr}{\int_{z=0}^{z=1} \int_{\varphi=0}^{\varphi=2\pi} \sigma_0 r \, d\varphi \, dz}$$

$$\Rightarrow R = \int_a^b \frac{dr}{\sigma_0 r \int_{z=0}^{z=1} \int_{\varphi=0}^{\varphi=2\pi} d\varphi \, dz}$$

18.3 Cylindrical Resistors

$$\Rightarrow R = \int_a^b \frac{dr}{\sigma_0 r (2\pi - 0)(l - 0)}$$

$$\Rightarrow R = \frac{1}{2\pi l \sigma_0} \int_a^b \frac{dr}{r}$$

$$\Rightarrow R = \frac{1}{2\pi l \sigma_0} [\ln r]_a^b$$

$$\Rightarrow R = \frac{1}{2\pi l \sigma_0} (\ln b - \ln a)$$

$$\Rightarrow R = \frac{\ln\left(\frac{b}{a}\right)}{2\pi l \sigma_0}$$

Choice (4) is the answer.

Method 2: The relation between capacitance, resistance, conductivity, and permittivity is as follows.

$$\varepsilon = \sigma R C$$

On the other hand, from previous chapters, we know that the capacitance of the cylinder between its inner and outer surface is as follows.

$$C = \frac{2\pi \varepsilon_0 l}{\ln\left(\frac{b}{a}\right)}$$

Therefore, the resistance of the cylinder can be calculated as follows.

$$R = \frac{\varepsilon_0}{\sigma_0 C} = \frac{\varepsilon_0}{\sigma_0 \left(\frac{2\pi \varepsilon_0 l}{\ln\left(\frac{b}{a}\right)}\right)}$$

$$\Rightarrow R = \frac{\ln\left(\frac{b}{a}\right)}{2\pi l \sigma_0}$$

Choice (4) is the answer (Fig. 18.9).

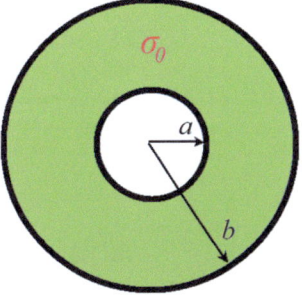

Fig. 18.9 A cylindrical resistor

> **Notes**
>
> In this problem, the relations below have been used.
>
> $$\int \frac{dx}{x} = \ln x + c$$
>
> $$\ln a - \ln b = \ln\left(\frac{a}{b}\right)$$

18.13. The cylindrical resistor illustrated in Fig. 18.10 can be considered a system of two parallel resistors.

The resistance of the left-side resistor can be calculated as follows.

$$R_1 = \int \frac{dr}{\int \sigma_1 dS}$$

$$\Rightarrow R_1 = \int \frac{dr}{\sigma_1 r \int_{z=0}^{z=l} \int_{\varphi=0}^{\varphi=\pi} d\varphi dz}$$

$$\Rightarrow R_1 = \int \frac{dr}{\sigma_1 r [\varphi]_0^\pi [z]_0^l}$$

$$\Rightarrow R_1 = \int \frac{dr}{\sigma_1 r \pi l}$$

$$\Rightarrow R_1 = \frac{1}{\pi \sigma_1 l} \int_{r=a}^{r=b} \frac{dr}{r}$$

$$\Rightarrow R_1 = \frac{1}{\pi \sigma_1 l} [\ln r]_a^b$$

$$\Rightarrow R_1 = \frac{1}{\pi \sigma_1 l} (\ln b - \ln a)$$

$$\Rightarrow R_1 = \frac{\ln\left(\frac{b}{a}\right)}{\pi \sigma_1 l}$$

Likewise, the resistance of the right-side resistor is as follows.

$$\Rightarrow R_2 = \frac{\ln\left(\frac{b}{a}\right)}{\pi \sigma_2 l}$$

Thus, the equivalent resistance of the system can be calculated as follows.

18.3 Cylindrical Resistors

$$R = \frac{R_1 R_2}{R_1 + R_2}$$

$$\Rightarrow R = \frac{\frac{\ln\left(\frac{b}{a}\right)}{\pi \sigma_1 l} \times \frac{\ln\left(\frac{b}{a}\right)}{\pi \sigma_2 l}}{\frac{\ln\left(\frac{b}{a}\right)}{\pi \sigma_1 l} + \frac{\ln\left(\frac{b}{a}\right)}{\pi \sigma_2 l}}$$

$$\Rightarrow R = \frac{\frac{\ln\left(\frac{b}{a}\right)}{\pi l} \times \frac{\ln\left(\frac{b}{a}\right)}{\pi \sigma_1 \sigma_2 l}}{\frac{\ln\left(\frac{b}{a}\right)}{\pi l}\left(\frac{\sigma_1 + \sigma_2}{\sigma_1 \sigma_2}\right)}$$

$$\Rightarrow R = \frac{\frac{\ln\left(\frac{b}{a}\right)}{\pi l}}{\frac{\sigma_1 + \sigma_2}{1}}$$

$$\Rightarrow R = \frac{\ln\left(\frac{b}{a}\right)}{\pi l (\sigma_1 + \sigma_2)}$$

Choice (1) is the answer.

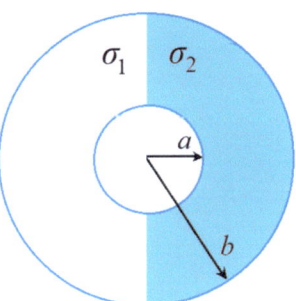

Fig. 18.10 A cylindrical resistor including two different electrical conductivities

> **Notes**
>
> In this problem, the relations below have been used.
>
> $$\int \frac{dx}{x} = \ln x + c$$
>
> $$\ln a - \ln b = \ln\left(\frac{a}{b}\right)$$

18.14. Based on the information given in the problem, we have:

$$\sigma = \frac{\sigma_0}{1+\frac{a}{r}}$$

$$l = 1 \, m$$

The electrical resistance between the internal and external surface of the coaxial cable with the nonuniform conductivity can be calculated as follows.

$$R = \int \frac{dr}{\int \sigma dS}$$

$$\Rightarrow R = \int_a^b \frac{dr}{\int_{z=0}^{z=1} \int_{\varphi=0}^{\varphi=2\pi} \left(\frac{\sigma_0}{1+\frac{a}{r}}\right) r d\varphi dz}$$

$$\Rightarrow R = \int_a^b \frac{dr}{\left(\frac{\sigma_0 r^2}{r+a}\right) \int_{z=0}^{z=1} \int_{\varphi=0}^{\varphi=2\pi} d\varphi dz}$$

$$\Rightarrow R = \int_a^b \left(\frac{r+a}{\sigma_0 r^2}\right) \frac{dr}{(2\pi-0)(1-0)}$$

$$\Rightarrow R = \frac{1}{2\pi\sigma_0} \int_a^b \left(\frac{1}{r}+\frac{a}{r^2}\right) dr$$

$$\Rightarrow R = \frac{1}{2\pi\sigma_0} \left[\ln r - \frac{a}{r}\right]_a^b$$

$$\Rightarrow R = \frac{1}{2\pi\sigma_0} \left(\ln b - \ln a - \left(\frac{a}{b}-1\right)\right)$$

$$\Rightarrow R = \frac{1}{2\pi\sigma_0} \left(\ln\left(\frac{b}{a}\right) + \frac{b-a}{b}\right)$$

Choice (3) is the answer (Fig. 18.11).

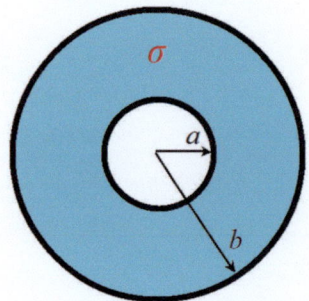

Fig. 18.11 A cylindrical resistor with nonuniform electrical conductivity

18.3 Cylindrical Resistors

> **Notes**
>
> In this problem, the relations below have been used.
>
> $$\int x^n dx = \frac{x^{n+1}}{n+1}$$
>
> $$\ln a - \ln b = \ln\left(\frac{a}{b}\right)$$

18.16. Based on the information given in the problem, we have:

$$\sigma = \frac{\sigma_0}{r}$$

$$l = 1 \ m$$

The electrical resistance between the internal and external surface of the coaxial cable with the nonuniform conductivity can be calculated as follows.

$$R = \int \frac{dr}{\int \sigma dS}$$

$$\Rightarrow R = \int_a^b \frac{dr}{\int_{z=0}^{z=1}\int_{\varphi=0}^{\varphi=2\pi} \left(\frac{\sigma_0}{r}\right) r d\varphi dz}$$

$$\Rightarrow R = \int_a^b \frac{dr}{\sigma_0 \int_{z=0}^{z=1}\int_{\varphi=0}^{\varphi=2\pi} d\varphi dz}$$

$$\Rightarrow R = \int_a^b \left(\frac{1}{\sigma_0}\right) \frac{dr}{(2\pi - 0)(1 - 0)}$$

$$\Rightarrow R = \frac{1}{2\pi\sigma_0} \int_a^b dr$$

$$\Rightarrow R = \frac{1}{2\pi\sigma_0} [r]_a^b$$

$$\Rightarrow R = \frac{b - a}{2\pi\sigma_0}$$

Based on Ohm's law (study General Physics II), the current in the resistor can be calculated as follows.

$$I = \frac{V_0}{R}$$

$$\Rightarrow I = \frac{2\pi\sigma_0 V_0}{b-a}$$

Choice (1) is the answer (Fig. 18.12).

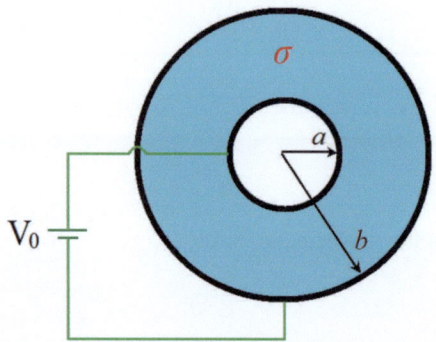

Fig. 18.12 A cylindrical resistor with nonuniform electrical conductivity connected to a voltage source

---

18.17. Based on the information given in the problem, we have:

$$\sigma_2 = 2\sigma_1$$

$$r_1 = 0.02 \ m$$

$$r_2 = 0.0225 \ m$$

$$r_3 = 0.025 \ m$$

$$V_0 = 100 \ V$$

As can be noticed, the circuit includes two series resistors that each of them is part of a cylindrical resistor. Hence, we need to use the following relation to calculate the resistance of each part.

$$R = \int \frac{dr}{\int \sigma dS}$$

For the first part, we have:

$$R_1 = \int \frac{dr}{\int \sigma_1 dS}$$

18.3 Cylindrical Resistors

$$\Rightarrow R_1 = \int \frac{dr}{\sigma_1 r \int\limits_{z=0}^{z=l} \int\limits_{\varphi=0}^{\varphi=\alpha} d\varphi dz}$$

$$\Rightarrow R_1 = \int \frac{dr}{\sigma_1 r [\varphi]_0^\alpha [z]_0^l}$$

$$\Rightarrow R_1 = \int \frac{dr}{\sigma_1 r \alpha l}$$

$$\Rightarrow R_1 = \frac{1}{\alpha \sigma_1 l} \int\limits_{r=r_1}^{r=r_2} \frac{dr}{r}$$

$$\Rightarrow R_1 = \frac{1}{\alpha \sigma_1 l} [\ln r]_{r_1}^{r_2}$$

$$\Rightarrow R_1 = \frac{1}{\alpha \sigma_1 l} (\ln r_2 - \ln r_1)$$

$$\Rightarrow R_1 = \frac{\ln\left(\frac{r_2}{r_1}\right)}{\alpha \sigma_1 l}$$

Likewise, for the second part, we have:

$$R_2 = \frac{\ln\left(\frac{r_3}{r_2}\right)}{\alpha \sigma_2 l}$$

Now, by putting the quantities in their parameters, we have:

$$R_1 = \frac{\ln\left(\frac{2.25}{2}\right)}{\alpha \sigma_1 l} = \frac{0.117}{\alpha \sigma_1 l}$$

$$R_2 = \frac{\ln\left(\frac{2.5}{2.25}\right)}{\alpha \sigma_2 l} = \frac{0.105}{\alpha \sigma_2 l} \xrightarrow{\sigma_2 = 2\sigma_1} R_2 = \frac{0.052}{\alpha \sigma_1 l}$$

From General Physics II, we know that the electric potential of a resistor in a circuit with two series resistors can be calculated as follows.

$$V_1 = \frac{R_1}{R_1 + R_2} V_0$$

$$V_2 = \frac{R_2}{R_1 + R_2} V_0$$

Therefore:

$$V_1 = \frac{\frac{0.117}{a\sigma_1 l}}{\frac{0.117}{a\sigma_1 l} + \frac{0.052}{a\sigma_1 l}} \times 100 \Rightarrow V_1 = \frac{0.117}{0.117 + 0.052} \times 100 \Rightarrow V_1 = 69.23\ V$$

$$V_2 = \frac{\frac{0.052}{a\sigma_1 l}}{\frac{0.117}{a\sigma_1 l} + \frac{0.052}{a\sigma_1 l}} \times 100 \Rightarrow V_2 = \frac{0.052}{0.117 + 0.052} \times 100 \Rightarrow V_2 = 30.77\ V$$

Choice (2) is the answer (Fig. 18.13).

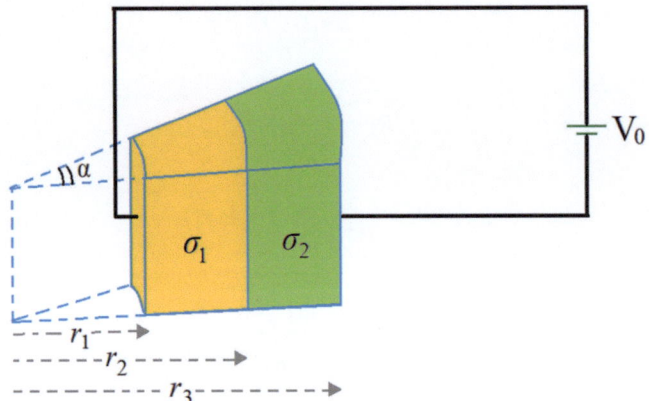

Fig. 18.13 A cylindrical resistor with different electrical conductivities connected to a voltage source

---

Notes

In this problem, the relations below have been used.

$$\int \frac{dx}{x} = \ln x$$

$$\ln a - \ln b = \ln\left(\frac{a}{b}\right)$$

18.4 Spherical Resistors

18.18. **Method 1**: The resistance of the resistor can be calculated as follows.

$$R = \int \frac{dR}{\int \sigma_0 dS}$$

18.4 Spherical Resistors

$$\Rightarrow R = \int \frac{dR}{\int_{\varphi=0}^{2\pi} \int_{\theta=0}^{\pi} \sigma_0 R^2 \sin\theta d\theta d\varphi}$$

$$\Rightarrow R = \int \frac{dR}{\sigma_0 R^2 \int_{\varphi=0}^{2\pi} \int_{\theta=0}^{\pi} \sin\theta d\theta d\varphi}$$

$$\Rightarrow R = \int \frac{dR}{\sigma_0 R^2 [-\cos\theta]_0^\pi [\varphi]_0^{2\pi}}$$

$$\Rightarrow R = \int \frac{dR}{\sigma_0 R^2 (\cos 0 - \cos\pi)(2\pi - 0)}$$

$$\Rightarrow R = \frac{1}{4\pi\sigma_0} \int_a^\infty \frac{dR}{R^2}$$

$$\Rightarrow R = \frac{1}{4\pi\sigma_0} \left[-\frac{1}{R}\right]_a^\infty$$

$$\Rightarrow R = \frac{1}{4\pi\sigma_0} \left(\frac{1}{a} - \frac{1}{\infty}\right)$$

$$\Rightarrow R = \frac{1}{4\pi\sigma_0 a}$$

Choice (4) is the answer.

Method 2: The relation between capacitance, resistance, conductivity, and permittivity is as follows.

$$\varepsilon = \sigma R C$$

On the other hand, from the previous chapters, we know that the capacitance of a single solid sphere with radius a is as follows.

$$C = 4\pi\varepsilon_0 a$$

Therefore, the resistance of the sphere can be calculated as follows (Fig. 18.14).

$$R = \frac{\varepsilon_0}{\sigma_0 C} = \frac{\varepsilon_0}{\sigma_0 (4\pi\varepsilon_0 a)}$$

$$\Rightarrow R = \frac{1}{4\pi\sigma_0 a}$$

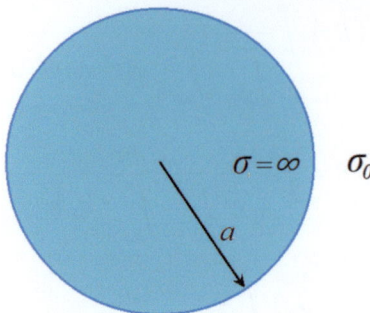

Fig. 18.14 spherical resistor

> **Notes**
> In this problem, the relations below have been used.
> $$\int \sin x \, dx = -\cos x$$
> $$\cos \pi = -1$$
> $$\cos 0 = 1$$
> $$\int x^n \, dx = \frac{x^{n+1}}{n+1}$$

18.19. **Method 1**: The electrical resistance between the internal and external surface of the hollow sphere with the uniform conductivity can be calculated as follows.

$$R = \int \frac{dR}{\int \sigma \, dS}$$

$$\Rightarrow R = \int \frac{dR}{\int_{\varphi=0}^{2\pi} \int_{\theta=0}^{\pi} \sigma_0 R^2 \sin\theta \, d\theta \, d\varphi}$$

$$\Rightarrow R = \int \frac{dR}{\sigma_0 R^2 \int_{\varphi=0}^{2\pi} \int_{\theta=0}^{\pi} \sin\theta \, d\theta \, d\varphi}$$

$$\Rightarrow R = \int \frac{dR}{\sigma_0 R^2 [-\cos\theta]_0^\pi [\varphi]_0^{2\pi}}$$

18.4 Spherical Resistors

$$\Rightarrow R = \int \frac{dR}{\sigma_0 R^2 (\cos 0 - \cos \pi)(2\pi - 0)}$$

$$\Rightarrow R = \frac{1}{4\pi\sigma_0} \int_a^b \frac{dR}{R^2}$$

$$\Rightarrow R = \frac{1}{4\pi\sigma_0} \left[-\frac{1}{R} \right]_a^b$$

$$\Rightarrow R = \frac{1}{4\pi\sigma_0} \left(\frac{1}{a} - \frac{1}{b} \right)$$

$$\Rightarrow R = \frac{b-a}{4\pi\sigma_0 ab}$$

Choice (2) is the answer.

Method 2: The relation between capacitance, resistance, conductivity, and permittivity is as follows.

$$\varepsilon = \sigma RC$$

On the other hand, from previous chapters, we know that the capacitance of the sphere between its inner and outer surface is as follows.

$$C = \frac{4\pi\varepsilon_0 ab}{b-a}$$

Therefore, the resistance of the cylinder is as follows.

$$R = \frac{\varepsilon_0}{\sigma_0 C} = \frac{\varepsilon_0}{\sigma_0 \left(\frac{4\pi\varepsilon_0 ab}{b-a} \right)}$$

$$\Rightarrow R = \frac{b-a}{4\pi\sigma_0 ab}$$

Choice (2) is the answer (Fig. 18.15).

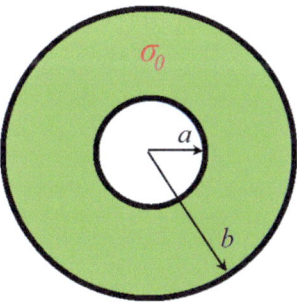

Fig. 18.15 A spherical resistor

> **Notes**
>
> In this problem, the relations below have been used.
>
> $$\int \sin x \, dx = -\cos x$$
>
> $$\cos \pi = -1$$
>
> $$\cos 0 = 1$$
>
> $$\int x^n \, dx = \frac{x^{n+1}}{n+1}$$

18.20. The spherical resistor illustrated in Fig. 18.16 can be considered a system of two parallel resistors.

The resistance of the left-side resistor can be calculated as follows.

$$R_1 = \int \frac{dR}{\int \sigma_1 dS}$$

$$\Rightarrow R_1 = \int \frac{dR}{\int_{\varphi=0}^{\pi} \int_{\theta=0}^{\pi} \sigma_1 R^2 \sin\theta \, d\theta \, d\varphi}$$

$$\Rightarrow R_1 = \int \frac{dR}{\sigma_1 R^2 \int_{\varphi=0}^{\pi} \int_{\theta=0}^{\pi} \sin\theta \, d\theta \, d\varphi}$$

$$\Rightarrow R_1 = \int \frac{dR}{\sigma_1 R^2 [-\cos\theta]_0^{\pi} [\varphi]_0^{\pi}}$$

$$\Rightarrow R_1 = \int \frac{dR}{\sigma_1 R^2 (\cos 0 - \cos \pi)(\pi - 0)}$$

$$\Rightarrow R_1 = \frac{1}{2\pi\sigma_1} \int_a^b \frac{dR}{R^2}$$

$$\Rightarrow R_1 = \frac{1}{2\pi\sigma_1} \left[-\frac{1}{R}\right]_a^b$$

$$\Rightarrow R_1 = \frac{1}{2\pi\sigma_1} \left(\frac{1}{a} - \frac{1}{b}\right)$$

18.4 Spherical Resistors

$$\Rightarrow R_1 = \frac{b-a}{2\pi\sigma_1 ab}$$

Likewise, the resistance of the right-side resistor is as follows.

$$\Rightarrow R_2 = \frac{b-a}{2\pi\sigma_2 ab}$$

Thus, the equivalent resistance of the system can be calculated as follows.

$$R = \frac{R_1 R_2}{R_1 + R_2}$$

$$\Rightarrow R = \frac{\frac{b-a}{2\pi\sigma_1 ab} \times \frac{b-a}{2\pi\sigma_2 ab}}{\frac{b-a}{2\pi\sigma_1 ab} + \frac{b-a}{2\pi\sigma_2 ab}}$$

$$\Rightarrow R = \frac{\frac{b-a}{2\pi ab} \times \frac{b-a}{2\pi\sigma_1\sigma_2 ab}}{\frac{b-a}{2\pi ab}\left(\frac{\sigma_1+\sigma_2}{\sigma_1\sigma_2}\right)}$$

$$\Rightarrow R = \frac{\frac{b-a}{2\pi ab}}{\frac{\sigma_1+\sigma_2}{1}}$$

$$\Rightarrow R = \frac{b-a}{2\pi ab(\sigma_1+\sigma_2)}$$

Choice (4) is the answer.

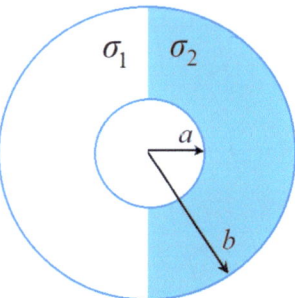

Fig. 18.16 A spherical resistor including different conductivities

> **Notes**
>
> In this problem, the relations below have been used.
>
> $$\int \sin x\, dx = -\cos x$$
>
> $$\cos \pi = -1$$
>
> $$\cos 0 = 1$$
>
> $$\int x^n\, dx = \frac{x^{n+1}}{n+1}$$

18.21. Based on the information given in the problem, we have:

$$\sigma = \frac{\sigma_0}{R^2}$$

The electrical resistance between the internal and external surface of the hollow sphere with the nonuniform conductivity can be calculated as follows.

$$R = \int \frac{dR}{\int \sigma dS}$$

$$\Rightarrow R = \int \frac{dR}{\int_{\varphi=0}^{2\pi} \int_{\theta=0}^{\pi} \frac{\sigma_0}{R^2} R^2 \sin\theta\, d\theta\, d\varphi}$$

$$\Rightarrow R = \int \frac{dR}{\sigma_0 \int_{\varphi=0}^{2\pi} \int_{\theta=0}^{\pi} \sin\theta\, d\theta\, d\varphi}$$

$$\Rightarrow R = \int \frac{dR}{\sigma_0 [-\cos\theta]_0^{\pi} [\varphi]_0^{2\pi}}$$

$$\Rightarrow R = \int \frac{dR}{\sigma_0 (\cos 0 - \cos \pi)(2\pi - 0)}$$

$$\Rightarrow R = \frac{1}{4\pi\sigma_0} \int_a^b dR$$

18.4 Spherical Resistors

$$\Rightarrow R = \frac{1}{4\pi\sigma_0}[R]_a^b$$

$$\Rightarrow R = \frac{b-a}{4\pi\sigma_0}$$

Choice (2) is the answer (Fig. 18.17).

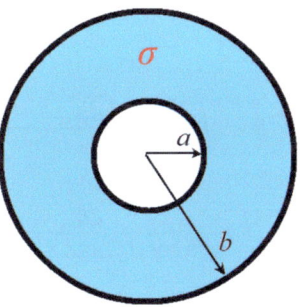

Fig. 18.17 A spherical resistor including nonuniform conductivity

> **Notes**
> In this problem, the relations below have been used.
>
> $$\int \sin x \, dx = -\cos x$$
>
> $$\cos \pi = -1$$
>
> $$\cos 0 = 1$$

18.22. Based on the information given in the problem, we have:

$$\sigma = \frac{\sigma_0}{R}$$

The electrical resistance between the internal and external surface of the hollow sphere with the nonuniform conductivity can be calculated as follows.

$$R = \int \frac{dR}{\int \sigma dS}$$

$$\Rightarrow R = \int \frac{dR}{\int_{\varphi=0}^{2\pi} \int_{\theta=0}^{\pi} \frac{\sigma_0}{R} R^2 \sin\theta \, d\theta \, d\varphi}$$

$$\Rightarrow R = \int \frac{dR}{\sigma_0 R \int_{\varphi=0}^{2\pi} \int_{\theta=0}^{\pi} \sin\theta \, d\theta \, d\varphi}$$

$$\Rightarrow R = \int \frac{dR}{\sigma_0 R [-\cos\theta]_0^{\pi} [\varphi]_0^{2\pi}}$$

$$\Rightarrow R = \int \frac{dR}{\sigma_0 R (\cos 0 - \cos\pi)(2\pi - 0)}$$

$$\Rightarrow R = \frac{1}{4\pi\sigma_0} \int_a^b \frac{dR}{R}$$

$$\Rightarrow R = \frac{1}{4\pi\sigma_0} [\ln R]_a^b$$

$$\Rightarrow R = \frac{1}{4\pi\sigma_0} (\ln b - \ln a)$$

$$\Rightarrow R = \frac{\ln\left(\frac{b}{a}\right)}{4\pi\sigma_0}$$

According to Ohm's law (study General Physics II), the current in the resistor can be calculated as follows.

$$I = \frac{V_0}{R}$$

$$\Rightarrow I = \frac{4\pi\sigma_0 V_0}{\ln\left(\frac{b}{a}\right)}$$

Moreover, the volume current density can be calculated as follows.

$$J = \frac{I}{S} = \frac{I}{4\pi R^2}$$

$$\Rightarrow J = \frac{\sigma_0 V_0}{R^2 \ln\left(\frac{b}{a}\right)}$$

Choice (1) is the answer (Fig. 18.18).

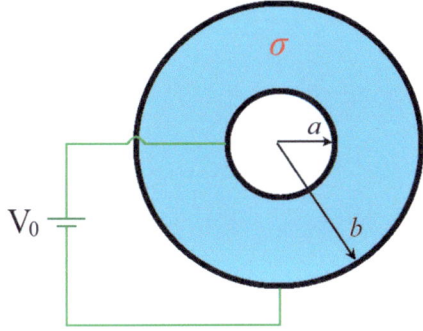

Fig. 18.18 A spherical resistor including nonuniform conductivity connected to a voltage source

> **Notes**
>
> In this problem, the relations below have been used.
>
> $$\int \sin x\, dx = -\cos x$$
>
> $$\cos \pi = -1$$
>
> $$\cos 0 = 1$$
>
> $$\int \frac{dx}{x} = \ln x$$
>
> $$\ln a - \ln b = \ln\left(\frac{a}{b}\right)$$

References

1. Rahmani-Andebili, M., General Physics II – Practice Problems, Methods, and Solutions, Springer Nature, 2025.
2. Rahmani-Andebili, M., General Physics I – Practice Problems, Methods, and Solutions, Springer Nature, 2025.
3. Rahmani-Andebili, M., Mathematics of Engineering and Science – Practice Problems, Methods, and Solutions, Springer Nature, 2024.
4. Rahmani-Andebili, M., Differential Equations – Practice Problems, Methods, and Solutions, Springer Nature, 2022.
5. Rahmani-Andebili, M., Calculus III – Practice Problems, Methods, and Solutions, Springer Nature, 2023.
6. Rahmani-Andebili, M., Calculus II – Practice Problems, Methods, and Solutions, Springer Nature, 2023.
7. Rahmani-Andebili, M., Calculus I (2nd Ed.) – Practice Problems, Methods, and Solutions, Springer Nature, 2023.
8. Rahmani-Andebili, M., Precalculus (2nd Ed.) – Practice Problems, Methods, and Solutions, Springer Nature, 2024.

Magnetic Field and Magnetic Flux: Part A

Abstract

In this chapter, the basic and advanced problems of magnetic flux and field due to a linear, surface, and volume currents are studied. Herein, different types of problems and exercises are presented that are categorized as follows.

- **Problems with detailed solution**: They have been designed to teach students the subjects in detail. Moreover, they have been categorized in different levels based on their difficulty levels (easy, normal, and hard) and calculation amounts (small, normal, and large).
- **Partially solved exercises**: They have been designed to encourage students to practice problems while guiding them through the problem-solving procedure and hinting the required formulas.
- **Exercises with final answer**: They have been designed to encourage students to practice more by themselves while hinting them by the final answer as well as to help instructors to give tests or quizzes.

19.1 Magnetic Field Due to a Linear Current

Problem

19.1. Figure 19.1 shows two parallel and long current-carrying wires, where the resultant magnetic field intensity at point p is zero. Calculate the distance between the wires if $I_1 = 4\ A$, $I_2 = 8\ A$, and $d_1 = 0.1\ m$ [1–8].

Difficulty level   ○ Easy   ● Normal   ○ Hard
Calculation amount ○ Small   ● Normal   ○ Large
1) $0.2\ m$
2) $0.3\ m$
3) $0.4\ m$
4) $0.5\ m$

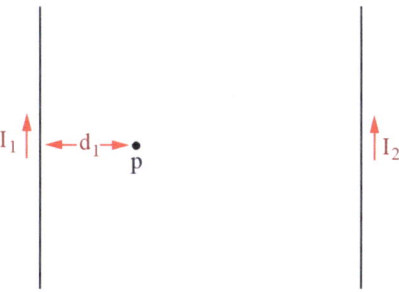

Fig. 19.1 Two parallel current-carrying wires

Problem

19.2. The resultant magnetic field intensity at point p, illustrated in Fig. 19.2, is zero. Calculate the distance between the wires if $I_1 = 3\ A$, $I_2 = 1\ A$, and $d_2 = 0.1\ m$.

Difficulty level   ○ Easy   ● Normal   ○ Hard
Calculation amount   ○ Small   ● Normal   ○ Large

1) $0.2\ m$
2) $0.3\ m$
3) $0.4\ m$
4) $0.5\ m$

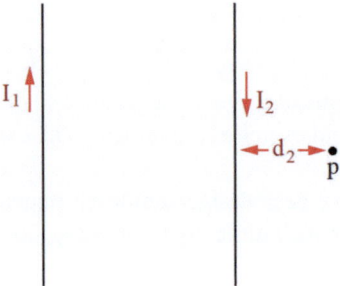

Fig. 19.2 Two parallel current-carrying wires

Partially Solved Exercise

19.3. As is shown in Fig. 19.3, two parallel and long wires carry the currents of $I_1 = I_2 = 2\pi\ A$ at the same direction. The distance between them is 1 m. Calculate the resultant magnetic field intensity at point p which is on the same plane but one meter away from the second wire.

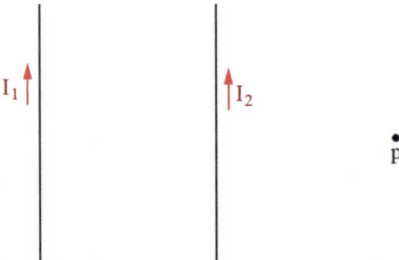

Fig. 19.3 Two parallel current-carrying wires

Solution

Figure 19.4 illustrates the magnetic field intensity of each wire at point p based on the right-hand rule. As can be seen, the field intensities are in the same direction (inward). Therefore:

$$H_{tot} = H_2 + H_1$$

Moreover, the magnitude of magnetic field intensity of the wires can be calculated as follows.

19.1 Magnetic Field Due to a Linear Current

$$H_1 = \frac{I_1}{2\pi r_1} = \frac{(\quad)}{2\pi \times (\quad)} = (\quad) \, A/m$$

$$H_2 = \frac{I_2}{2\pi r_2} = \frac{(\quad)}{2\pi \times (\quad)} = (\quad) \, A/m$$

Thus:

$$H_{tot} = (\quad) + (\quad) \Rightarrow H_{tot} = 1.5 \, A/m$$

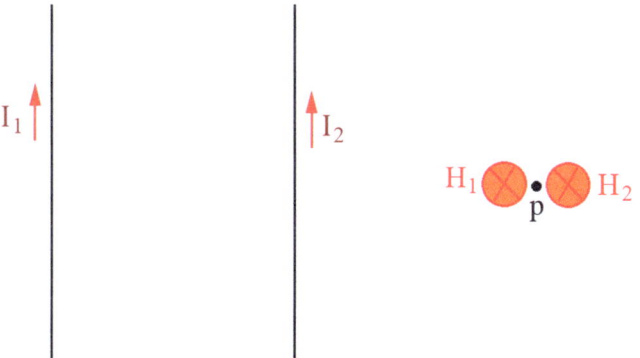

Fig. 19.4 Two parallel current-carrying wires

Problem

19.4. Two parallel wires carry the currents in opposite directions as is illustrated in Fig. 19.5. Calculate the magnitude of resultant magnetic field intensity at point p if $d = 2 \, m$, $I_1 = 3 \, A$, and $I_2 = 4 \, A$.
Difficulty level   ○ Easy   ○ Normal   ● Hard
Calculation amount   ○ Small   ● Normal   ○ Large

1) $H = \dfrac{5}{2\sqrt{2}\pi} \, A/m$
2) $H = \dfrac{5}{2\sqrt{2}} \, A/m$
3) $H = \dfrac{1}{2\sqrt{2}\pi} \, A/m$
4) $H = \dfrac{5}{\sqrt{2}\pi} \, A/m$

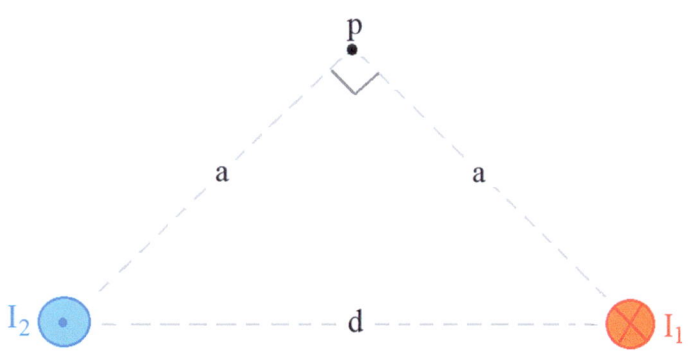

Fig. 19.5 Two parallel current-carrying wires

Problem

19.5. Two concentric rings with radiuses $r_1 = r$ and $r_2 = 2r$ generate equal magnetic field intensity at their centers. What is the relation between the currents of rings? (Fig. 19.6)

Difficulty level   ● Easy   ○ Normal   ○ Hard
Calculation amount ● Small ○ Normal   ○ Large

1) $I_1 = I_2$
2) $I_1 = \dfrac{1}{2} I_2$
3) $I_1 = 2 I_2$
4) $I_1 = \sqrt{2} I_2$

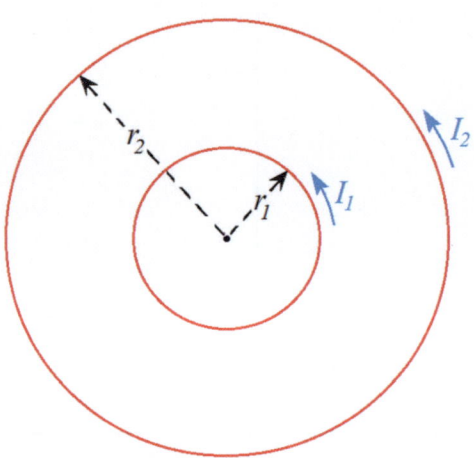

Fig. 19.6 Two concentric rings carrying different currents

Exercise

19.6. Calculate the magnetic field intensity in free space at the center of a circle carrying a current. Herein, assume that radius and current of the circle are infinite.

Final Answer

$H = 0.5 \, A/m$

Problem

19.7. Calculate the magnitude of magnetic field intensity at the center of the circular arc shown in Fig. 19.7.

Difficulty level   ○ Easy   ● Normal   ○ Hard
Calculation amount ● Small ○ Normal   ○ Large

1) $\dfrac{I\theta}{2\pi R}$
2) $\dfrac{I}{2R}$
3) $\dfrac{I\theta}{4R}$
4) $\dfrac{I\theta}{4\pi R}$

19.1 Magnetic Field Due to a Linear Current

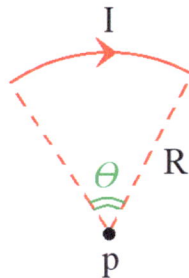

Fig. 19.7 A circular arc carrying a current

Partially Solved Exercise

19.8. Calculate the magnitude of magnetic field intensity at point p in the circuit shown in Fig. 19.8.

Solution

As we know, the magnitude of magnetic field intensity (A/m) at the center of a circular arc with angle α, radius r, and current I can be calculated as follows.

$$H = \frac{\alpha}{2\pi} \frac{I}{2r}$$

It should be noted that the two straight pieces of wire do not generate any magnetic field intensity at point p.

Therefore:

$$H = \frac{(\ \ )}{(\ \ )} \times \frac{(\ \ )}{(\ \ )}$$

$$\Rightarrow H = \frac{I}{8R}$$

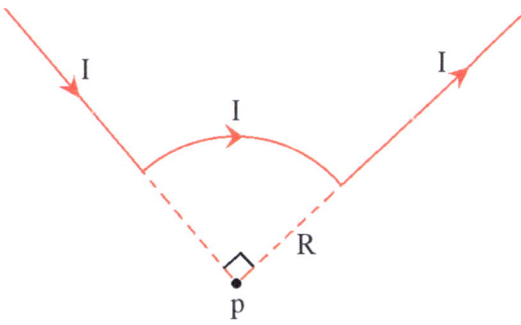

Fig. 19.8 A circuit carrying a current

Problem

19.9. In the circuit shown in Fig. 19.9, calculate the magnitude and direction of magnetic field intensity at point p.
Difficulty level   ○ Easy   ● Normal   ○ Hard
Calculation amount   ○ Small   ● Normal   ○ Large

1) $\dfrac{I(a+b)}{4ab}$, ⊙

2) $\dfrac{I(a+b)}{4ab}$, ⊗

3) $\dfrac{I(a-b)}{4ab}$, ⊙

4) $\dfrac{I(a-b)}{4ab}$, ⊗

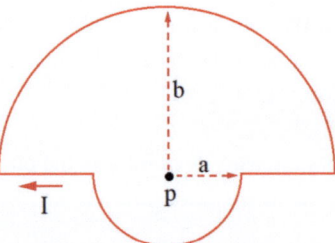

Fig. 19.9 A circuit carrying a current

Partially Solved Exercise

19.10. Calculate the magnitude and direction of magnetic field intensity at point p in the circuit shown in Fig. 19.10.

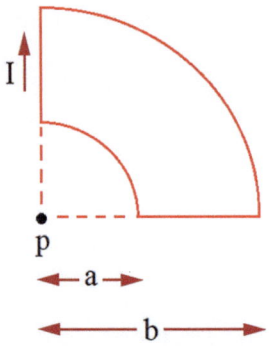

Fig. 19.10 A circuit carrying a current

Final Answer

The two straight pieces of wire do not generate any magnetic field intensity at point p. The direction of magnetic field intensity can be calculated based on the right-hand rule. As is shown in Fig. 19.11, the quarter-circle with radius a generates outward magnetic field intensity while the quarter-circle with radius b generates inward magnetic field intensity.

Moreover, the magnitude of magnetic field intensity (A/m) at the center of a circular arc with angle α, radius r, and current I can be calculated as follows.

$$H = \frac{\alpha}{2\pi} \frac{I}{2r}$$

19.1 Magnetic Field Due to a Linear Current

Hence:

$$H_{tot} = H_1 - H_2$$

$$\Rightarrow H = \frac{(\ \ )}{(\ \ )} \times \frac{(\ \ )}{(\ \ )} - \frac{(\ \ )}{(\ \ )} \times \frac{(\ \ )}{(\ \ )}$$

$$\Rightarrow H = \frac{I}{8}\left(\frac{1}{a} - \frac{1}{b}\right)$$

$$\Rightarrow H = \frac{I(a-b)}{8ab}$$

The total magnetic field intensity will be outward, that is, $\odot$.

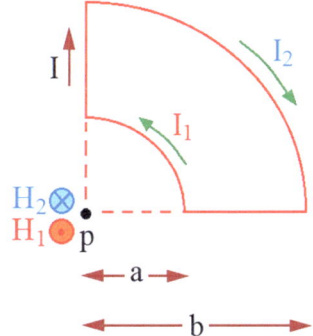

Fig. 19.11 Magnetic field intensities at point p

Problem

19.11. In Fig. 19.12, calculate the magnetic field intensity at the center if $I = 3\ A$ and $r = 0.1\ m$.
Difficulty level ○ Easy ● Normal ○ Hard
Calculation amount ○ Small ● Normal ○ Large
1) 3 A/m
2) 10 A/m
3) 20 A/m
4) 30 A/m

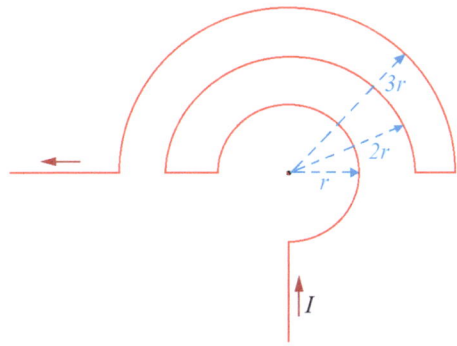

Fig. 19.12 A circuit carrying a current

Problem

19.12. In Fig. 19.13, the current radially enters a circle and leaves it. Calculate the magnitude and direction of magnetic field intensity at the center of the circle.

Difficulty level ○ Easy ○ Normal ● Hard
Calculation amount ○ Small ● Normal ○ Large

1) $\frac{1}{2}\frac{I}{2R}$, $\odot$
2) $\frac{1}{2}\frac{I}{2R}$, $\otimes$
3) $\frac{1}{4}\frac{I}{2R}$, $\otimes$
4) 0

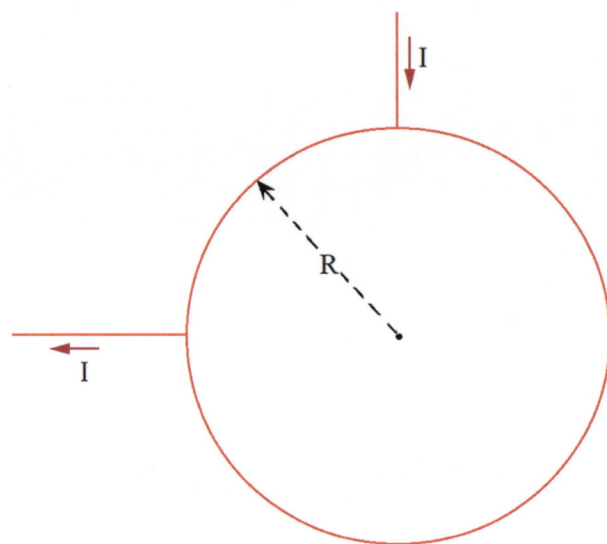

Fig. 19.13 A circuit carrying a current

19.2 Magnetic Field Due to a Surface Current

Problem

19.13. Figure 19.14 illustrates a uniform surface current with the surface current density $\vec{J} = J_0\hat{y}$ A/m^2 on the xy plane. Calculate the magnetic flux density on the top and bottom of the surface.

Difficulty level ○ Easy ● Normal ○ Hard
Calculation amount ● Small ○ Normal ○ Large

1) $\vec{B} = \begin{cases} \frac{1}{2}\mu_0 J_0\hat{x} & z>0 \\ -\frac{1}{2}\mu_0 J_0\hat{x} & z<0 \end{cases}$

2) $\vec{B} = \begin{cases} \mu_0 J_0\hat{x} & z>0 \\ -\mu_0 J_0\hat{x} & z<0 \end{cases}$

3) $\vec{B} = \begin{cases} \mu_0 J_0\hat{x} & z>0 \\ \mu_0 J_0\hat{x} & z<0 \end{cases}$

4) $\vec{B} = \begin{cases} -\frac{1}{2}\mu_0 J_0\hat{x} & z>0 \\ \frac{1}{2}\mu_0 J_0\hat{x} & z<0 \end{cases}$

19.2 Magnetic Field Due to a Surface Current

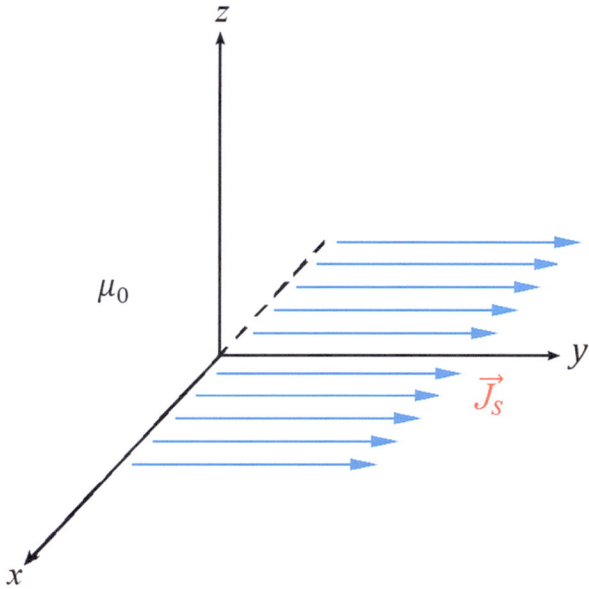

Fig. 19.14 A uniform surface current

Problem

19.14. In Fig. 19.15, the uniform surface current is on the xy plane. If the magnetic flux density at point p is zero, calculate the value of current I.

Difficulty level ○ Easy ● Normal ○ Hard
Calculation amount ○ Small ● Normal ○ Large

1) πJ_0
2) $\pi b J_0$
3) $0.5 \pi b J_0$
4) $2\pi(a-b)J_0$

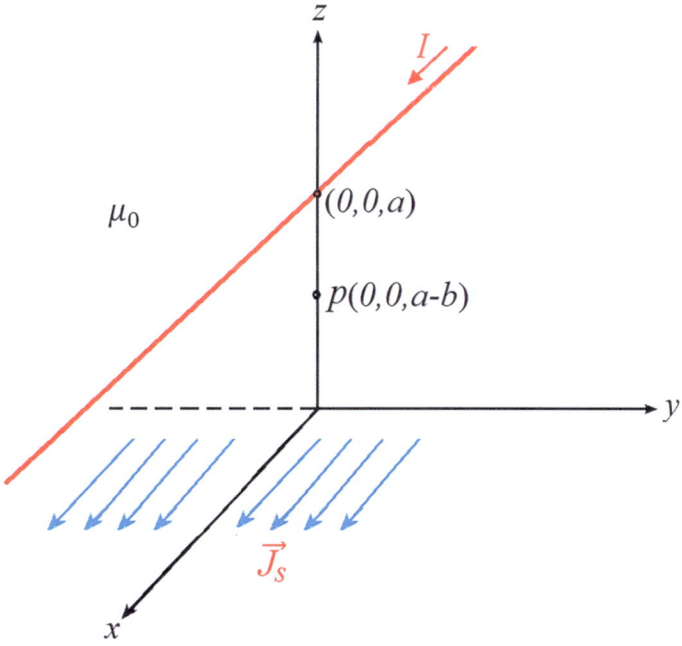

Fig. 19.15 A uniform surface current and a current-carrying wire

19.3 Magnetic Field Due to a Volume Current

Problem

19.15. In a free space, in the region $-a < z < a$, the uniform volume current density is $\vec{J} = J_0 \hat{x}$ A/m². Calculate the magnetic field intensity at point $p(0, 0, 2a)$ in the Cartesian coordinate system (Fig. 19.16).

Difficulty level   ○ Easy   ○ Normal   ● Hard
Calculation amount   ○ Small   ● Normal   ○ Large
1) $-2J_0 a\hat{y}$
2) $J_0 a\hat{y}$
3) $-J_0 a\hat{y}$
4) $2J_0 a\hat{y}$

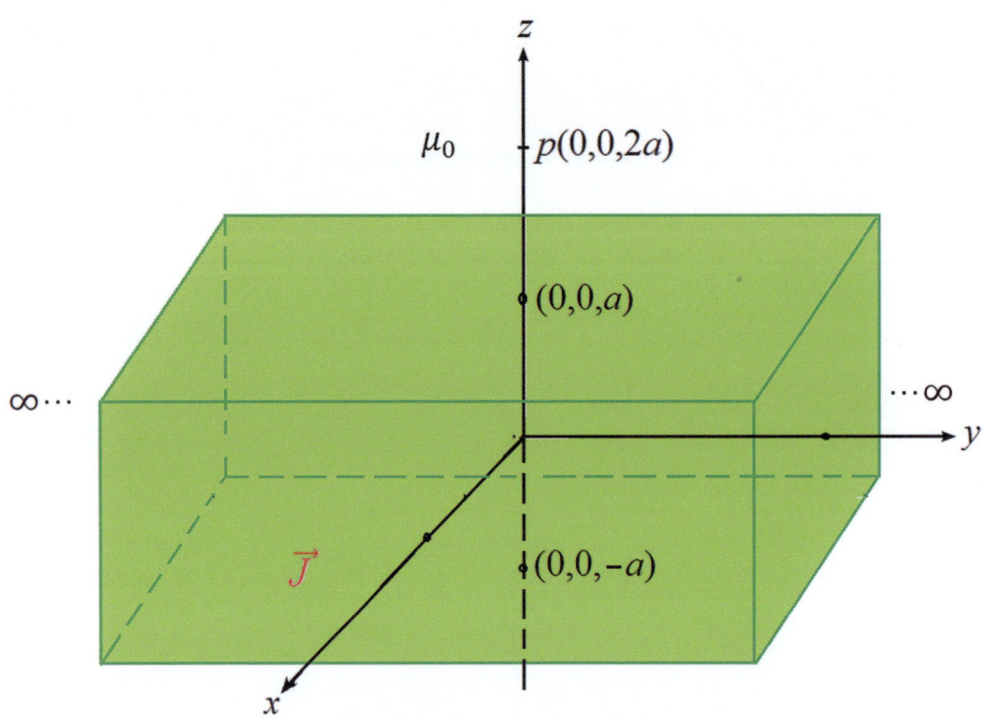

Fig. 19.16 A uniform volume current

19.4 Magnetic Flux

Problem

19.16. In a free space on the *xoy* plane, a surface current with surface current density $\vec{J_s} = -J_0\hat{x}$ is available. Calculate the magnetic flux passing through an arbitrary square with the unity side length on the *xoz* plane.

Difficulty level   ○ Easy   ○ Normal   ● Hard
Calculation amount   ○ Small   ● Normal   ○ Large

1) $\frac{\mu_0 J_0}{2}$
2) $\mu_0 J_0$
3) $2\mu_0 J_0$
4) $4\mu_0 J_0$

Exercise

19.17. In Problem 19.16, calculate the magnetic flux passing through a square with the unity side length on the *zoy* plane.

Final Answer

$\varphi = 0$

References

1. Rahmani-Andebili, M., General Physics II – Practice Problems, Methods, and Solutions, Springer Nature, 2025.
2. Rahmani-Andebili, M., General Physics I – Practice Problems, Methods, and Solutions, Springer Nature, 2025.
3. Rahmani-Andebili, M., Mathematics of Engineering and Science – Practice Problems, Methods, and Solutions, Springer Nature, 2024.
4. Rahmani-Andebili, M., Differential Equations – Practice Problems, Methods, and Solutions, Springer Nature, 2022.
5. Rahmani-Andebili, M., Calculus III – Practice Problems, Methods, and Solutions, Springer Nature, 2023.
6. Rahmani-Andebili, M., Calculus II – Practice Problems, Methods, and Solutions, Springer Nature, 2023.
7. Rahmani-Andebili, M., Calculus I (2nd Ed.) – Practice Problems, Methods, and Solutions, Springer Nature, 2023.
8. Rahmani-Andebili, M., Precalculus (2nd Ed.) – Practice Problems, Methods, and Solutions, Springer Nature, 2024.

Magnetic Field and Magnetic Flux: Part B

Abstract

In this chapter, the problems of the nineteenth chapter are fully solved, in detail, step-by-step, and with different methods.

20.1 Magnetic Field Due to a Linear Current

20.1. Based on the information given in the problem, we have [1–8]:

$$I_1 = 4\ A$$

$$I_2 = 8\ A$$

$$d_1 = 0.1\ m$$

$$H_p = 0$$

Figure 20.1 illustrates the magnetic field intensity of each wire at point p based on the right-hand rule. As can be seen, the field intensities are in opposite directions. Therefore:

$$H_p = H_2 - H_1$$

Moreover, the magnitude of magnetic field intensity of each wire can be calculated as follows.

$$H_1 = \frac{I_1}{2\pi d_1} = \frac{4}{2\pi \times 0.1} = \frac{20}{\pi}\ A/m$$

$$H_2 = \frac{I_2}{2\pi d_2} = \frac{8}{2\pi d_2} = \frac{4}{\pi d_2}\ A/m$$

Hence:

$$\frac{4}{\pi d_2} - \frac{20}{\pi} = 0$$

$$\Rightarrow \frac{4}{d_2} = 20 \Rightarrow d_2 = 0.2\ m$$

Thus:

$$\Rightarrow d = d_2 + d_1 = 0.2 + 0.1$$

$$\Rightarrow d = 0.3 \, m$$

Choice (2) is the answer.

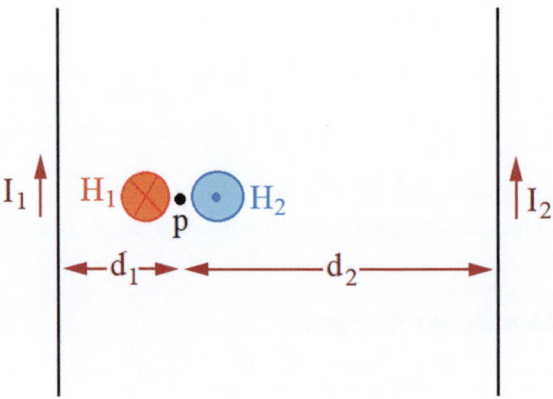

Fig. 20.1 Two parallel current-carrying wires

20.2. Based on the information given in the problem, we have:

$$I_1 = 3 \, A$$

$$I_2 = 1 \, A$$

$$d_2 = 0.1 \, m$$

$$H_p = 0$$

Figure 20.2 illustrates the magnetic field intensity of each wire at point p based on the right-hand rule. As can be seen, the field intensities are in opposite directions. Therefore:

$$H_p = H_2 - H_1$$

Moreover, the magnitude of magnetic field intensity of each wire can be calculated as follows.

$$H_1 = \frac{I_1}{2\pi d_1} = \frac{3}{2\pi \times d_1} = \frac{3}{2\pi d_1} \, A/m$$

$$H_2 = \frac{I_2}{2\pi d_2} = \frac{1}{2\pi \times 0.1} = \frac{5}{\pi} \, A/m$$

Hence:

$$\frac{5}{\pi} - \frac{3}{2\pi d_1} = 0$$

$$\Rightarrow 5 = \frac{3}{2d_1} \Rightarrow d_1 = 0.3 \, m$$

Thus:

$$\Rightarrow d = d_1 - d_2 = 0.3 - 0.1$$

$$\Rightarrow d = 0.2 \, m$$

Choice (1) is the answer.

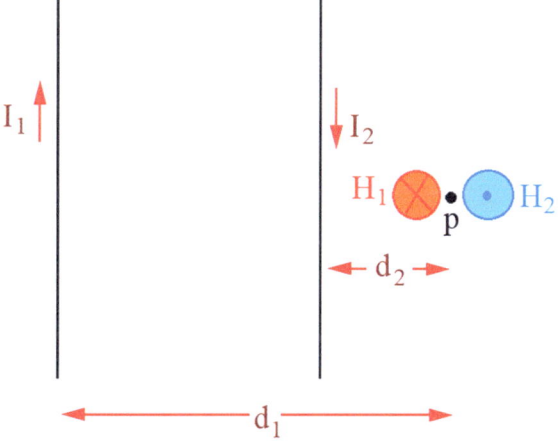

Fig. 20.2 Two parallel current-carrying wires

20.4. Based on the information given in the problem, we have:

$$I_1 = 3 \, A$$

$$I_2 = 4 \, A$$

$$d = 2 \, m$$

Figure 20.3 illustrates the magnetic field intensity of each wire at point p based on the right-hand rule. The distance of the point from each wire can be calculated as follows.

$$\cos 45° = \frac{a}{d}$$

$$\Rightarrow \frac{\sqrt{2}}{2} = \frac{a}{d} \Rightarrow a = \frac{d}{\sqrt{2}}$$

The magnitude of magnetic field intensity of each wire at point p can be calculated as follows.

$$H_1 = \frac{I_1}{2\pi a} = \frac{I_1}{2\pi\left(\frac{d}{\sqrt{2}}\right)} = \frac{\sqrt{2}}{2\pi d}I_1$$

$$H_2 = \frac{I_2}{2\pi a} = \frac{I_2}{2\pi\left(\frac{d}{\sqrt{2}}\right)} = \frac{\sqrt{2}}{2\pi d}I_2$$

The angle between the vectors of magnetic field intensity is 90°. The magnitude of resultant magnetic field intensity can be calculated by using the Pythagorean formula as follows.

$$H = \sqrt{H_1^2 + H_2^2}$$

$$\Rightarrow H = \frac{\sqrt{2}}{2\pi d}\sqrt{I_1^2 + I_2^2}$$

$$\Rightarrow H = \frac{\sqrt{2}}{2\pi \times 2}\sqrt{3^2 + 4^2}$$

$$\Rightarrow H = \frac{5}{2\sqrt{2}\pi} \; A/m$$

Choice (1) is the answer.

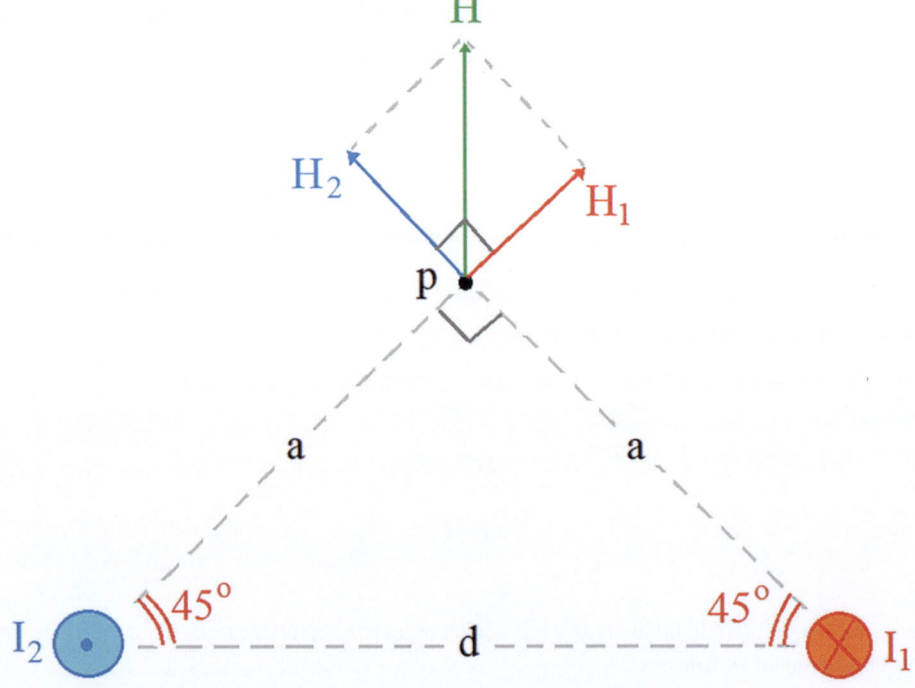

Fig. 20.3 The magnetic field intensity of two parallel current-carrying wires at point p

20.5. Based on the information given in the problem, we have:

$$r_1 = r$$

$$r_2 = 2r$$

$$H_1 = H_2$$

As we know, the magnitude of magnetic field intensity (A/m) at the center of a ring with radius *r* and current *I* is calculated as follows.

$$H = \frac{I}{2r}$$

Therefore, for this problem, we have:

$$\frac{I_1}{2r} = \frac{I_2}{2(2r)}$$

$$\Rightarrow I_1 = \frac{1}{2} I_2$$

Choice (2) is the answer (Fig. 20.4).

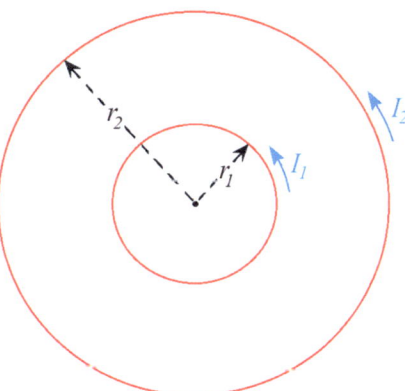

Fig. 20.4 Two concentric rings carrying different currents

20.7. The magnitude of magnetic field intensity (A/m) at the center of a circular arc with angle α, radius *r*, and current *I* can be calculated as follows.

$$H = \frac{\alpha}{2\pi} \frac{I}{2r}$$

Hence:

$$H = \frac{\theta I}{4\pi R}$$

Choice (4) is the answer (Fig. 20.5).

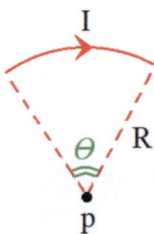

Fig. 20.5 A circular arc carrying a current

20.9. The two straight pieces of wire do not generate any magnetic field intensity at point *p*. The direction of magnetic field intensity can be calculated based on the right-hand rule. Both semicircles generate inward magnetic field intensity shown in Fig. 20.6. Thus, the total magnetic field intensity will be inward, that is, ⊗.

Moreover, the magnitude of magnetic field intensity (*A/m*) at the center of a circular arc with angle α, radius *r*, and current *I* can be calculated as follows.

$$H = \frac{\alpha}{2\pi} \frac{I}{2r}$$

Hence:

$$H_{tot} = H_1 + H_2$$

$$H = \frac{\pi}{2\pi} \frac{I}{2a} + \frac{\pi}{2\pi} \frac{I}{2b} \Rightarrow H = \frac{I}{4}\left(\frac{1}{a} + \frac{1}{b}\right)$$

$$\Rightarrow H = \frac{I(a+b)}{4ab}$$

Choice (2) is the answer.

20.1 Magnetic Field Due to a Linear Current

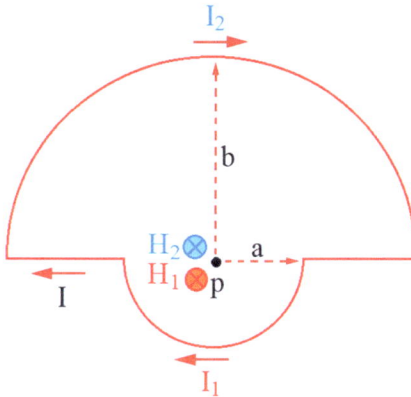

Fig. 20.6 Magnetic field intensities at point p

20.11. Based on the information given in the problem, we have:

$$I = 3\,A$$

$$r = 0.1\,m$$

As we know, the magnitude of magnetic field intensity (A/m) at the center of a circular arc with angle α, radius r, and current I can be calculated as follows.

$$H = \frac{\alpha}{2\pi}\frac{I}{2r}$$

The straight pieces of wire do not generate any magnetic field intensity at the center. Based on the right-hand rule, the direction of magnetic field intensity of the three-quarter circle with radius r and the semicircle with radius $3r$ is outward, while the direction of magnetic field intensity of the three-quarter circle with radius $2r$ is inward.

Therefore:

$$H_{tot} = H_1 - H_2 + H_3$$

$$H = \frac{\frac{3\pi}{2}}{2\pi}\frac{I}{2r} - \frac{\pi}{2\pi}\frac{I}{2(2r)} + \frac{\pi}{2\pi}\frac{I}{2(3r)}$$

$$H = \left(\frac{3}{8} - \frac{1}{8} + \frac{1}{12}\right)\frac{I}{r}$$

$$H = \left(\frac{1}{3}\right)\frac{3}{0.1}$$

$$\Rightarrow H = 10\,A/m$$

Choice (2) is the answer (Fig. 20.7).

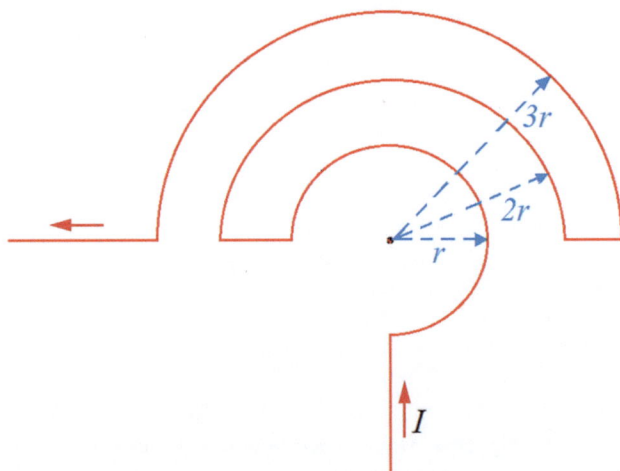

Fig. 20.7 Three magnetic field intensities at point p

20.12. First, the current of each part of the circle needs to be calculated as follows. Based on the current division rule, we can write:

$$I_1 = \frac{R_2}{R} I$$

$$I_2 = \frac{R_1}{R} I$$

where, R is the total resistance of the circle, R_1 is the resistance of the quarter-circle, and R_2 is the resistance of the rest of the circle. Therefore:

$$I_1 = \frac{3}{4} I$$

$$I_2 = \frac{1}{4} I$$

The two straight pieces of wire do not generate any magnetic field intensity at point p. The direction of magnetic field intensity can be calculated based on the right-hand rule. As is shown in Fig. 20.8, the quarter-circle generates outward magnetic field intensity while the other part of the circle generates inward magnetic field intensity.

Moreover, the magnitude of magnetic field intensity (A/m) at the center of a circular arc with angle α, radius r, and current I can be calculated as follows.

$$H = \frac{\alpha}{2\pi} \frac{I}{2r}$$

Hence:

$$H_{tot} = H_1 - H_2$$

20.2 Magnetic Field Due to a Surface Current

$$H = \frac{\frac{\pi}{2}}{2\pi}\frac{I_1}{2R} - \frac{\frac{3\pi}{2}}{2\pi}\frac{I_2}{2R}$$

$$\Rightarrow H = \frac{I_1}{8R} - \frac{3I_2}{8R}$$

$$\Rightarrow H = \frac{\frac{3}{4}I}{8R} - \frac{3 \times \frac{1}{4}I}{8R}$$

$$\Rightarrow H = \frac{3I}{32R} - \frac{3I}{32R}$$

$$\Rightarrow H = 0$$

Choice (4) is the answer.

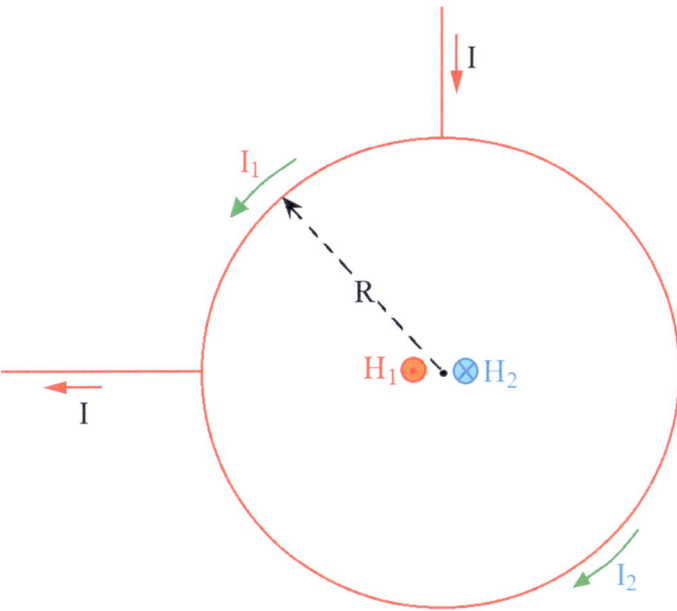

Fig. 20.8 Magnetic field intensities at the center of the circle

20.2 Magnetic Field Due to a Surface Current

20.13. The magnetic flux density of a uniform surface current density $\vec{J_s}$ can be calculated as follows.

$$\vec{B} = \frac{1}{2}\mu_0 \vec{J_s} \times \hat{n}$$

where, $\hat{n}$ is the normal vector of the $\vec{J_s}$.

Therefore, the magnetic flux density on the top and bottom of the surface is as follows.

$$\vec{B} = \begin{cases} \frac{1}{2}\mu_0 \vec{J_s} \times \hat{z} & z > 0 \\ -\frac{1}{2}\mu_0 \vec{J_s} \times \hat{z} & z < 0 \end{cases}$$

$$\Rightarrow \vec{B} = \begin{cases} \frac{1}{2}\mu_0 J_0 \hat{y} \times \hat{z} & z > 0 \\ -\frac{1}{2}\mu_0 J_0 \hat{y} \times \hat{z} & z < 0 \end{cases}$$

$$\Rightarrow \vec{B} = \begin{cases} \frac{1}{2}\mu_0 J_0 \hat{x} & z > 0 \\ -\frac{1}{2}\mu_0 J_0 \hat{x} & z < 0 \end{cases}$$

Choice (1) is the answer (Fig. 20.9).

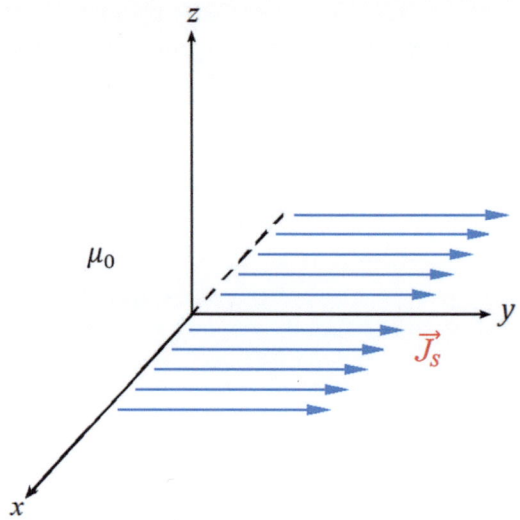

Fig. 20.9 A uniform surface current

20.14. Based on the information given in the problem, we have:

$$\vec{J_s} = J_0 \hat{x}$$

$$B_p = 0 \ T$$

The magnetic flux density of a straight wire at a distance r can be calculated as follows.

$$B = \frac{\mu_0 I}{2\pi r}$$

20.2 Magnetic Field Due to a Surface Current

Moreover, the magnetic flux density of a uniform surface current density $\vec{J_s}$ can be calculated as follows.

$$\vec{B} = \frac{1}{2}\mu_0 \vec{J_s} \times \hat{n}$$

where, $\hat{n}$ is the normal vector of the $\vec{J_s}$.

Thus, in this problem, for $z > 0$, we can write:

$$\vec{B}_1 = \frac{1}{2}\mu_0 J_0 \hat{x} \times \hat{z} \Rightarrow \vec{B}_1 = -\frac{1}{2}\mu_0 J_0 \hat{y}$$

The magnetic flux densities at point p is as follows.

$$\vec{B}_{1,p} = -\frac{1}{2}\mu_0 J_0 \hat{y}$$

$$\vec{B}_{2,p} = \frac{\mu_0 I}{2\pi b}\hat{y}$$

Therefore:

$$\vec{B}_{1,p} + \vec{B}_{2,p} = 0$$

$$\Rightarrow -\frac{1}{2}\mu_0 J_0 \hat{y} + \frac{\mu_0 I}{2\pi b}\hat{y} = 0$$

$$\Rightarrow \frac{1}{2}\mu_0 J_0 = \frac{\mu_0 I}{2\pi b}$$

$$\Rightarrow I = \pi b J_0$$

Choice (2) is the answer (Fig. 20.10).

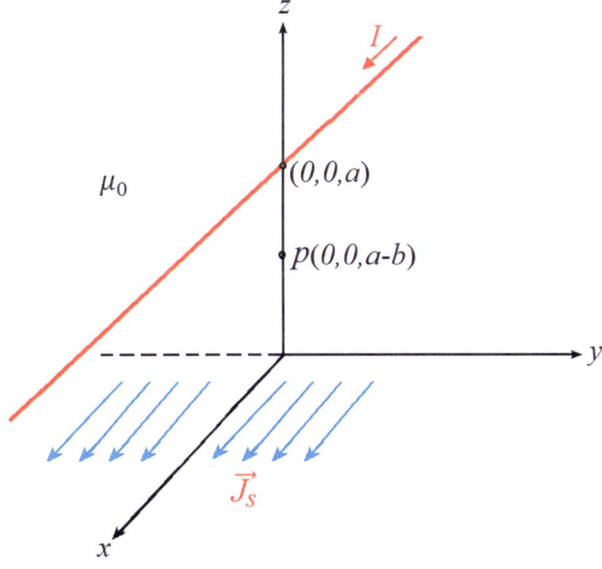

Fig. 20.10 A uniform surface current and a current-carrying wire

20.3 Magnetic Field Due to a Volume Current

20.15. Based on the information given in the problem, we have:

$$-a < z < a$$

$$\vec{J} = J_0 \hat{x} \; A/m^2$$

The magnetic flux density of a uniform surface current density $\vec{J_s}$ can be calculated as follows.

$$\vec{H} = \frac{1}{2} \vec{J_s} \times \hat{n}$$

where, $\hat{n}$ is the normal vector of the $\vec{J_s}$.

The region $-a < z < a$ can be considered as many differential layers with the depth dz and the uniform surface current density $\vec{J_s} = J_0 dz \hat{x}$.

The magnetic flux density of each differential layer can be calculated as follows.

$$d\vec{H} = \frac{1}{2} J_0 dz \hat{x} \times \hat{z} = -\frac{1}{2} J_0 dz \hat{y}$$

Now, for the whole region, we need to integrate it for the given boundaries as follows.

$$\Rightarrow \vec{H} = \int_{-a}^{a} -\frac{1}{2} J_0 dz \hat{y}$$

$$\Rightarrow \vec{H} = -\frac{1}{2} J_0 \hat{y} \int_{-a}^{a} dz$$

$$\Rightarrow \vec{H} = -\frac{1}{2} J_0 \hat{y} (a - (-a))$$

$$\Rightarrow \vec{H} = -J_0 a \hat{y}$$

Choice (3) is the answer (Fig. 20.11).

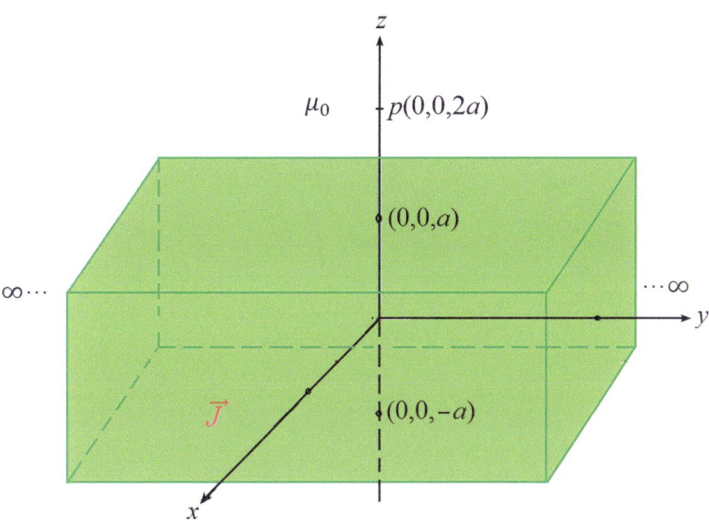

Fig. 20.11 A uniform volume current

20.4 Magnetic Flux

20.16. Based on the information given in the problem, we have:

$$\vec{J_s} = -J_0 \hat{x}$$

$$S = 1 \ m^2$$

The magnetic flux density of a uniform surface current density $\vec{J_s}$ can be calculated as follows.

$$\vec{B} = \frac{1}{2}\mu_0 \vec{J_s} \times \hat{n}$$

where, $\hat{n}$ is the normal vector of the $\vec{J_s}$.

In addition, the magnetic flux can be calculated as follows.

$$\varphi = \int \vec{B} \cdot \vec{dS}$$

Therefore, for this problem for $z > 0$, we can write:

$$\vec{B} = -\frac{1}{2}\mu_0 J_0 \hat{x} \times \hat{z}$$

$$\Rightarrow \vec{B} = \frac{1}{2}\mu_0 J_0 \hat{y}$$

As is illustrated in Fig. 20.12, the square can be in an arbitrary location on the *xoz* plane since the magnetic flux density is uniform in that region. Therefore:

$$\varphi = \int \frac{1}{2}\mu_0 J_0 \hat{y} \cdot dS\hat{y}$$

$$\Rightarrow \varphi = \frac{1}{2}\mu_0 J_0 \int dS$$

$$\Rightarrow \varphi = \frac{1}{2}\mu_0 J_0 \times 1$$

$$\Rightarrow \varphi = \frac{1}{2}\mu_0 J_0$$

Choice (1) is the answer.

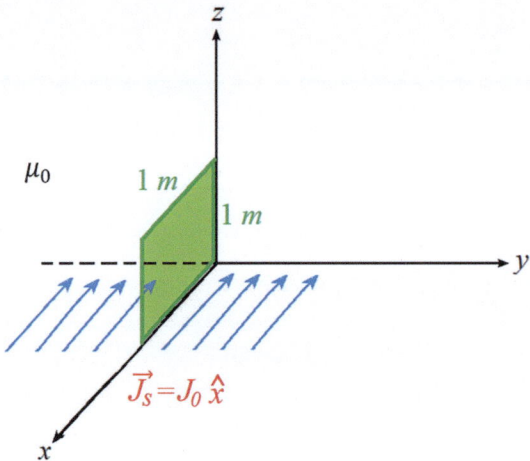

Fig. 20.12 The magnetic flux passing through the square due to a uniform surface current

References

1. Rahmani-Andebili, M., General Physics II – Practice Problems, Methods, and Solutions, Springer Nature, 2025.
2. Rahmani-Andebili, M., General Physics I – Practice Problems, Methods, and Solutions, Springer Nature, 2025.
3. Rahmani-Andebili, M., Mathematics of Engineering and Science – Practice Problems, Methods, and Solutions, Springer Nature, 2024.
4. Rahmani-Andebili, M., Differential Equations – Practice Problems, Methods, and Solutions, Springer Nature, 2022.
5. Rahmani-Andebili, M., Calculus III – Practice Problems, Methods, and Solutions, Springer Nature, 2023.
6. Rahmani-Andebili, M., Calculus II – Practice Problems, Methods, and Solutions, Springer Nature, 2023.
7. Rahmani-Andebili, M., Calculus I (2nd Ed.) – Practice Problems, Methods, and Solutions, Springer Nature, 2023.
8. Rahmani-Andebili, M., Precalculus (2nd Ed.) – Practice Problems, Methods, and Solutions, Springer Nature, 2024.

21
Electromagnetic Force and Torque: Part A

Abstract
In this chapter, the basic and advanced problems of electromagnetic force and torque are studied. Herein, different types of problems and exercises are presented that are categorized as follows.

- **Problems with detailed solution**: They have been designed to teach students the subjects in detail. Moreover, they have been categorized in different levels based on their difficulty levels (easy, normal, and hard) and calculation amounts (small, normal, and large).
- **Partially solved exercises**: They have been designed to encourage students to practice problems while guiding them through the problem-solving procedure and hinting the required formulas.
- **Exercises with final answer**: They have been designed to encourage students to practice more by themselves while hinting them by the final answer as well as to help instructors to give tests or quizzes.

21.1 Electromagnetic Force

Problem

21.1. An electron with the velocity of $\vec{v} = 10^6 \hat{x} + 1.5 \times 10^6 \hat{y}$ m/s enters an environment with the magnetic field of $\vec{B} = 0.06\hat{x} - 0.3\hat{y}$ T. Calculate the magnetic force exerted on the electron. Herein, assume that the magnitude of charge of an electron is 1.6×10^{-19} C [1–8].

Difficulty level ● Easy ○ Normal ○ Hard
Calculation amount ● Small ○ Normal ○ Large

1) $-6.24 \times 10^{-14} \hat{z}$ N
2) $6.24 \times 10^{-14} \hat{z}$ N
3) $6.24 \times 10^{-14} \hat{y}$ N
4) $-6.24 \times 10^{-14} \hat{y}$ N

Partially Solved Exercise

21.2. Calculate the magnetic force exerted on a proton that with the velocity of $\vec{v} = 10^6 \hat{z}$ m/s enters an environment with the magnetic field of $\vec{B} = 2\hat{y}$ T. The magnitude of charge of a proton is 1.6×10^{-19} C.

Solution

As we know, the magnetic force exerted on a moving charge with the velocity of $\vec{v}$ in an environment with the magnetic field of $\vec{B}$ can be calculated as follows.

$$\vec{F} = q\vec{v} \times \vec{B}$$

Therefore:

$$\vec{F} = (\qquad)(\qquad) \times (\qquad)$$

$$\Rightarrow \vec{F} = (\qquad)(\qquad)$$

$$\Rightarrow \vec{F} = -3.2 \times 10^{-13}\hat{x}\ N$$

Problem

21.3. In Fig. 21.1, the wire is placed on xy plane. Calculate the force that the wire exerts on the point charge when it passes through the origin. The velocity vector of point charge is $v = -v_0\hat{y}$.

Difficulty level ○ Easy ● Normal ○ Hard
Calculation amount ○ Small ● Normal ○ Large

1) $-\dfrac{qv_0\mu_0 I}{4a}\hat{x}$
2) $-\dfrac{qv_0\mu_0 I}{8a}\hat{z}$
3) $-\dfrac{qv_0\mu_0 I}{8a}\hat{y}$
4) $-\dfrac{qv_0\mu_0 I}{2a}\hat{x}$

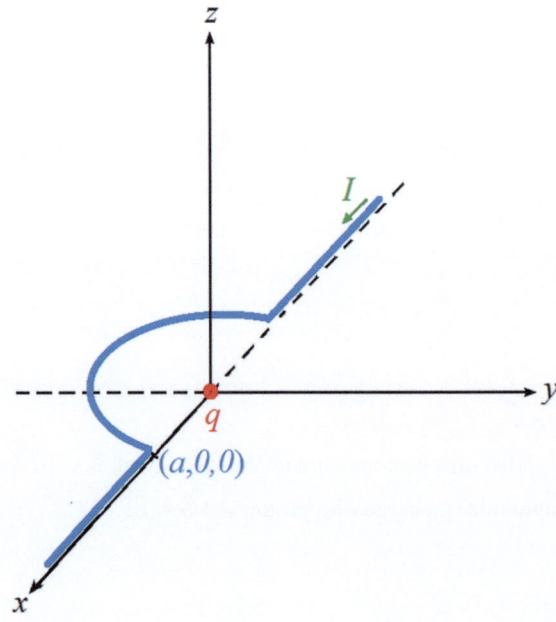

Fig. 21.1 A moving point charge and a wire with the specific shape placed on xy plane

Problem

21.4. Two long wires carry the currents I_1 and I_2 in the same direction in a free space shown in Fig. 21.2. The distance between the wires is d. Determine if the force between them is attractive or repulsive. In addition, calculate the magnetic force exerted on per unit length of them.

Difficulty level ○ Easy ● Normal ○ Hard
Calculation amount ○ Small ● Normal ○ Large

1) $\dfrac{\mu_0 I_1 I_2}{2\pi d}$, repulsive
2) $\dfrac{\mu_0 I_1}{2\pi d}$, attractive
3) $\dfrac{\mu_0 I_2}{2\pi d}$, repulsive
4) $\dfrac{\mu_0 I_1 I_2}{2\pi d}$, attractive

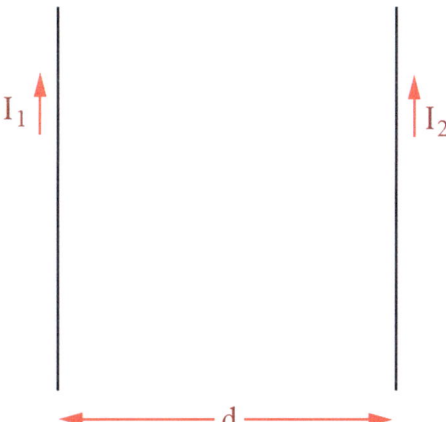

Fig. 21.2 Two parallel wires carrying currents in the same direction

Problem

21.5. Solve Problem 21.4 while assuming that the currents are in opposite directions (Fig. 21.3).

Difficulty level ○ Easy ● Normal ○ Hard
Calculation amount ○ Small ● Normal ○ Large

1) $\dfrac{\mu_0 I_1 I_2}{2\pi d}$, repulsive
2) $\dfrac{\mu_0 I_1}{2\pi d}$, attractive
3) $\dfrac{\mu_0 I_2}{2\pi d}$, repulsive
4) $\dfrac{\mu_0 I_1 I_2}{2\pi d}$, attractive

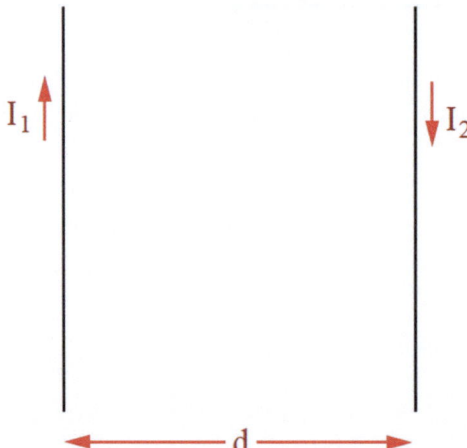

Fig. 21.3 Two parallel wires carrying currents in opposite directions

Exercise

21.6. Two long wires carry the currents $I_1 = 1\ A$ and $I_2 = \pi\ A$ in a free space. The distance between the wires is $0.5\ m$. Calculate the magnetic force that each of the wires exerts on two meters of the other one.

Final Answer

$F = 2\mu_0\ N$

Problem

21.7. In Fig. 21.4, the wire and the square loop are on xy plane. Calculate the force exerted on the loop by the wire.
Difficulty level   ○ Easy   ○ Normal   ● Hard
Calculation amount   ○ Small   ○ Normal   ● Large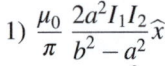

1) $\dfrac{\mu_0}{\pi}\dfrac{2a^2 I_1 I_2}{b^2 - a^2}\widehat{x}$

2) $-\dfrac{\mu_0}{\pi}\dfrac{2a^2 I_1 I_2}{b^2 - a^2}\widehat{x}$

3) $\dfrac{\mu_0}{\pi} I_1 I_2 \ln\left(\dfrac{b+a}{b-a}\right)\widehat{y}$

4) $-\dfrac{\mu_0}{\pi} I_1 I_2 \ln\left(\dfrac{b+a}{b-a}\right)\widehat{y}$

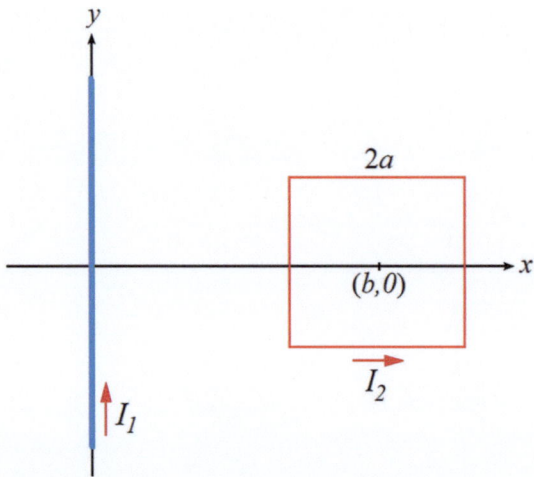

Fig. 21.4 A wire and a square loop placed on xy plane

21.2 Electromagnetic Torque

Problem

21.8. Consider a ring with radius a that is placed in free space on $z = 0$ plane on the origin. Calculate the torque that is exerted on the ring if its current is $\vec{I} = I_0\hat{\varphi}$ and the magnetic field intensity in the area is $\vec{H} = H_0\hat{\varphi}$ in the cylindrical coordinates system (Fig. 21.5).

Difficulty level   ○ Easy   ○ Normal   ● Hard
Calculation amount   ● Small ○ Normal   ○ Large

1) $-\mu_0 H_0 \pi a^2 I_0 \hat{r}$
2) $-\mu_0 H_0 \pi a^2 I_0 \hat{\varphi}$
3) $\mu_0 H_0 \pi a^2 I_0 \hat{\varphi}$
4) $\mu_0 H_0 \pi a^2 I_0 \hat{r}$

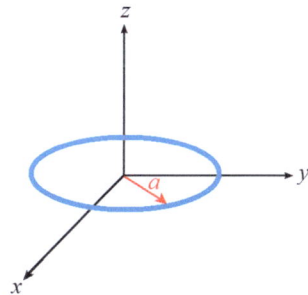

Fig. 21.5 A ring placed in free space on *xy* plane at the origin in presence of a magnetic field

Exercise

21.9. In Problem 21.8, calculate the torque that is exerted on the ring if $a = 2$ m, $\vec{I} = 3\hat{\varphi}$ A, and $\vec{H} = 4\hat{\varphi}$ A/m.

Final Answer

$\vec{\tau} = -48\pi\mu_0 \hat{r}$

Partially Solved Exercise

21.10. In Fig. 21.6, calculate the torque that is exerted on the ring if its current and the magnetic field intensity in the area are as follows in the cylindrical coordinates system.

$$\vec{I} = I_0\hat{\varphi}$$

$$\vec{H} = H_0\hat{z}$$

Solution

As we know, the torque acting on a ring (magnetic dipole) can be calculated as follows.

$$\vec{\tau} = \vec{m} \times \vec{B}$$

where, $\vec{B}$ is the magnetic flux density in the area and $\vec{m}$ is the magnetic moment of the magnetic dipole that can be calculated as follows.

$$\vec{m} = SI\hat{n} = \pi a^2 I \hat{n}$$

Herein, a, I, and $\hat{n}$ are the radius, current, and normal vector of the magnetic dipole. The normal vector of a magnetic dipole is determined based on the right-hand rule where if the thumb finger is in the current direction, the other four fingers show the $\hat{n}$.

Hence, for this problem, we have:

$$\vec{m} =$$

$$\vec{B} =$$

$$\Rightarrow \vec{\tau} = (\qquad) \times (\qquad)$$

$$\Rightarrow \vec{\tau} = 0$$

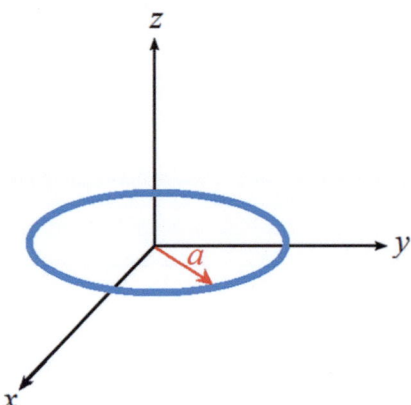

Fig. 21.6 A ring placed in free space on *xy* plane at the origin in presence of a magnetic field

References

1. Rahmani-Andebili, M., General Physics II – Practice Problems, Methods, and Solutions, Springer Nature, 2025.
2. Rahmani-Andebili, M., General Physics I – Practice Problems, Methods, and Solutions, Springer Nature, 2025.
3. Rahmani-Andebili, M., Mathematics of Engineering and Science – Practice Problems, Methods, and Solutions, Springer Nature, 2024.
4. Rahmani-Andebili, M., Differential Equations – Practice Problems, Methods, and Solutions, Springer Nature, 2022.
5. Rahmani-Andebili, M., Calculus III – Practice Problems, Methods, and Solutions, Springer Nature, 2023.
6. Rahmani-Andebili, M., Calculus II – Practice Problems, Methods, and Solutions, Springer Nature, 2023.
7. Rahmani-Andebili, M., Calculus I (2nd Ed.) – Practice Problems, Methods, and Solutions, Springer Nature, 2023.
8. Rahmani-Andebili, M., Precalculus (2nd Ed.) – Practice Problems, Methods, and Solutions, Springer Nature, 2024.

22. Electromagnetic Force and Torque: Part B

Abstract

In this chapter, the problems of the twenty first chapter are fully solved, in detail, step-by-step, and with different methods.

22.1 Electromagnetic Force

22.1. Based on the information given in the problem, we have [1–8]:

$$\vec{v} = 10^6 \hat{x} + 1.5 \times 10^6 \hat{y} \ m/s$$

$$\vec{B} = 0.06 \hat{x} - 0.3 \hat{y} \ T$$

$$q = -1.6 \times 10^{-19} \ C$$

As we know, the magnetic force exerted on a moving charge with the velocity of $\vec{v}$ in an environment with the magnetic field of $\vec{B}$ can be calculated as follows.

$$\vec{F} = q \vec{v} \times \vec{B}$$

Therefore:

$$\vec{F} = \left(-1.6 \times 10^{-19}\right) \left(10^6 \hat{x} + 1.5 \times 10^6 \hat{y}\right) \times (0.06 \hat{x} - 0.3 \hat{y})$$

$$\Rightarrow \vec{F} = \left(-1.6 \times 10^{-19}\right) \left(-39 \times 10^4 \hat{z}\right)$$

$$\Rightarrow \vec{F} = 6.24 \times 10^{-14} \hat{z} \ N$$

Choice (2) is the answer.

Notes

In this problem, the relation below has been used.

$$a \times b = (a_1 \hat{x} + a_2 \hat{y}) \times (b_1 \hat{x} + b_2 \hat{y}) = (a_1 b_2 - a_2 b_1) \hat{z}$$

22.3. As we know, the magnetic force exerted on a moving charge with the velocity of $\vec{v}$ in an environment with the magnetic field of $\vec{B}$ can be calculated as follows.

$$\vec{F} = q\vec{v} \times \vec{B}$$

Moreover, the magnitude of magnetic flux density (T) at the center of a circular arc with angle α, radius r, and current I can be calculated as follows.

$$B = \frac{\alpha}{2\pi} \frac{\mu_0 I}{2r}$$

Also, the direction of magnetic flux density can be determined by using right-hand rule. For this problem, the direction is $\hat{z}$.

In addition, it should be noted that the straight parts of the wire do not generate any magnetic flux density at the origin.

Thus:

$$\vec{F} = q(-v_0 \hat{y}) \times \frac{\pi}{2\pi} \frac{\mu_0 I}{2a} \hat{z}$$

$$\Rightarrow \vec{F} = -\frac{q v_0 \mu_0 I}{4a} \hat{y} \times \hat{z}$$

$$\Rightarrow \vec{F} = -\frac{q v_0 \mu_0 I}{4a} \hat{x}$$

Choice (1) is the answer (Fig. 22.1).

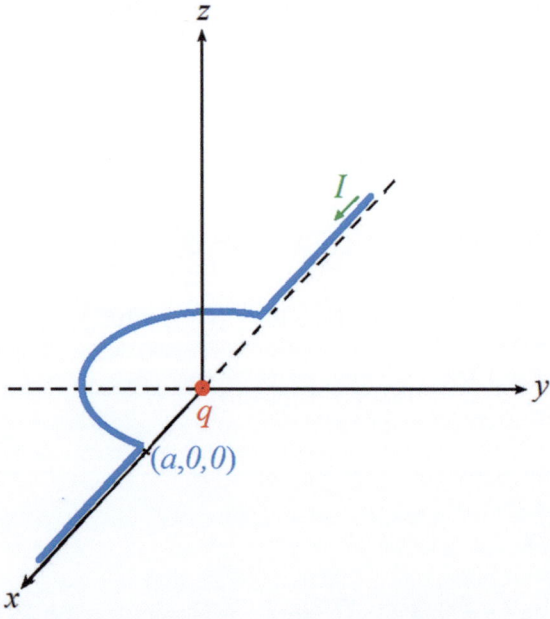

Fig. 22.1 A moving point charge and a wire with the specific shape placed on *xy* plane

22.4. As we know, the magnetic force exerted on a current-carrying wire with the current and length of I and l in an environment with the magnetic field of $\vec{B}$ can be calculated as follows.

$$\vec{F} = I\vec{l} \times \vec{B} = BIl \sin\theta$$

where, θ is the angle between the $\vec{l}$ and $\vec{B}$.

The magnitude of magnetic flux density of the first wire on the place of the second wire can be calculated as follows. As can be seen in Fig. 22.2 (a), it is inward.

$$B_1 = \frac{\mu_0 I_1}{2\pi d}$$

Thus, the magnitude of magnetic force that the first wire exerts on per unit length of the second one can be calculated as follows. Herein, $\theta = 90°$, as can be noticed from Fig. 22.2 (a).

$$F_{12} = \left(\frac{\mu_0 I_1}{2\pi d}\right)(I_2) \sin 90°$$

$$\Rightarrow F_{12} = \frac{\mu_0 I_1 I_2}{2\pi d}$$

The direction of magnetic force that the first wire exerts on the second one is determined based on the right-hand rule shown in Fig. 22.2 (a). As can be seen, it is toward the first wire.

On the other hand, the magnitude of magnetic flux density of the second wire on the place of the first one can be calculated as follows. As can be seen in Fig. 22.2 (b), it is outward.

$$B_2 = \frac{\mu_0 I_2}{2\pi d}$$

Hence, the magnitude of magnetic force that the second wire exerts on per unit length of the first one can be calculated as follows. Herein, $\theta = 90°$, as can be noticed from Fig. 22.2 (b).

$$F_{21} = \left(\frac{\mu_0 I_2}{2\pi d}\right)(I_1) \sin 90°$$

$$\Rightarrow F_{21} = \frac{\mu_0 I_1 I_2}{2\pi d}$$

The direction of magnetic force that the second wire exerts on the first one is determined based on the right-hand rule shown in Fig. 22.2 (b). As can be seen, it is toward the second wire.

As can be noticed from Figs. 22.2 (a) and (b), the wires attract each other. Moreover, we have:

$$F_{12} = F_{21} = \frac{\mu_0 I_1 I_2}{2\pi d}$$

Choice (4) is the answer.

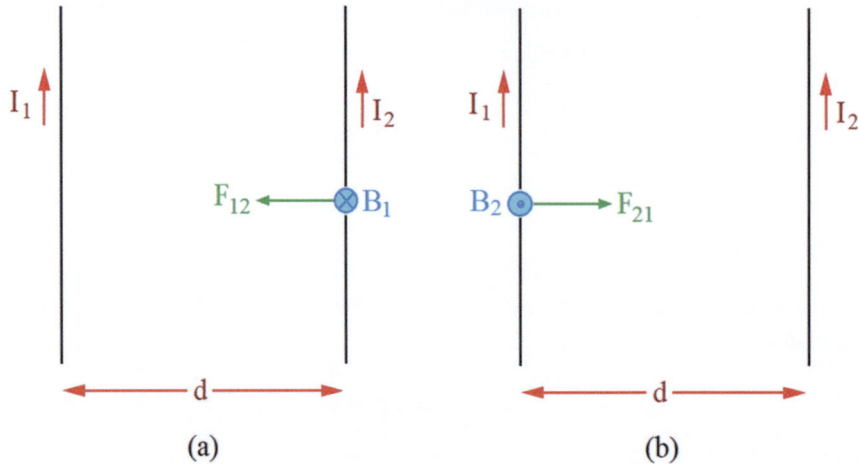

Fig. 22.2 (a) The force that the first wire exerts on the second one. (b) The force that the second wire exerts on the first one

---

22.5. As we know, the magnetic force exerted on a current-carrying wire with the current and length of I and l in an environment with the magnetic field of $\vec{B}$ can be calculated as follows.

$$\vec{F} = I\vec{l} \times \vec{B} = BIl\sin\theta$$

where, θ is the angle between the $\vec{l}$ and $\vec{B}$.

The magnitude of magnetic flux density of the first wire on the place of the second wire can be calculated as follows. As can be seen in Fig. 22.3 (a), it is inward.

$$B_1 = \frac{\mu_0 I_1}{2\pi d}$$

Thus, the magnitude of magnetic force that the first wire exerts on per unit length of the second one can be calculated as follows. Herein, $\theta = 90°$, as can be noticed from Fig. 22.3 (a).

$$F_{12} = \left(\frac{\mu_0 I_1}{2\pi d}\right)(I_2)\sin 90°$$

$$\Rightarrow F_{12} = \frac{\mu_0 I_1 I_2}{2\pi d}$$

The direction of magnetic force that the first wire exerts on the second one is determined based on the right-hand rule shown in Fig. 22.3 (a). As can be seen, it is away from the first wire.

On the other hand, the magnitude of magnetic flux density of the second wire on the place of the first one can be calculated as follows. As can be seen in Fig. 22.3 (b), it is inward.

$$B_2 = \frac{\mu_0 I_2}{2\pi d}$$

Hence, the magnitude of magnetic force that the second wire exerts on per unit length of the first one can be calculated as follows. Herein, $\theta = 90°$, as can be noticed from Fig. 22.3 (b).

$$F_{21} = \left(\frac{\mu_0 I_2}{2\pi d}\right)(I_1) \sin 90°$$

$$\Rightarrow F_{21} = \frac{\mu_0 I_1 I_2}{2\pi d}$$

The direction of magnetic force that the second wire exerts on the first one is determined based on the right-hand rule shown in Fig. 22.3 (b). As can be seen, it is away from the second wire.

As can be noticed from Figs. 22.3 (a) and (b), the wires repel each other. Moreover, we have:

$$F_{12} = F_{21} = \frac{\mu_0 I_1 I_2}{2\pi d}$$

Choice (1) is the answer.

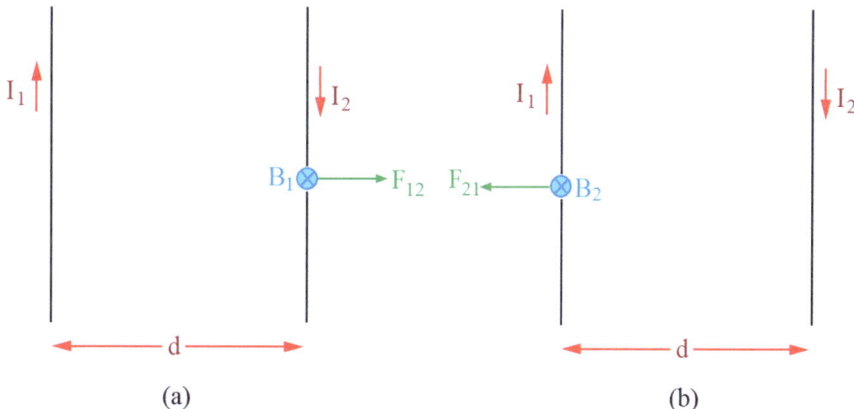

Fig. 22.3 (a) The force that the first wire exerts on the second one. (b) The force that the second wire exerts on the first one

---

22.7. The magnetic force exerted on a current-carrying wire with the current and length of I and l in an environment with the magnetic field of $\vec{B}$ can be calculated as follows.

$$\vec{F} = I\vec{l} \times \vec{B} = BIl \sin \theta$$

Herein, θ is the angle between the $\vec{l}$ and $\vec{B}$ and the direction of force can be determined by using right-hand rule.

Moreover, the magnitude of magnetic flux density of a current-carrying at a distance r can be calculated as follows. In addition, the direction of magnetic flux density can be determined by using right-hand rule.

$$B = \frac{\mu_0 I_1}{2\pi d}$$

The direction of magnetic flux density generated by the straight wire for all four sides of the square is inward.

Due to the symmetry, the force on the first and second sides of the square cancels each other. However, the force on the third and fourth sides of the square can be calculated as follows.

The force on the third side of the square:

$$F_3 = B_1 I_2 l \sin\theta$$

$$\Rightarrow F_3 = \frac{\mu_0 I_1}{2\pi(b-a)} I_2 (2a) \sin 90°$$

$$\Rightarrow F_3 = \frac{\mu_0 I_1 I_2 a}{\pi(b-a)}$$

By using right-hand rule, the direction of force is obtained $\hat{x}$. Thus:

$$\vec{F_3} = \frac{\mu_0 I_1 I_2 a}{\pi(b-a)} \hat{x}$$

Likewise, for the force on the fourth side of the square, we have:

$$F_4 = B_1 I_2 l \sin\theta$$

$$\Rightarrow F_4 = \frac{\mu_0 I_1}{2\pi(b+a)} I_2 (2a) \sin 90°$$

$$\Rightarrow F_4 = \frac{\mu_0 I_1 I_2 a}{\pi(b+a)}$$

By using right-hand rule, the direction of force is obtained $-\hat{x}$. Hence:

$$\vec{F_4} = -\frac{\mu_0 I_1 I_2 a}{\pi(b+a)} \hat{x}$$

The net force exerted on the loop by the wire can be calculated as follows.

$$\vec{F_{net}} = \vec{F_3} + \vec{F_4}$$

$$\Rightarrow \vec{F_{net}} = \frac{\mu_0 I_1 I_2 a}{\pi(b-a)} \hat{x} + \left(-\frac{\mu_0 I_1 I_2 a}{\pi(b+a)} \hat{x}\right)$$

$$\Rightarrow \vec{F_{net}} = \frac{\mu_0 I_1 I_2 a}{\pi} \left(\frac{1}{b-a} - \frac{1}{b+a}\right) \hat{x}$$

$$\Rightarrow \vec{F_{net}} = \frac{\mu_0 I_1 I_2 a}{\pi} \left(\frac{b+a-(b-a)}{b^2-a^2}\right) \hat{x}$$

$$\Rightarrow \vec{F_{net}} = \frac{\mu_0}{\pi} \frac{2a^2 I_1 I_2}{b^2-a^2} \hat{x}$$

Choice (1) is the answer (Fig. 22.4).

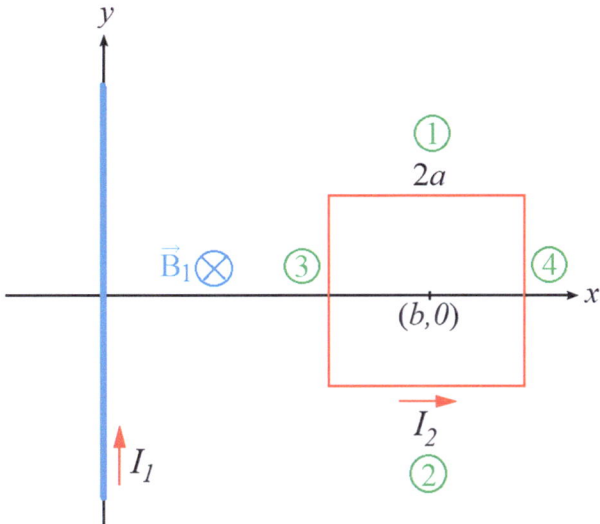

Fig. 22.4 A wire and a square loop placed on xy plane

22.2 Electromagnetic Torque

22.8. Based on the information given in the problem, we have:

$$\vec{I} = I_0 \hat{\varphi}$$

$$\vec{H} = H_0 \hat{\varphi}$$

The torque acting on a ring, called magnetic dipole, can be calculated as follows.

$$\vec{\tau} = \vec{m} \times \vec{B}$$

where, $\vec{B}$ is the magnetic flux density in the area and $\vec{m}$ is the magnetic moment of the magnetic dipole that can be calculated as follows.

$$\vec{m} = SI\hat{n} = \pi a^2 I \hat{n}$$

Herein, a, I, and $\hat{n}$ are the radius, current, and normal vector of the magnetic dipole. The normal vector of a magnetic dipole is determined based on the right-hand rule where if the thumb finger is in the current direction, the other four fingers show the $\hat{n}$.

Hence, for this problem, we have:

$$\vec{m} = \pi a^2 I_0 \hat{z}$$

$$\vec{B} = \mu_0 H_0 \hat{\varphi}$$

$$\Rightarrow \vec{\tau} = \pi a^2 I_0 \hat{z} \times \mu_0 H_0 \hat{\varphi}$$

$$\Rightarrow \vec{\tau} = -\mu_0 H_0 \pi a^2 I_0 \hat{r}$$

Choice (1) is the answer (Fig. 22.5).

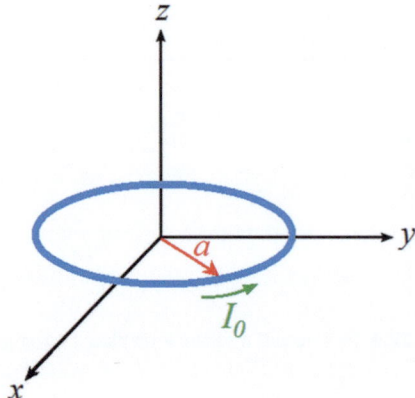

Fig. 22.5 A ring placed in free space on *xy* plane at the origin in presence of a magnetic field

Notes

In this problem, the relations below have been used.

$$\hat{r} \times \hat{\varphi} = \hat{z} \quad \text{or} \quad \hat{\varphi} \times \hat{r} = -\hat{z}$$

$$\hat{\varphi} \times \hat{z} = \hat{r} \quad \text{or} \quad \hat{z} \times \hat{\varphi} = -\hat{r}$$

$$\hat{z} \times \hat{r} = \hat{\varphi} \quad \text{or} \quad \hat{r} \times \hat{z} = -\hat{\varphi}$$

References

1. Rahmani-Andebili, M., General Physics II – Practice Problems, Methods, and Solutions, Springer Nature, 2025.
2. Rahmani-Andebili, M., General Physics I – Practice Problems, Methods, and Solutions, Springer Nature, 2025.
3. Rahmani-Andebili, M., Mathematics of Engineering and Science – Practice Problems, Methods, and Solutions, Springer Nature, 2024.
4. Rahmani-Andebili, M., Differential Equations – Practice Problems, Methods, and Solutions, Springer Nature, 2022.
5. Rahmani-Andebili, M., Calculus III – Practice Problems, Methods, and Solutions, Springer Nature, 2023.
6. Rahmani-Andebili, M., Calculus II – Practice Problems, Methods, and Solutions, Springer Nature, 2023.
7. Rahmani-Andebili, M., Calculus I (2nd Ed.) – Practice Problems, Methods, and Solutions, Springer Nature, 2023.
8. Rahmani-Andebili, M., Precalculus (2nd Ed.) – Practice Problems, Methods, and Solutions, Springer Nature, 2024.

23. Ampere's Circuital Law and Magnetic Energy: Part A

Abstract

In this chapter, the basic and advanced problems of magnetic energy and Ampere's circuital law for uniform and nonuniform current densities are studied. Herein, different types of problems and exercises are presented that are categorized as follows.

- *Problems with detailed solution*: They have been designed to teach students the subjects in detail. Moreover, they have been categorized in different levels based on their difficulty levels (easy, normal, and hard) and calculation amounts (small, normal, and large).
- *Partially solved exercises*: They have been designed to encourage students to practice problems while guiding them through the problem-solving procedure and hinting the required formulas.
- *Exercises with final answer*: They have been designed to encourage students to practice more by themselves while hinting them by the final answer as well as to help instructors to give tests or quizzes.

23.1 Ampere's Circuital Law

Problem

23.1. A long conducting cylinder with radius a and length l carries a uniform current I. Calculate the magnetic field intensity in the region $0 < r < a$ [1–8] (Fig. 23.1).

Difficulty level   ○ Easy   ● Normal   ○ Hard
Calculation amount   ○ Small   ● Normal   ○ Large

1) $\vec{H} = \dfrac{Ia^2 r}{2\pi} \hat{\varphi}$
2) $\vec{H} = \dfrac{Ir}{2\pi a} \hat{\varphi}$
3) $\vec{H} = \dfrac{Ir}{2\pi a^2} \hat{\varphi}$
4) $\vec{H} = \dfrac{I}{2\pi a} \hat{\varphi}$

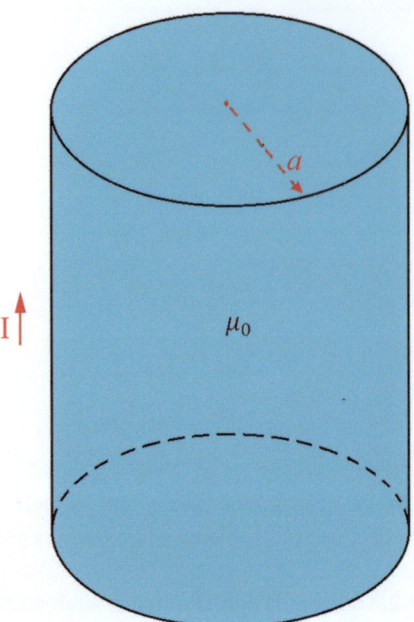

Fig. 23.1 A long conducting cylinder carrying a uniform current

Problem

23.2. In Problem 23.1, calculate the magnetic field intensity in the region $r > a$.

Difficulty level   ○ Easy   ● Normal   ○ Hard
Calculation amount   ○ Small   ● Normal   ○ Large

1) $\vec{H} = \dfrac{I}{2\pi a}\hat{\varphi}$
2) $\vec{H} = \dfrac{Ia}{2\pi r}\hat{\varphi}$
3) $\vec{H} = \dfrac{Ir}{2\pi}\hat{\varphi}$
4) $\vec{H} = \dfrac{I}{2\pi r}\hat{\varphi}$

Problem

23.3. As is illustrated in Fig. 23.2, a uniform current I is passing through the long hollow cylinder made by nonmagnetic material. Calculate the magnitude of magnetic flux density at $r = 1.5a$.

Difficulty level   ○ Easy   ○ Normal   ○ Hard
Calculation amount   ○ Small   ○ Normal   ○ Large

1) $B = \dfrac{\mu_0 I}{3\pi r}$
2) $B = \dfrac{\mu_0 I}{4\pi r}$
3) $B = \dfrac{5\mu_0 I}{18\pi r}$
4) $B = \dfrac{5\mu_0 I}{36\pi r}$

23.1 Ampere's Circuital Law

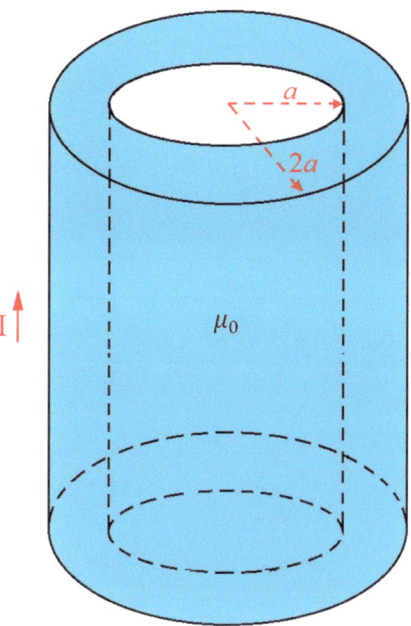

Fig. 23.2 A long conducting hollow cylinder carrying a current

Problem

23.4. Figure 23.3 shows a long straight wire that carries an outward current $I_0 = 2\,A$ and a long hollow cylinder with radiuses $a = 3\pi$ and $b = 4\pi$. Calculate the value of the integral below on the given contour (dashed line). Herein, assume that $\mu_r = 10$.

$$\alpha = \oint \vec{B} \cdot d\vec{l}$$

Difficulty level ○ Easy ○ Normal ● Hard
Calculation amount ○ Small ● Normal ○ Large
1) $18\mu_0$
2) $-18\mu_0$
3) $22\mu_0$
4) $4.5\mu_0$

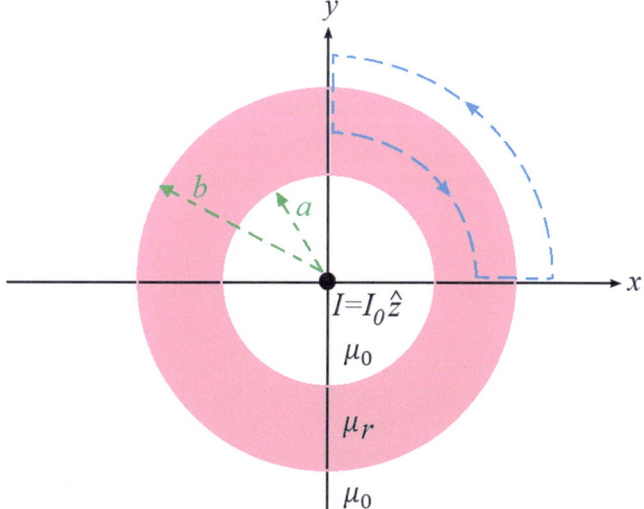

Fig. 23.3 A long straight wire carrying an outward current and a long hollow cylinder

Problem

23.5. As is shown in Fig. 23.4, the nonuniform volume current density inside a long solid cylinder with radius a is as follows.

$$\vec{J} = J_0 \frac{r}{a} \hat{z}$$

Calculate the magnetic flux density inside the cylinder.

Difficulty level   ○ Easy   ○ Normal   ● Hard
Calculation amount   ○ Small   ○ Normal   ● Large

1) $\vec{B} = \frac{\mu_0 J_0 r^2}{2a^2} \hat{\varphi}$
2) $\vec{B} = \frac{\mu_0 J_0 r^2}{3a} \hat{\varphi}$
3) $\vec{B} = \frac{3\mu_0 J_0 r}{a^2} \hat{\varphi}$
4) $\vec{B} = \frac{3\mu_0 J_0 r^2}{a} \hat{\varphi}$

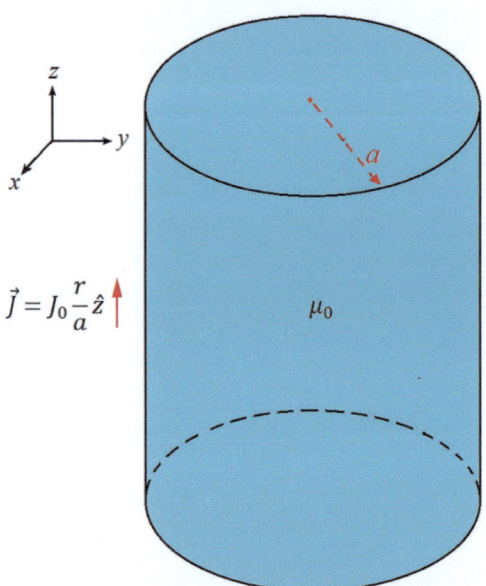

Fig. 23.4 A long cylinder carrying a nonuniform current

23.2 Magnetic Energy and Magnetic Energy Density

Problem

23.6. A long conducting cylinder with radius a and length l carries a current I. Calculate the value of the following term.

$$\frac{U(0 < r < a)}{U(r = a)}$$

Herein, $U(0 < r < a)$ and $U(r = a)$ are the magnetic energies stored in the cylinder in the regions $0 < r < a$ and $r = a$, respectively (Fig. 23.5).

Difficulty level   ○ Easy   ○ Normal   ● Hard
Calculation amount   ○ Small   ○ Normal   ● Large

1) 0
2) 1
3) $\left(\dfrac{r}{a}\right)^3$
4) $\left(\dfrac{r}{a}\right)^4$

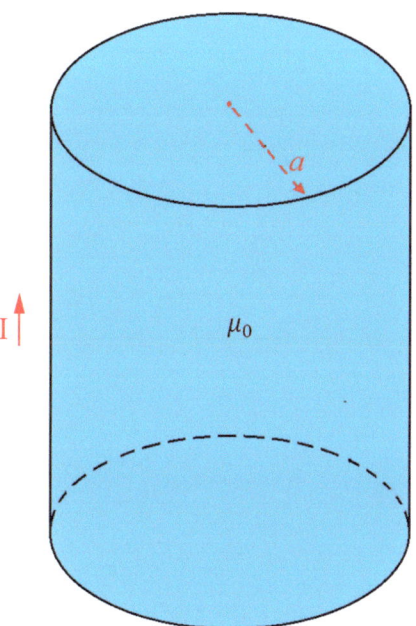

Fig. 23.5 A long conducting cylinder carrying a uniform current

Problem

23.7. Calculate the magnitude of an electric field so that its energy density is equal to the energy density generated by 1 T magnetic field. Herein, assume that $\varepsilon_0 = \dfrac{1}{36\pi} \times 10^{-9}$ F/m and $\mu_0 = 4\pi \times 10^{-7}$ H/m.

Difficulty level   ○ Easy   ○ Normal   ○ Hard
Calculation amount   ○ Small   ○ Normal   ○ Large

1) 1.5×10^8 V/m
2) 1.5×10^{-8} V/m
3) 3×10^{-8} V/m
4) 3×10^8 V/m

References

1. Rahmani-Andebili, M., General Physics II – Practice Problems, Methods, and Solutions, Springer Nature, 2025.
2. Rahmani-Andebili, M., General Physics I – Practice Problems, Methods, and Solutions, Springer Nature, 2025.
3. Rahmani-Andebili, M., Mathematics of Engineering and Science – Practice Problems, Methods, and Solutions, Springer Nature, 2024.
4. Rahmani-Andebili, M., Differential Equations – Practice Problems, Methods, and Solutions, Springer Nature, 2022.
5. Rahmani-Andebili, M., Calculus III – Practice Problems, Methods, and Solutions, Springer Nature, 2023.
6. Rahmani-Andebili, M., Calculus II – Practice Problems, Methods, and Solutions, Springer Nature, 2023.
7. Rahmani-Andebili, M., Calculus I (2nd Ed.) – Practice Problems, Methods, and Solutions, Springer Nature, 2023.
8. Rahmani-Andebili, M., Precalculus (2nd Ed.) – Practice Problems, Methods, and Solutions, Springer Nature, 2024.

Ampere's Circuital Law and Magnetic Energy: Part B

Abstract
In this chapter, the problems of the twenty third chapter are fully solved, in detail, step-by-step, and with different methods.

24.1 Ampere's Circuital Law

24.1. To calculate the magnetic field intensity in the cylinder, we need to use Ampere's circuital law as follows [1–8].

$$I_{in} = \oint \frac{\vec{B}}{\mu_0} \cdot \vec{dl}$$

For the region $0 < r < a$, we have:

$$I_{in} = \oint \vec{H} \cdot \vec{dl}$$

$$\Rightarrow I_{(0<r<a)} = \int_{\varphi=0}^{\varphi=2\pi} H_\varphi \widehat{\varphi} \cdot r d\varphi \widehat{\varphi}$$

$$\Rightarrow \left(\frac{r}{a}\right)^2 I = H_\varphi r \int_{\varphi=0}^{\varphi=2\pi} d\varphi$$

$$\Rightarrow \left(\frac{r}{a}\right)^2 I = H_\varphi r (2\pi - 0)$$

$$\Rightarrow H_\varphi = \frac{Ir}{2\pi a^2}$$

$$\Rightarrow \vec{H} = \frac{Ir}{2\pi a^2} \widehat{\varphi}$$

Choice (3) is the answer (Fig. 24.1).

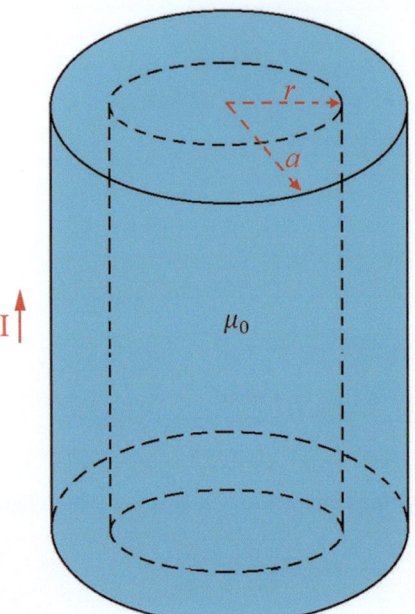

Fig. 24.1 Applying Ampere's circuital law on the long conducting cylinder carrying a uniform current

24.2. Applying Ampere's circuital law for the region $r > a$:

$$\Rightarrow I_{in} = \oint \frac{\vec{B}}{\mu_0} \cdot \vec{dl}$$

$$\Rightarrow I_{in} = \oint \vec{H} \cdot \vec{dl}$$

$$\Rightarrow I = \int_{\varphi=0}^{\varphi=2\pi} H_\varphi \widehat{\varphi} \cdot r d\varphi \widehat{\varphi}$$

$$\Rightarrow I = H_\varphi r \int_{\varphi=0}^{\varphi=2\pi} d\varphi$$

$$\Rightarrow I = H_\varphi r (2\pi - 0)$$

$$\Rightarrow H_\varphi = \frac{I}{2\pi r}$$

$$\Rightarrow \vec{H} = \frac{I}{2\pi r} \widehat{\varphi}$$

Choice (4) is the answer (Fig. 24.2).

24.1 Ampere's Circuital Law

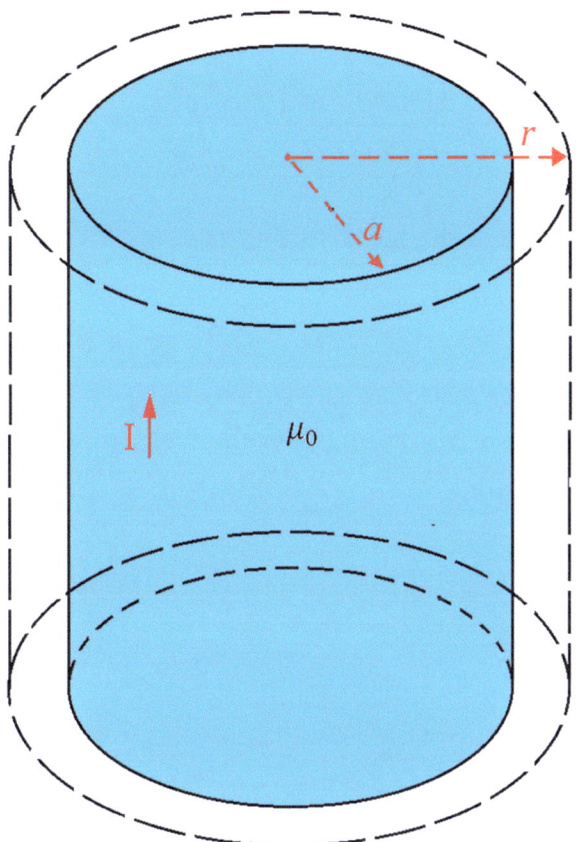

Fig. 24.2 Applying Ampere's circuital law on the long conducting cylinder carrying a current

24.3. Based on the information given in the problem, we have:

$$r = 1.5a$$

Ampere's circuital law presents a relation between the current and magnetic field intensity created by it as follows.

$$I_{in} = \oint \frac{\vec{B}}{\mu_0} \cdot \vec{dl}$$

Hence, for this problem, we can write as follows.

$$\mu_0 I_r = \oint \vec{B} \cdot \vec{dl}$$

$$\Rightarrow \mu_0 \frac{\pi(r^2 - a^2)}{\pi\left((2a)^2 - a^2\right)} I = \int_{\varphi=0}^{\varphi=2\pi} B_\varphi \widehat{\varphi} \cdot r d\varphi \widehat{\varphi}$$

For $r = 1.5a$, we have:

$$\Rightarrow \mu_0 \frac{\pi\left((1.5a)^2 - a^2\right)}{\pi\left((2a)^2 - a^2\right)} I = \int_{\varphi=0}^{\varphi=2\pi} 1.5 B_\varphi d\varphi$$

$$\Rightarrow \mu_0 \frac{1.25}{3} I = 1.5 B_\varphi (2\pi - 0)$$

$$\Rightarrow B_\varphi = \frac{1.25 \mu_0 I}{9\pi r}$$

$$\vec{B} = \frac{5\mu_0 I}{36\pi r} \hat{\varphi}$$

$$\Rightarrow B = \frac{5\mu_0 I}{36\pi r}$$

Choice (4) is the answer (Fig. 24.3).

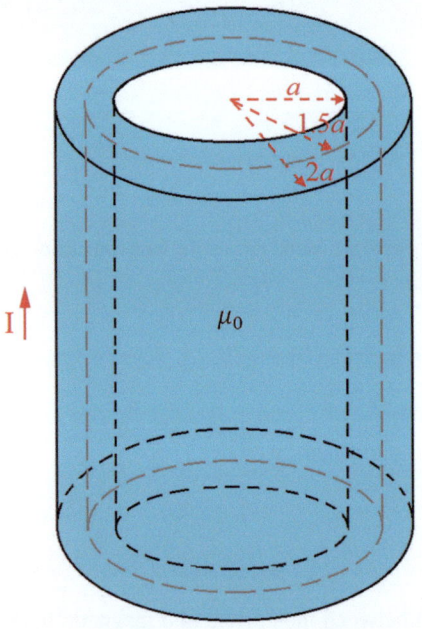

Fig. 24.3 Applying Ampere's circuital law on the long hollow cylinder carrying a current

---

24.4. Based on the information given in the problem, we have:

$$a = 3\pi$$

$$b = 4\pi$$

$$I_0 = 2 \text{ A}$$

$$\mu_r = 10$$

24.1 Ampere's Circuital Law

As we know, the magnetic flux density at a distance r from a long straight wire can be calculated as follows.

$$B = \frac{\mu_0 \mu_r I}{2\pi r}$$

Hence, according to Fig. 24.4, the magnetic flux densities in the magnetic material and free space are as follows.

$$\vec{B_1} = \frac{\mu_0 \times 10 \times 2}{2\pi r}\hat{\varphi} = \frac{10\mu_0}{\pi r}\hat{\varphi}$$

$$\vec{B_3} = \frac{\mu_0 \times 2}{2\pi r}\hat{\varphi} = \frac{\mu_0}{\pi r}\hat{\varphi}$$

Now, the integral on the given contour can be calculated as follows.

$$\alpha = \oint \vec{B}.\vec{dl}$$

$$\alpha = \oint \vec{B_1}.\vec{dl_1} + \oint \vec{B_2}.\vec{dl_2} + \oint \vec{B_3}.\vec{dl_3} + \oint \vec{B_4}.\vec{dl_4}$$

$$\Rightarrow \alpha = \int_0^{\frac{\pi}{2}} \frac{10\mu_0}{\pi r}\hat{\varphi}.(-rd\varphi\hat{\varphi}) + 0 + \int_0^{\frac{\pi}{2}} \frac{\mu_0}{\pi r}\hat{\varphi}.(rd\varphi\hat{\varphi}) + 0$$

$$\Rightarrow \alpha = -\frac{10\mu_0}{\pi}\int_0^{\frac{\pi}{2}} d\varphi + \frac{\mu_0}{\pi}\int_0^{\frac{\pi}{2}} d\varphi$$

$$\Rightarrow \alpha = -\frac{10\mu_0}{\pi}\left(\frac{\pi}{2} - 0\right) + \frac{\mu_0}{\pi}\left(\frac{\pi}{2} - 0\right)$$

$$\Rightarrow \alpha = -4.5\mu_0$$

Choice (4) is the answer.

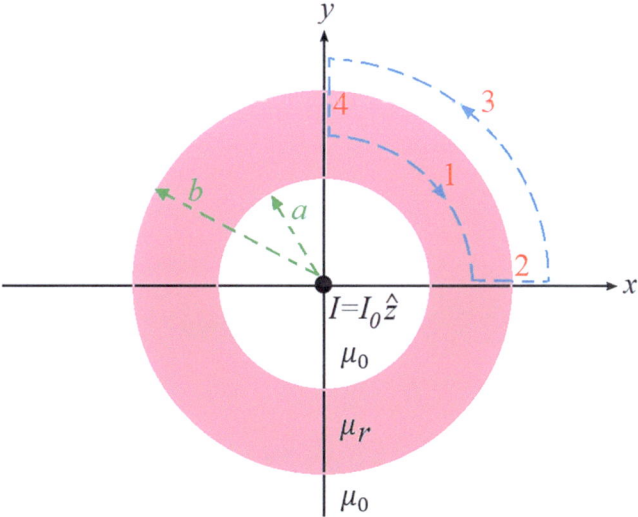

Fig. 24.4 A long straight wire carrying an outward current and a long hollow cylinder

24.5. Based on the information given in the problem, we have:

$$\vec{J} = J_0 \frac{r}{a} \hat{z}$$

As we know, the relation between the current and volume current density is as follows.

$$I = \int \vec{J} \cdot \vec{dS} \qquad (24.1)$$

In addition, Ampere's circuital law presents a relation between the current and magnetic field intensity created by it as follows.

$$I_{in} = \oint \frac{\vec{B}}{\mu_0} \cdot \vec{dl} \qquad (24.2)$$

By applying (24.1) for $r < a$, we have:

$$I = \int_{\varphi=0}^{\varphi=2\pi} \int_{r=0}^{r=r} J_0 \frac{r}{a} \hat{z} \cdot r dr d\varphi \hat{z}$$

$$\Rightarrow I = \frac{J_0}{a} \int_{\varphi=0}^{\varphi=2\pi} \int_{r=0}^{r=r} r^2 dr d\varphi$$

$$\Rightarrow I = \frac{J_0}{a} \left[\frac{r^3}{3}\right]_0^r \left[\varphi\right]_0^{2\pi}$$

$$\Rightarrow I = \frac{J_0}{a} \left(\frac{r^3}{3} - 0\right)(2\pi - 0)$$

$$\Rightarrow I = \frac{2\pi J_0}{3a} r^3 \qquad (24.3)$$

By applying (24.2) for $r < a$, we have:

$$\mu_0 I = \oint \vec{B} \cdot \vec{dl}$$

$$\Rightarrow \mu_0 I = \int_{\varphi=0}^{\varphi=2\pi} B_\varphi \hat{\varphi} \cdot r d\varphi \hat{\varphi}$$

$$\Rightarrow \mu_0 I = B_\varphi r \int_{\varphi=0}^{\varphi=2\pi} d\varphi$$

$$\Rightarrow \mu_0 I = B_\varphi r (2\pi - 0)$$

24.1 Ampère's Circuital Law

$$\Rightarrow B_\varphi = \frac{\mu_0 I}{2\pi r}$$

$$\Rightarrow \vec{B} = \frac{\mu_0 I}{2\pi r}\hat{\varphi} \qquad (24.4)$$

Solving (24.3) and (24.4):

$$\vec{B} = \frac{\mu_0}{2\pi r} \times \frac{2\pi J_0}{3a} r^3 \hat{\varphi}$$

$$\vec{B} = \frac{\mu_0 J_0 r^2}{3a}\hat{\varphi}$$

Choice (2) is the answer (Fig. 24.5).

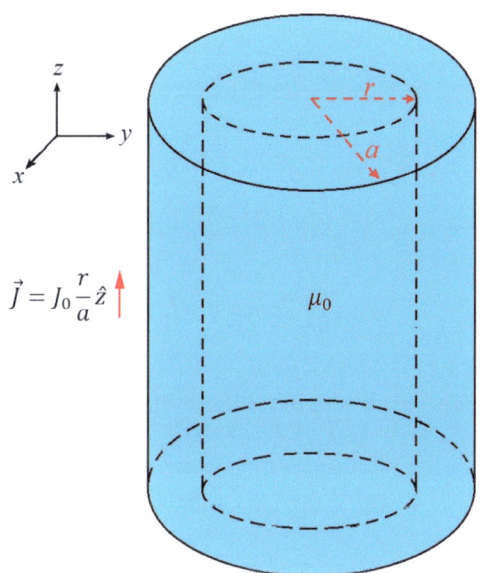

Fig. 24.5 Applying Ampere's circuital law on the long cylinder carrying a nonuniform current

> **Notes**
> In this problem, the relation below has been used.
>
> $$\int x^n dx = \frac{x^{n+1}}{n+1}$$

24.2 Magnetic Energy and Magnetic Energy Density

24.6. As we know, magnetic energy in an area can be calculated as follows.

$$U = \frac{1}{2}\int \mu_0 H^2 dV = \frac{1}{2}\int \frac{B^2}{\mu_0} dV = \frac{1}{2}\int \mu_0 BH dV$$

As can be noticed, we need to calculate the magnetic field intensity in the cylinder by using Ampere's circuital law as follows.

$$I_{in} = \oint \frac{\vec{B}}{\mu_0} \cdot \vec{dl}$$

For the region $0 < r < a$, we have:

$$I_{in} = \oint \vec{H} \cdot \vec{dl}$$

$$\Rightarrow I_{(0<r<a)} = \int_{\varphi=0}^{\varphi=2\pi} H_\varphi \widehat{\varphi} \cdot r d\varphi \widehat{\varphi}$$

$$\Rightarrow \left(\frac{r}{a}\right)^2 I = H_\varphi r \int_{\varphi=0}^{\varphi=2\pi} d\varphi$$

$$\Rightarrow \left(\frac{r}{a}\right)^2 I = H_\varphi r (2\pi - 0)$$

$$\Rightarrow H_\varphi = \frac{Ir}{2\pi a^2}$$

$$\Rightarrow \vec{H} = \frac{Ir}{2\pi a^2} \widehat{\varphi}$$

Now, the magnetic energy for the region $0 < r < a$ can be calculated as follows.

$$U = \frac{1}{2}\int \mu_0 H^2 dV$$

$$\Rightarrow U(0<r<a) = \frac{1}{2}\mu_0 \int_0^l \int_0^{2\pi} \int_0^r \left(\frac{Ir}{2\pi a^2}\right)^2 r dr d\varphi dz$$

$$\Rightarrow U(0<r<a) = \frac{1}{2}\mu_0 \left(\frac{I}{2\pi a^2}\right)^2 \int_0^l \int_0^{2\pi} \int_0^r r^3 dr d\varphi dz$$

$$\Rightarrow U(0<r<a) = \frac{1}{2}\mu_0 \left(\frac{I}{2\pi a^2}\right)^2 \left[\frac{r^4}{4}\right]_0^r \left[\varphi\right]_0^{2\pi} \left[z\right]_0^l$$

$$\Rightarrow U(0<r<a) = \frac{1}{2}\mu_0 \left(\frac{I}{2\pi a^2}\right)^2 \left(\frac{r^4}{4} - 0\right)(2\pi - 0)(l - 0)$$

24.2 Magnetic Energy and Magnetic Energy Density

$$\Rightarrow U(0<r<a) = \mu_0 \frac{I^2 l}{16\pi} \left(\frac{r}{a}\right)^4$$

The magnetic energy for the region $r = a$ can be calculated by using $r = a$ in the previous equation as follows.

$$U(r=a) = \mu_0 \frac{I^2 l}{16\pi}$$

Therefore:

$$\frac{U(0<r<a)}{U(r=a)} = \left(\frac{r}{a}\right)^4$$

Choice (4) is the answer (Fig. 24.6).

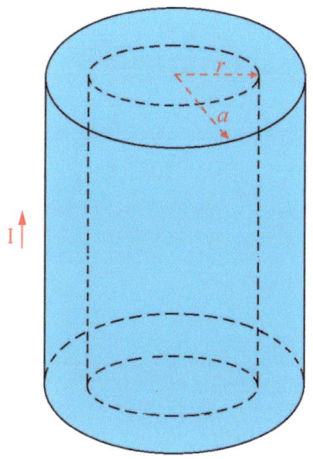

Fig. 24.6 Applying Ampere's circuital law on the long conducting cylinder carrying a uniform current

---

Notes

In this problem, the relation below has been used.

$$\int x^n dx = \frac{x^{n+1}}{n+1}$$

24.7. Based on the information given in the problem, we have:

$$u_B = u_E$$

$$B = 1\ T$$

As we know, the magnetic energy density and electric energy density are calculated as follows.

$$u_B = \frac{1}{2}\frac{B^2}{\mu_0}$$

$$u_E = \frac{1}{2}\varepsilon_0 E^2$$

Therefore:

$$\frac{1}{2}\frac{(1)^2}{\mu_0} = \frac{1}{2}\varepsilon_0 E^2$$

$$\Rightarrow E^2 = \frac{1}{\mu_0 \varepsilon_0}$$

$$\Rightarrow E = \frac{1}{\sqrt{\mu_0 \varepsilon_0}}$$

$$\Rightarrow E = \frac{1}{\sqrt{4\pi \times 10^{-7} \times \frac{1}{36\pi} \times 10^{-9}}}$$

$$\Rightarrow E = 3 \times 10^8 \; V/m$$

Choice (4) is the answer.

> **Notes**
>
> As can be noticed, for an equal energy density, a much stronger electric field is needed. Moreover, it should be mentioned the value of $\frac{1}{\sqrt{\mu_0 \varepsilon_0}}$ is equal to the speed of light. In other words:
>
> $$c = \frac{1}{\sqrt{\mu_0 \varepsilon_0}} = 3 \times 10^8$$

References

1. Rahmani-Andebili, M., General Physics II – Practice Problems, Methods, and Solutions, Springer Nature, 2025.
2. Rahmani-Andebili, M., General Physics I – Practice Problems, Methods, and Solutions, Springer Nature, 2025.
3. Rahmani-Andebili, M., Mathematics of Engineering and Science – Practice Problems, Methods, and Solutions, Springer Nature, 2024.
4. Rahmani-Andebili, M., Differential Equations – Practice Problems, Methods, and Solutions, Springer Nature, 2022.
5. Rahmani-Andebili, M., Calculus III – Practice Problems, Methods, and Solutions, Springer Nature, 2023.
6. Rahmani-Andebili, M., Calculus II – Practice Problems, Methods, and Solutions, Springer Nature, 2023.
7. Rahmani-Andebili, M., Calculus I (2nd Ed.) – Practice Problems, Methods, and Solutions, Springer Nature, 2023.
8. Rahmani-Andebili, M., Precalculus (2nd Ed.) – Practice Problems, Methods, and Solutions, Springer Nature, 2024.

Magnetic Vector Potential: Part A

Abstract

In this chapter, the basic and advanced problems of magnetic vector potential are studied. Herein, different types of problems and exercises are presented that are categorized as follows.

- **Problems with detailed solution**: They have been designed to teach students the subjects in detail. Moreover, they have been categorized in different levels based on their difficulty levels (easy, normal, and hard) and calculation amounts (small, normal, and large).
- **Partially solved exercises**: They have been designed to encourage students to practice problems while guiding them through the problem-solving procedure and hinting the required formulas.
- **Exercises with final answer**: They have been designed to encourage students to practice more by themselves while hinting them by the final answer as well as to help instructors to give tests or quizzes.

Problem

25.1. Which one of the following choices correctly presents the relation between magnetic vector potential ($\vec{A}$) and magnetic flux (φ) [1–8].

Difficulty level   ○ Easy   ● Normal   ○ Hard
Calculation amount   ● Small   ○ Normal   ○ Large

1) $\varphi = \oint \vec{A} \cdot d\vec{S}$
2) $\varphi = \int \frac{\mu_0 A}{4\pi R} dV$
3) $\varphi = \oint \nabla \cdot \vec{A} \cdot d\vec{S}$
4) $\varphi = \oint \vec{A} \cdot d\vec{l}$

Problem

25.2. The magnetic vector potential in one part of a space is $\vec{A} = \frac{\varphi_0}{2\pi r}\hat{r}$ in cylindrical coordinates system. Calculate the magnetic flux density in the region.

Difficulty level   ○ Easy   ● Normal   ○ Hard
Calculation amount   ○ Small   ● Normal   ○ Large

1) 0

2) $\dfrac{\varphi_0}{2\pi r^2}\hat{z}$

3) $-\dfrac{\varphi_0}{2\pi r^3 \sin\varphi}\hat{z}$

4) $-\dfrac{\varphi_0}{2\pi r^2}\hat{r}$

Partially Solved Exercise

25.3. The magnetic vector potential in one part of a space is as follows in spherical coordinates system. Calculate the magnetic flux density in the region.

$$\vec{A} = \frac{1}{R^2}\cos\theta \hat{R}$$

Solution

The relation between magnetic flux density and magnetic vector potential is as follows.

$$\vec{B} = \nabla \times \vec{A}$$

Moreover, the application of curl operator on a vector ($\vec{A}$) in spherical coordinates system is as follows.

$$\nabla \times \vec{A} = \frac{1}{R^2 \sin\theta}\begin{vmatrix} \hat{R} & R\hat{\theta} & R\sin\theta\hat{\varphi} \\ \dfrac{\partial}{\partial R} & \dfrac{\partial}{\partial \theta} & \dfrac{\partial}{\partial \varphi} \\ A_R & RA_\varphi & R\sin\theta A_\varphi \end{vmatrix}$$

$$= \frac{1}{R\sin\theta}\left(\frac{\partial(\sin\theta A_\varphi)}{\partial \theta} - \frac{\partial A_\theta}{\partial \varphi}\right)\hat{R} + \frac{1}{R}\left(\frac{1}{\sin\theta}\frac{\partial A_R}{\partial \varphi} - \frac{\partial(RA_\varphi)}{\partial R}\right)\hat{\theta} + \frac{1}{R}\left(\frac{\partial(RA_\theta)}{\partial R} - \frac{\partial A_R}{\partial \theta}\right)\hat{\varphi}$$

Therefore:

$$\vec{B} = \nabla \times \frac{1}{R^2}\cos\theta \hat{R}$$

$$\vec{B} = \frac{1}{R\sin\theta}\left(\frac{\partial(\quad)}{\partial \theta} - \frac{\partial(\quad)}{\partial \varphi}\right)\hat{R} + \frac{1}{R}\left(\frac{1}{\sin\theta}\frac{\partial}{\partial \varphi}(\quad) - \frac{\partial(\quad)}{\partial R}\right)\hat{\theta} + \frac{1}{R}\left(\frac{\partial(\quad)}{\partial R} - \frac{\partial}{\partial \theta}(\quad)\right)\hat{\varphi}$$

Hence:

$$\Rightarrow \vec{B} = (\quad)\hat{R} + (\quad)\hat{\theta} + (\quad)(\quad)\hat{\varphi}$$

$$\Rightarrow \vec{B} = \frac{1}{R^3}\sin\theta\hat{\varphi}$$

Problem

25.4. As is illustrated in Fig. 25.1, in a free space, there is a long cylindrical solenoid with radius a, current I, and number of turns per unit length n which is aligned with z-axis. Calculate the magnetic vector potential inside the solenoid at $r = 0.5a$.

Difficulty level   ○ Easy   ○ Normal   ● Hard
Calculation amount   ○ Small   ● Normal   ○ Large

1) $\mu_0 n I a \widehat{\varphi}$
2) $\frac{1}{4}\mu_0 n I a^2 \widehat{\varphi}$
3) $\frac{1}{4}\mu_0 n I a \widehat{\varphi}$
4) $\frac{1}{2}\mu_0 n I a \widehat{\varphi}$

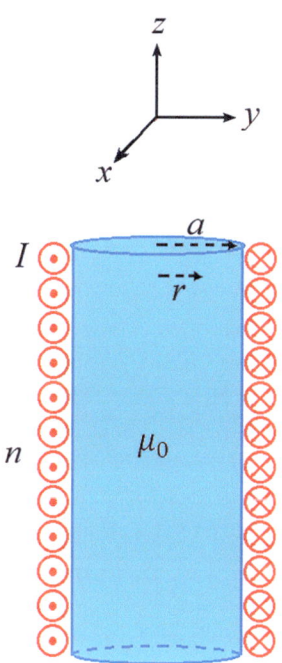

Fig. 25.1 A long cylindrical solenoid placed in free space and aligned with z-axis

Magnetic Vector Potential: Part B

26

Abstract

In this chapter, the problems of the twenty fifth chapter are fully solved, in detail, step-by-step, and with different methods.

26.1. As we know, the relation between magnetic flux and magnetic flux density is as follows [1–8].

$$\varphi = \int \vec{B} \cdot \vec{dS} \tag{26.1}$$

Moreover, the relation between magnetic flux density and magnetic vector potential is as follows.

$$\vec{B} = \nabla \times \vec{A} \tag{26.2}$$

In addition, the Stokes's theorem is as follows.

$$\int \left(\nabla \times \vec{F} \right) \cdot \vec{dS} = \oint \vec{F} \cdot \vec{dl} \tag{26.3}$$

Solving (26.1) and (26.2):

$$\varphi = \int \left(\nabla \times \vec{A} \right) \cdot \vec{dS} \tag{26.4}$$

By applying the Stokes's theorem in (26.4), we have:

$$\varphi = \oint \vec{A} \cdot \vec{dl}$$

Choice (4) is the answer.

26.2. The relation between magnetic flux density and magnetic vector potential is as follows.

$$\vec{B} = \nabla \times \vec{A}$$

Moreover, the application of curl operator on a vector ($\vec{A}$) in cylindrical coordinates system is as follows.

$$\nabla \times \vec{A} = \frac{1}{r}\begin{vmatrix} \hat{r} & r\hat{\varphi} & \hat{z} \\ \frac{\partial}{\partial r} & \frac{\partial}{\partial \varphi} & \frac{\partial}{\partial z} \\ A_r & rA_\varphi & A_z \end{vmatrix} = \left(\frac{1}{r}\frac{\partial A_z}{\partial \varphi} - \frac{\partial A_\varphi}{\partial z}\right)\hat{r} + \left(\frac{\partial A_r}{\partial z} - \frac{\partial A_z}{\partial r}\right)\hat{\varphi} + \frac{1}{r}\left(\frac{\partial (rA_\varphi)}{\partial r} - \frac{\partial A_r}{\partial \varphi}\right)\hat{z}$$

Therefore:

$$\vec{B} = \nabla \times \frac{\varphi_0}{2\pi r}\hat{r}$$

$$\Rightarrow \vec{B} = \left(\frac{1}{r}\frac{\partial}{\partial \varphi}(0) - \frac{\partial}{\partial z}(0)\right)\hat{r} + \left(\frac{\partial}{\partial z}\left(\frac{\varphi_0}{2\pi r}\right) - \frac{\partial}{\partial r}(0)\right)\hat{\varphi} + \frac{1}{r}\left(\frac{\partial}{\partial r}(r \times 0) - \frac{\partial}{\partial \varphi}\left(\frac{\varphi_0}{2\pi r}\right)\right)\hat{z}$$

$$\Rightarrow \vec{B} = 0\hat{r} + 0\hat{\varphi} + 0\hat{z}$$

$$\Rightarrow \vec{B} = 0$$

Choice (1) is the answer.

26.4. As we know, magnetic flux density inside a cylindrical solenoid is uniform and as follows.

$$\vec{B} = \mu n I \hat{n}$$

Herein, μ is the magnetic permeability of the material placed inside the solenoid, I is the current, n is the number of turns per unit length, and $\hat{n}$ is the direction of magnetic flux density. In this regard, based on the right-hand rule, the thumb finger shows $\hat{n}$ if the other four fingers are aligned with the current direction. For this problem, $\hat{n} = \hat{z}$.

On the other hand, we know that the relation between magnetic flux density and magnetic vector potential is as follows.

$$\vec{B} = \nabla \times \vec{A}$$

Therefore:

$$\mu_0 n I \hat{z} = \frac{1}{r}\begin{vmatrix} \hat{r} & r\hat{\varphi} & \hat{z} \\ \frac{\partial}{\partial r} & \frac{\partial}{\partial \varphi} & \frac{\partial}{\partial z} \\ A_r & rA_\varphi & A_z \end{vmatrix}$$

$$\Rightarrow \mu_0 n I \hat{z} = \left(\frac{1}{r}\frac{\partial A_z}{\partial \varphi} - \frac{\partial A_\varphi}{\partial z}\right)\hat{r} + \left(\frac{\partial A_r}{\partial z} - \frac{\partial A_z}{\partial r}\right)\hat{\varphi} + \frac{1}{r}\left(\frac{\partial (rA_\varphi)}{\partial r} - \frac{\partial A_r}{\partial \varphi}\right)\hat{z}$$

Herein, since the current of solenoid is uniform and in $\hat{\varphi}$ direction, $A_r = A_z = 0$ and $A_\varphi =$ Constant are assumed. Therefore:

$$\mu_0 n I \hat{z} = \left(\frac{1}{r}\frac{\partial}{\partial \varphi}(0) - \frac{\partial}{\partial z}(A_\varphi)\right)\hat{r} + \left(\frac{\partial}{\partial z}(0) - \frac{\partial}{\partial r}(0)\right)\hat{\varphi} + \frac{1}{r}\left(\frac{\partial}{\partial r}(rA_\varphi) - \frac{\partial}{\partial \varphi}(0)\right)\hat{z}$$

$$\Rightarrow \mu_0 n I \hat{z} = 0\hat{r} + 0\hat{\varphi} + \frac{1}{r}(A_\varphi - 0)\hat{z}$$

$$\Rightarrow \mu_0 n I \hat{z} = \frac{1}{r} A_\varphi \hat{z}$$

$$\Rightarrow A_\varphi = r\mu_0 n I$$

$$\Rightarrow \vec{A} = r\mu_0 n I \hat{\varphi}$$

For $r = \frac{a}{2}$, we have:

$$\vec{A} = \frac{1}{2}\mu_0 n I a \hat{\varphi}$$

Choice (4) is the answer (Fig. 26.1).

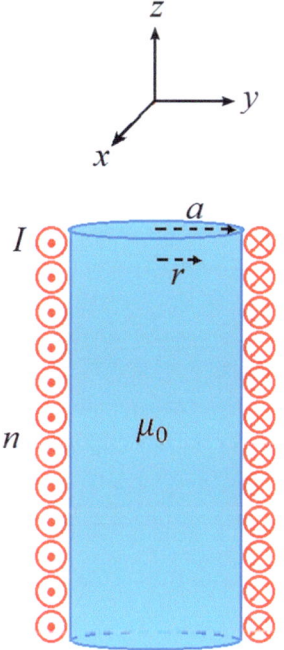

Fig. 26.1 A long cylindrical solenoid placed in free space and aligned with z-axis

References

1. Rahmani-Andebili, M., General Physics II – Practice Problems, Methods, and Solutions, Springer Nature, 2025.
2. Rahmani-Andebili, M., General Physics I – Practice Problems, Methods, and Solutions, Springer Nature, 2025.
3. Rahmani-Andebili, M., Mathematics of Engineering and Science – Practice Problems, Methods, and Solutions, Springer Nature, 2024.
4. Rahmani-Andebili, M., Differential Equations – Practice Problems, Methods, and Solutions, Springer Nature, 2022.
5. Rahmani-Andebili, M., Calculus III – Practice Problems, Methods, and Solutions, Springer Nature, 2023.
6. Rahmani-Andebili, M., Calculus II – Practice Problems, Methods, and Solutions, Springer Nature, 2023.
7. Rahmani-Andebili, M., Calculus I (2nd Ed.) – Practice Problems, Methods, and Solutions, Springer Nature, 2023.
8. Rahmani-Andebili, M., Precalculus (2nd Ed.) – Practice Problems, Methods, and Solutions, Springer Nature, 2024.

Magnetization: Part A

Abstract

In this chapter, the basic and advanced problems of magnetization as well as bound surface and volume current densities are studied. Herein, different types of problems and exercises are presented that are categorized as follows.

- *Problems with detailed solution*: They have been designed to teach students the subjects in detail. Moreover, they have been categorized in different levels based on their difficulty levels (easy, normal, and hard) and calculation amounts (small, normal, and large).
- *Partially solved exercises*: They have been designed to encourage students to practice problems while guiding them through the problem-solving procedure and hinting the required formulas.
- *Exercises with final answer*: They have been designed to encourage students to practice more by themselves while hinting them by the final answer as well as to help instructors to give tests or quizzes.

Problem

27.1. In Fig. 27.1, the current of inner and outer conductors of the coaxial cable are 2 A inward and 2 A outward, respectively. Calculate the value of the closed integral below on the contour. Herein, $\mu_r = 3$ [1–8].

$$\alpha = \oint \vec{M} \cdot \vec{dl}$$

Difficulty level   ○ Easy   ○ Normal   ● Hard
Calculation amount   ○ Small   ● Normal   ○ Large
1) -2
2) 2
3) $\frac{2}{3}$
4) $-\frac{2}{3}$

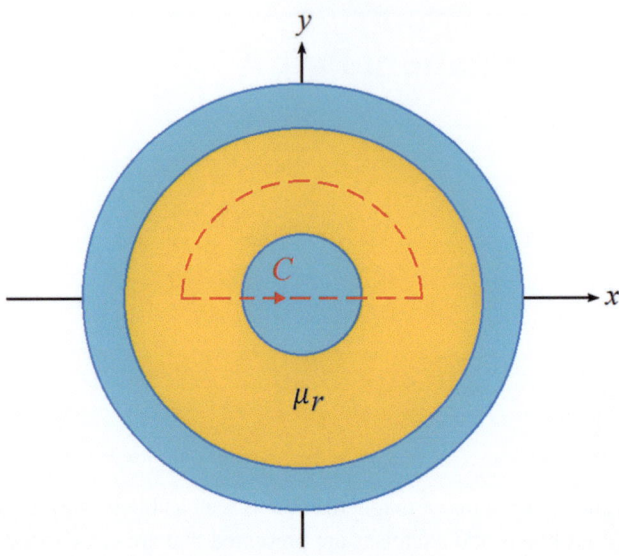

Fig. 27.1 A coaxial cable

Problem

27.2. The region $a < r < b$ in the cylindrical coordinates system is filled with a nonuniform magnetic material. A wire on the z-axis that carries the current I in the positive direction of z-axis generates the uniform magnetic flux density $\vec{B} = \frac{\mu_0 I}{2\pi a}\hat{\varphi}$ in the material. Calculate the bound surface current density $\vec{J_{bs}}$ on the surface $r = b$ (Fig. 27.2).

Difficulty level   ○ Easy   ○ Normal   ● Hard
Calculation amount   ○ Small   ○ Normal   ● Large

1) 0
2) $\dfrac{I(b-a)}{2\pi ab}\hat{z}$
3) $\dfrac{I(a+b)}{2\pi ab}\hat{z}$
4) $\dfrac{I(a-b)}{2\pi ab}\hat{z}$

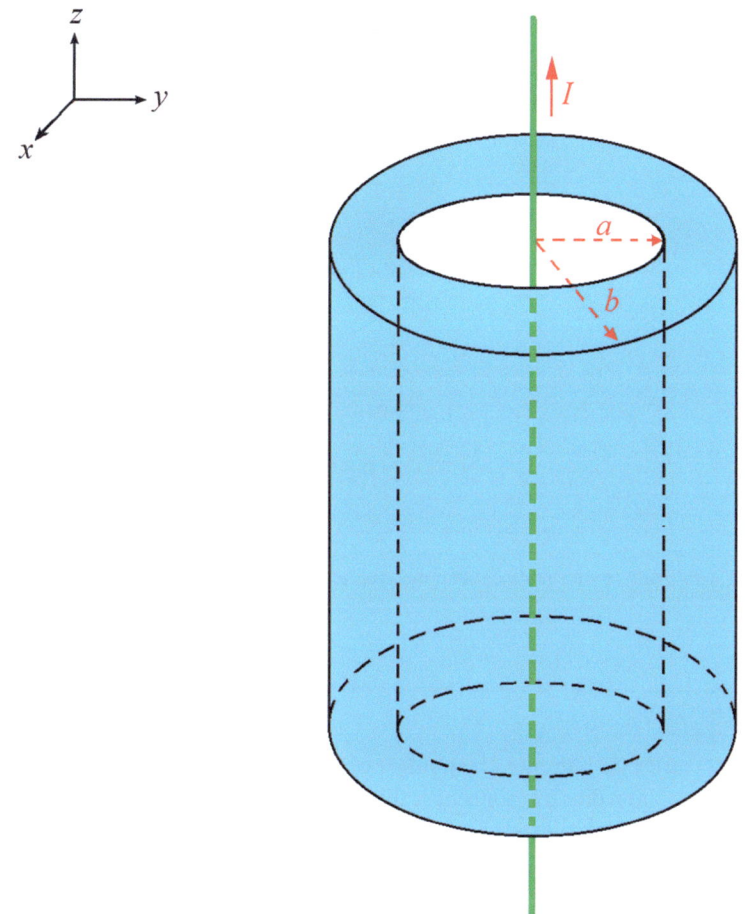

Fig. 27.2 A current-carrying wire aligned with the axis of a hollow cylinder with nonuniform magnetic material

Problem

27.3. Figure 27.3 illustrates a long cylindrical wire with radius a made from a nonmagnetic material and immersed in two symmetric regions with different magnetic permeabilities. Calculate the bound surface current density $\overrightarrow{J_{bs}}$ on the surface $r = 2a$ on xy plane.

Difficulty level   ○ Easy   ○ Normal   ● Hard
Calculation amount   ○ Small   ● Normal   ○ Large

1) $-\dfrac{2}{\pi a}\widehat{r}$
2) $-\dfrac{1}{\pi a}\widehat{r}$
3) $\dfrac{1}{\pi a}\widehat{r}$
4) $\dfrac{2}{\pi a}\widehat{r}$

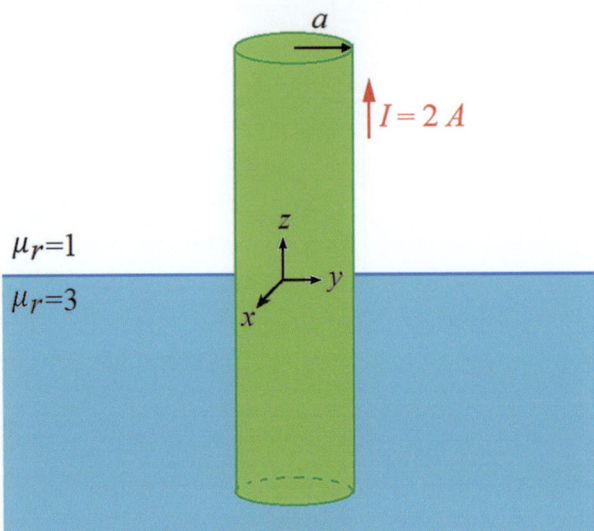

Fig. 27.3 A long cylindrical wire made from a nonmagnetic material and immersed in two symmetric regions with different magnetic permeabilities

Problem

27.4. For a long solenoid that carries the current I, the number of turns per unit length and radius are n and b. Moreover, a solid cylinder with permeability μ and radius a is concentrically placed inside the solenoid ($a < b$). Calculate the magnitude of magnetization vectors in the regions $r < a$ and $a < r < b$ (Fig. 27.4).

Difficulty level ○ Easy ○ Normal ● Hard
Calculation amount ● Small ○ Normal ○ Large

1) $M_{r<a} = \left(\dfrac{\mu}{\mu_0} - 1\right)nI, \quad M_{a<r<b} = \left(\dfrac{\mu}{\mu_0} + 1\right)nI$

2) $M_{r<a} = \left(\dfrac{\mu_0}{\mu} - 1\right)nI, \quad M_{a<r<b} = \left(\dfrac{\mu_0}{\mu} + 1\right)nI$

3) $M_{r<a} = \left(\dfrac{\mu}{\mu_0} - 1\right)nI, \quad M_{a<r<b} = 0$

4) $M_{r<a} = \left(\dfrac{\mu}{\mu_0} + 1\right)nI, \quad M_{a<r<b} = \dfrac{\mu}{\mu_0}nI$

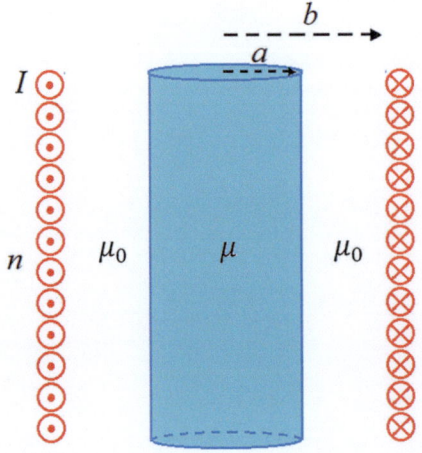

Fig. 27.4 A solid cylinder with permeability μ and radius a which is concentrically placed inside a solenoid with radius b

References

1. Rahmani-Andebili, M., General Physics II – Practice Problems, Methods, and Solutions, Springer Nature, 2025.
2. Rahmani-Andebili, M., General Physics I – Practice Problems, Methods, and Solutions, Springer Nature, 2025.
3. Rahmani-Andebili, M., Mathematics of Engineering and Science – Practice Problems, Methods, and Solutions, Springer Nature, 2024.
4. Rahmani-Andebili, M., Differential Equations – Practice Problems, Methods, and Solutions, Springer Nature, 2022.
5. Rahmani-Andebili, M., Calculus III – Practice Problems, Methods, and Solutions, Springer Nature, 2023.
6. Rahmani-Andebili, M., Calculus II – Practice Problems, Methods, and Solutions, Springer Nature, 2023.
7. Rahmani-Andebili, M., Calculus I (2nd Ed.) – Practice Problems, Methods, and Solutions, Springer Nature, 2023.
8. Rahmani-Andebili, M., Precalculus (2nd Ed.) – Practice Problems, Methods, and Solutions, Springer Nature, 2024.

Magnetization: Part B

Abstract

In this chapter, the problems of the twenty seventh chapter are fully solved, in detail, step-by-step, and with different methods.

28.1. Based on the information given in the problem, we have [1–8]:

$$I_{inward} = I_{outward} = 2\ A$$

$$\mu_r = 3$$

$$\alpha = \oint_C \vec{M} \cdot \vec{dl}$$

As we know, if the magnetic material is uniform, $\vec{M} = (\mu_r - 1)\vec{H}$ is used for the magnetization vector; otherwise, $\vec{M} = \dfrac{\vec{B}}{\mu_0} - \vec{H}$.

Therefore, for this problem, we can write as follows.

$$\alpha = \oint_C (\mu_r - 1)\vec{H} \cdot \vec{dl}$$

$$\Rightarrow \alpha = (3-1)\oint_C \vec{H} \cdot \vec{dl} = 2\oint_C \vec{H} \cdot \vec{dl} \tag{28.1}$$

On the other hand, according to Ampere's circuital law we have the following relation.

$$I_{in} = \oint_C \vec{H} \cdot \vec{dl} \tag{28.2}$$

As can be noticed from Fig. 28.1, the contour encloses half of the inner conductor. Moreover, since the direction of current of inner conductor is outward, based on right-hand rule, the direction of four fingers is in the opposite direction of the contour. Thus:

$$I_{in} = -0.5 I_{inward} = -0.5 \times 2 = -1\ A \tag{28.3}$$

Solving (28.2) and (28.3):

$$\oint_C \vec{H}.\vec{dl} = -1 \tag{28.4}$$

Solving (28.1) and (28.4):

$$\alpha = 2 \times (-1)$$

$$\Rightarrow \alpha = -2$$

Choice (1) is the answer.

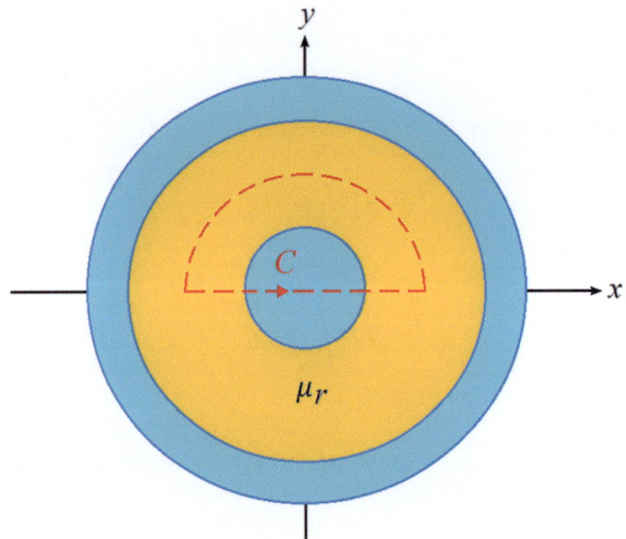

Fig. 28.1 A coaxial cable

28.2. Based on the information given in the problem, we have:

$$\vec{B} = \frac{\mu_0 I}{2\pi a}\hat{\varphi} \tag{28.5}$$

According to Ampere's circuital law, we have the following relation.

$$I_{in} = \oint_C \vec{H}.\vec{dl}$$

Therefore, for the region $a < r < b$, we have:

$$\Rightarrow I = \int_{\varphi=0}^{\varphi=2\pi} H_\varphi \hat{\varphi}.rd\varphi\hat{\varphi}$$

28 Magnetization: Part B

$$\Rightarrow I = H_\varphi r \int_{\varphi=0}^{\varphi=2\pi} d\varphi$$

$$\Rightarrow I = H_\varphi r (2\pi - 0)$$

$$\Rightarrow H_\varphi = \frac{I}{2\pi r}$$

$$\Rightarrow \vec{H} = \frac{I}{2\pi r}\hat{\varphi} \qquad (28.6)$$

Since the magnetic material is nonuniform, the relation below needs to be used instead of $\vec{M} = (\mu_r - 1)\vec{H}$ for the magnetization vector.

$$\vec{M} = \frac{\vec{B}}{\mu_0} - \vec{H} \qquad (28.7)$$

Solving (28.5), (28.6) and (28.7):

$$\vec{M} = \frac{I}{2\pi a}\hat{\varphi} - \frac{I}{2\pi r}\hat{\varphi}$$

$$\Rightarrow \vec{M} = \frac{I}{2\pi}\left(\frac{1}{a} - \frac{1}{r}\right)\hat{\varphi} \qquad (28.8)$$

As we know, the bound surface current density can be calculated as follows.

$$\vec{J}_{bs} = \vec{M} \times \hat{n}$$

Thus, for this problem, the bound surface current density on the surface $r = b$ is calculated as follows.

$$\vec{J}_{bs} = \vec{M} \times \hat{r}\Big|_{r=b}$$

$$\Rightarrow \vec{J}_{bs} = \frac{I}{2\pi}\left(\frac{1}{a} - \frac{1}{r}\right)\hat{\varphi} \times \hat{r}\Big|_{r=b} = \frac{I}{2\pi}\left(\frac{1}{a} - \frac{1}{b}\right)(-\hat{z})$$

$$\Rightarrow \vec{J}_{bs} = \frac{I(a-b)}{2\pi ab}\hat{z}$$

Choice (4) is the answer (Fig. 28.2).

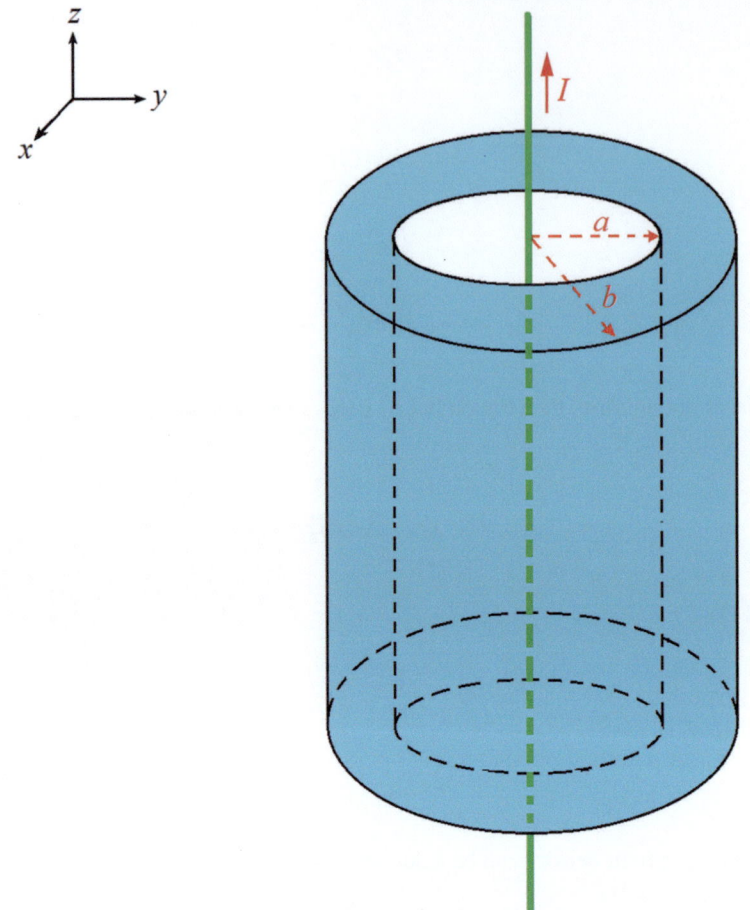

Fig. 28.2 A current-carrying wire aligned with the axis of a hollow cylinder with nonuniform magnetic material

28.3. According to Ampere's circuital law, we have the following relation.

$$I_{in} = \oint_C \vec{H} \cdot \vec{dl}$$

For the region $r > a$, we have:

$$\Rightarrow 2 = \int_{\varphi=0}^{\varphi=2\pi} H_\varphi \widehat{\varphi} \cdot r d\varphi \widehat{\varphi}$$

$$\Rightarrow 2 = H_\varphi r \int_{\varphi=0}^{\varphi=2\pi} d\varphi$$

$$\Rightarrow 2 = H_\varphi r (2\pi - 0)$$

$$\Rightarrow H_\varphi = \frac{1}{\pi r}$$

$$\Rightarrow \vec{H} = \frac{1}{\pi r} \widehat{\varphi} \tag{28.9}$$

As we know, the relation below can be used for the magnetization vector if the magnetic material is uniform.

$$\vec{M} = (\mu_r - 1)\vec{H} \qquad (28.10)$$

Solving (28.9) and (28.10):

$$\vec{M} = (3-1)\frac{1}{\pi r}\hat{\varphi}$$

$$\Rightarrow \vec{M} = \frac{2}{\pi r}\hat{\varphi} \qquad (28.11)$$

As we know, the bound surface current density can be calculated as follows.

$$\vec{J}_{bs} = \vec{M} \times \hat{n}$$

Thus, for this problem, the bound surface current density on the surface $r = 2a$ is calculated as follows.

$$\vec{J}_{bs} = \vec{M} \times \hat{z}\bigg|_{r=2a}$$

$$\Rightarrow \vec{J}_{bs} = \frac{2}{\pi r}\hat{\varphi} \times \hat{z}\bigg|_{r=2a}$$

$$\Rightarrow \vec{J}_{bs} = \frac{1}{\pi a}\hat{r}$$

Choice (3) is the answer (Fig. 28.3).

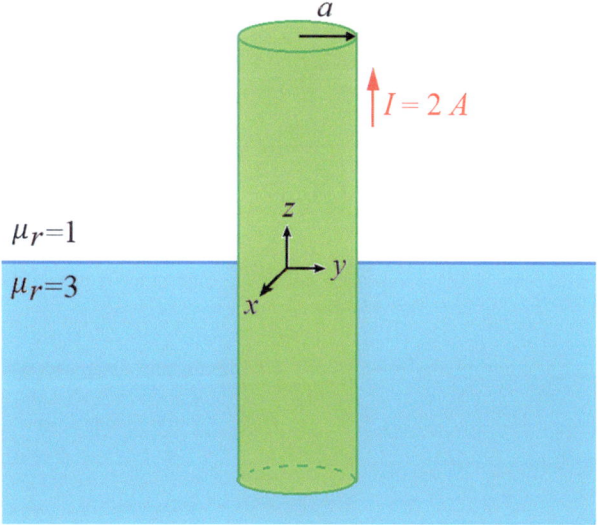

Fig. 28.3 A long cylindrical wire made from a nonmagnetic material and immersed in two symmetric regions with different magnetic permeabilities

28.4. As we know, magnitude of magnetic field intensity inside a cylindrical solenoid is uniform and as follows.

$$H = nI$$

Herein, I is the current and n is the number of turns per unit length.

As we know, for the magnetization vector, the relation below can be used if the magnetic material is uniform.

$$\vec{M} = (\mu_r - 1)\vec{H}$$

Therefore, for the region $r < a$, we have:

$$M_{r<a} = \left(\frac{\mu}{\mu_0} - 1\right)nI$$

Also, for the region $a < r < b$, we have:

$$M_{a<r<b} = (1 - 1)nI$$

$$\Rightarrow M_{a<r<b} = 0$$

Choice (3) is the answer (Fig. 28.4).

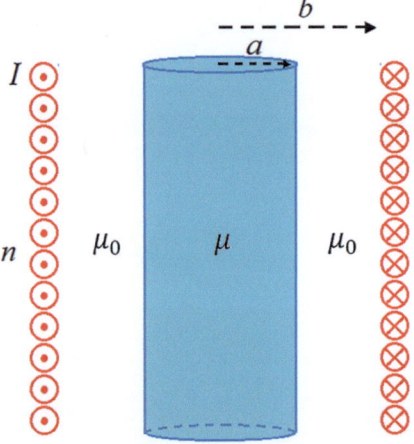

Fig. 28.4 A solid cylinder with permeability μ and radius a which is concentrically placed inside a solenoid with radius b

References

1. Rahmani-Andebili, M., General Physics II – Practice Problems, Methods, and Solutions, Springer Nature, 2025.
2. Rahmani-Andebili, M., General Physics I – Practice Problems, Methods, and Solutions, Springer Nature, 2025.
3. Rahmani-Andebili, M., Mathematics of Engineering and Science – Practice Problems, Methods, and Solutions, Springer Nature, 2024.
4. Rahmani-Andebili, M., Differential Equations – Practice Problems, Methods, and Solutions, Springer Nature, 2022.
5. Rahmani-Andebili, M., Calculus III – Practice Problems, Methods, and Solutions, Springer Nature, 2023.
6. Rahmani-Andebili, M., Calculus II – Practice Problems, Methods, and Solutions, Springer Nature, 2023.
7. Rahmani-Andebili, M., Calculus I (2nd Ed.) – Practice Problems, Methods, and Solutions, Springer Nature, 2023.
8. Rahmani-Andebili, M., Precalculus (2nd Ed.) – Practice Problems, Methods, and Solutions, Springer Nature, 2024.

Boundary Conditions and Method of Image Current in Magnetostatics: Part A

Abstract

In this chapter, the basic and advanced problems of boundary conditions in magnetic field and method of image current in magnetostatics are studied. Herein, different types of problems and exercises are presented that are categorized as follows.

- *Problems with detailed solution*: They have been designed to teach students the subjects in detail. Moreover, they have been categorized in different levels based on their difficulty levels (easy, normal, and hard) and calculation amounts (small, normal, and large).
- *Partially solved exercises*: They have been designed to encourage students to practice problems while guiding them through the problem-solving procedure and hinting the required formulas.
- *Exercises with final answer*: They have been designed to encourage students to practice more by themselves while hinting them by the final answer as well as to help instructors to give tests or quizzes.

29.1 Boundary Conditions in Magnetic Field

Problem

29.1. On the boundary of two environments, a surface current with the surface current density $\vec{J}_s = 5\hat{y}$ A/m is available. Calculate the magnetic field intensity in the second region if $\vec{H}_1 = 4\hat{x} - 10\hat{y} + 6\hat{z}$ [1–8] (Fig. 29.1).

Difficulty level   ○ Easy   ○ Normal   ● Hard
Calculation amount   ○ Small   ● Normal   ○ Large

1) $\vec{H}_2 = 1.6\hat{x} - 10\hat{y} + \hat{z}$
2) $\vec{H}_2 = 1.2\hat{x} - 12\hat{y} + 2\hat{z}$
3) $\vec{H}_2 = 1.6\hat{x} - 15\hat{y} + 6\hat{z}$
4) $\vec{H}_2 = 4\hat{x} - 15\hat{y} + 6\hat{z}$

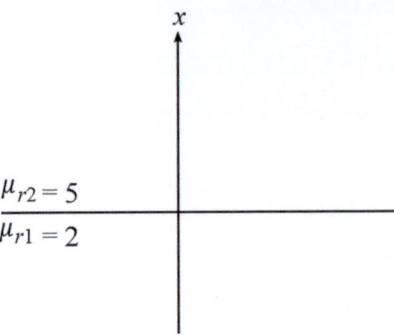

Fig. 29.1 Studying the boundary conditions for the magnetic field intensity

Partially Solved Exercise

29.2. On the boundary of two environments, a surface current with the surface current density $\vec{J}_s = (\hat{x} + 2\hat{y})$ A/m is available. Determine the magnetic flux density in the second region if $\vec{H}_1 = 2\hat{x} - 3\hat{y} + \hat{z}$.

Solution

As we know, the normal component of magnetic flux density on the boundary of two regions is continuous. In other words:

$$B_{1n} = B_{2n} \tag{29.1}$$

Moreover, the tangent component of magnetic field intensity on the boundary of two regions is discontinuous and the amount of discontinuity is equal to the free surface current density. Herein, $\hat{n}$ is the unit normal vector of the first region. In other words:

$$\hat{n} \times (\vec{H}_{2t} - \vec{H}_{1t}) = \vec{J}_s \tag{29.2}$$

Therefore, from (29.1), we can write:

$$\mu_0 \mu_{r1} H_{1z} = \mu_0 \mu_{r2} H_{2z}$$

$$\Rightarrow \mu_{r1} H_{1z} = \mu_{r2} H_{2z}$$

$$\Rightarrow (\quad) \times (\quad) = (\quad) H_{2z}$$

$$\Rightarrow H_{2z} = (\quad)$$

$$\Rightarrow \vec{H}_{2n} = (\quad)\hat{z} \tag{29.3}$$

Moreover, from (29.2), we can write the following relation. Herein, it should be noticed that the normal vector of the first region is $-\hat{z}$. Hence:

$$-\hat{z} \times (\vec{H}_{2t} - (\qquad)) = \hat{x} + 2\hat{y}$$

$$\Rightarrow -\hat{z} \times (\vec{H}_{2t} - (\qquad)) = -\hat{z} \times (\hat{y} - 2\hat{x})$$

29.1 Boundary Conditions in Magnetic Field

$$\Rightarrow \vec{H}_{2t} - (\quad) = \hat{y} - 2\hat{x}$$

$$\Rightarrow \vec{H}_{2t} = \quad \quad (29.4)$$

Therefore, from (29.3) and (29.4), we have (Fig. 29.2):

$$\vec{H}_2 = \vec{H}_{2n} + \vec{H}_{2t}$$

$$\Rightarrow \vec{H}_2 =$$

$$\Rightarrow \vec{B}_2 = \mu_0(-4\hat{y} + 3\hat{z})$$

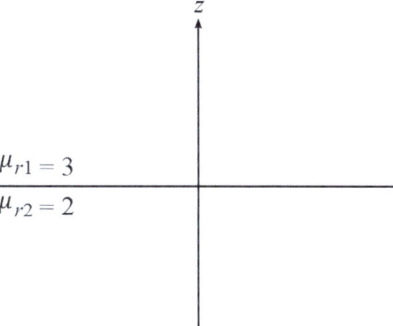

Fig. 29.2 Studying the boundary conditions for the magnetic flux density

---

Problem

29.3. On the boundary of two environments, the surface current density is $\vec{J}_s = 9\hat{y}$ A/m. Calculate the magnetic flux density in the first region if $\vec{H}_2 = 3\hat{x} + 8\hat{z}$ (Fig. 29.3).

Difficulty level ○ Easy ○ Normal ● Hard
Calculation amount ○ Small ● Normal ○ Large

1) $\vec{B}_1 = 24\mu_0(\hat{x} - \hat{z})$
2) $\vec{B}_1 = 24\mu_0(2\hat{x} + \hat{z})$
3) $\vec{B}_1 = 24\mu_0(-2\hat{x} + \hat{z})$
4) $\vec{B}_1 = 24\mu_0(-\hat{x} + \hat{z})$

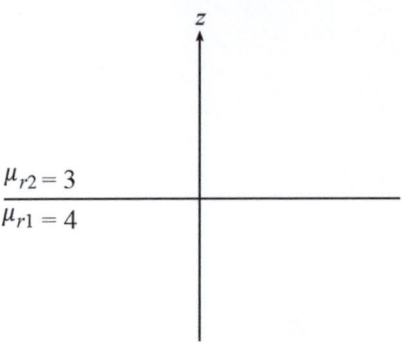

Fig. 29.3 Studying the boundary conditions for the magnetic flux density

Problem

29.4. On the boundary of the two environments, shown in Fig. 29.4, a bound surface current density $\vec{J}_{sb}$ is available. Determine the relation exists between the magnetization vectors on the boundary.

Difficulty level ○ Easy ○ Normal ● Hard
Calculation amount ● Small ○ Normal ○ Large

1) $\vec{J}_{sb} = \hat{n} \times \vec{M}_1 - \hat{n} \times \vec{M}_2$
2) $\vec{J}_{sb} = -\hat{n} \times \vec{M}_1 + \hat{n} \times \vec{M}_2$
3) $\vec{J}_{sb} = -\hat{n} \times \vec{M}_1 - \hat{n} \times \vec{M}_2$
4) $\vec{J}_{sb} = \hat{n} \times \vec{M}_1 + \hat{n} \times \vec{M}_2$

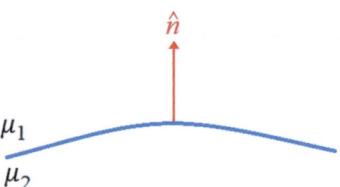

Fig. 29.4 Studying the boundary conditions for the magnetization vector

29.2 Method of Image Current in Magnetostatics

Problem

29.5. In Fig. 29.5, calculate the force exerted on the per unit length of the wire by the region placed at $z < 0$.

Difficulty level ○ Easy ● Normal ○ Hard
Calculation amount ● Small ○ Normal ○ Large

1) $\dfrac{\mu_0 I^2}{4\pi d}\hat{z}$
2) $\dfrac{\mu_0 I^2}{4\pi d}\hat{z}$
3) $\dfrac{\mu_0 I^2}{2\pi d}\hat{z}$
4) $\dfrac{\mu_0 I^2}{2\pi d}\hat{z}$

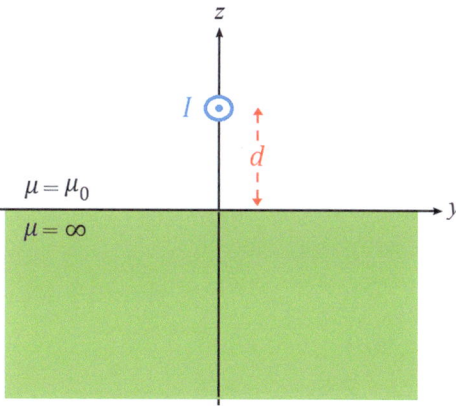

Fig. 29.5 A current-carrying wire placed in front of a large region with infinite magnetic permeability

References

1. Rahmani-Andebili, M., General Physics II - Practice Problems, Methods, and Solutions, Springer Nature, 2025.
2. Rahmani-Andebili, M., General Physics I – Practice Problems, Methods, and Solutions, Springer Nature, 2025.
3. Rahmani-Andebili, M., Mathematics of Engineering and Science – Practice Problems, Methods, and Solutions, Springer Nature, 2024.
4. Rahmani-Andebili, M., Differential Equations – Practice Problems, Methods, and Solutions, Springer Nature, 2022.
5. Rahmani-Andebili, M., Calculus III – Practice Problems, Methods, and Solutions, Springer Nature, 2023.
6. Rahmani-Andebili, M., Calculus II – Practice Problems, Methods, and Solutions, Springer Nature, 2023.
7. Rahmani-Andebili, M., Calculus I (2nd Ed.) – Practice Problems, Methods, and Solutions, Springer Nature, 2023.
8. Rahmani-Andebili, M., Precalculus (2nd Ed.) – Practice Problems, Methods, and Solutions, Springer Nature, 2024.

Boundary Conditions and Method of Image Current in Magnetostatics: Part B

Abstract

In this chapter, the problems of the twenty ninth chapter are fully solved, in detail, step-by-step, and with different methods.

30.1 Boundary Conditions in Magnetic Field

30.1. Based on the information given in the problem, we have [1–8]:

$$\vec{J}_s = 5\hat{y} \ A/m$$

$$\vec{H}_1 = 4\hat{x} - 10\hat{y} + 6\hat{z}$$

As we know, the normal component of magnetic flux density on the boundary of two regions is continuous. In other words:

$$B_{1n} = B_{2n} \tag{30.1}$$

Moreover, the tangent component of magnetic field intensity on the boundary of two regions is discontinuous and the amount of discontinuity is equal to the free surface current density. Herein, $\hat{n}$ is the unit normal vector of the first region. In other words:

$$\hat{n} \times \left(\vec{H}_{2t} - \vec{H}_{1t}\right) = \vec{J}_s \tag{30.2}$$

Therefore, from (30.1), we can write:

$$\mu_0 \mu_{r1} H_{1x} = \mu_0 \mu_{r2} H_{2x}$$

$$\Rightarrow \mu_{r1} H_{1x} = \mu_{r2} H_{2x}$$

$$\Rightarrow 2 \times 4 = 5 H_{2x}$$

$$\Rightarrow H_{2x} = 1.6$$

$$\Rightarrow \vec{H}_{2n} = 1.6\hat{x} \tag{30.3}$$

Moreover, from (30.2), we can write:

$$\hat{x} \times \left(\vec{H}_{2t} - (-10\hat{y} + 6\hat{z})\right) = 5\hat{y}$$

$$\Rightarrow \hat{x} \times \left(\vec{H}_{2t} - (-10\hat{y} + 6\hat{z})\right) = -5\hat{x} \times \hat{y}$$

$$\Rightarrow \vec{H}_{2t} - (-10\hat{y} + 6\hat{z}) = -5\hat{z}$$

$$\Rightarrow \vec{H}_{2t} = -10\hat{y} + \hat{z} \tag{30.4}$$

Therefore, from (30.3) and (30.4), we have:

$$\vec{H}_2 = \vec{H}_{2n} + \vec{H}_{2t}$$

$$\Rightarrow \vec{H}_2 = 1.6\hat{x} - 10\hat{y} + \hat{z}$$

Choice (1) is the answer (Fig. 30.1).

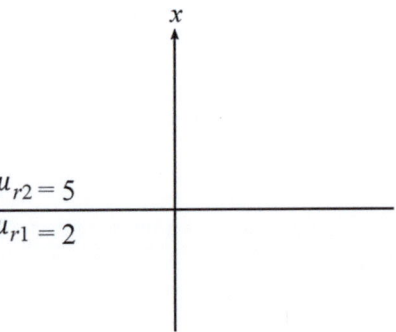

Fig. 30.1 Studying the boundary conditions for the magnetic field intensity

30.3. Based on the information given in the problem, we have:

$$\vec{J}_s = 9\hat{y} \; A/m$$

$$\vec{H}_2 = 3\hat{x} + 8\hat{z}$$

As we know, the normal component of magnetic flux density on the boundary of two regions is continuous. In other words:

$$B_{1n} = B_{2n} \tag{30.5}$$

30.1 Boundary Conditions in Magnetic Field

Moreover, the tangent component of magnetic field intensity on the boundary of two regions is discontinuous and the amount of discontinuity is equal to the free surface current density. Herein, $\hat{n}$ is the unit normal vector of the first region. In other words:

$$\hat{n} \times \left(\vec{H}_{2t} - \vec{H}_{1t}\right) = \vec{J}_s \tag{30.6}$$

Therefore, from (30.5), we can write:

$$\mu_0 \mu_{r1} H_{1z} = \mu_0 \mu_{r2} H_{2z}$$

$$\Rightarrow \mu_{r1} H_{1z} = \mu_{r2} H_{2z}$$

$$\Rightarrow 4 H_{1z} = 3 \times 8$$

$$\Rightarrow H_{1z} = 6$$

$$\Rightarrow \vec{H}_{1n} = 6\hat{z} \tag{30.7}$$

Moreover, from (30.6), we can write:

$$\hat{z} \times \left(3\hat{x} - \vec{H}_{1t}\right) = 9\hat{y}$$

$$\Rightarrow \hat{z} \times \left(3\hat{x} - \vec{H}_{1t}\right) = 9\hat{z} \times \hat{x}$$

$$\Rightarrow 3\hat{x} - \vec{H}_{1t} = 9\hat{x}$$

$$\Rightarrow \vec{H}_{1t} = -6\hat{x} \tag{30.8}$$

Therefore, from (30.7) and (30.8), we have:

$$\vec{H}_1 = \vec{H}_{1n} + \vec{H}_{1t}$$

$$\Rightarrow \vec{H}_1 = -6\hat{x} + 6\hat{z}$$

$$\Rightarrow \vec{B}_1 = 24\mu_0(-\hat{x} + \hat{z})$$

Choice (4) is the answer (Fig. 30.2).

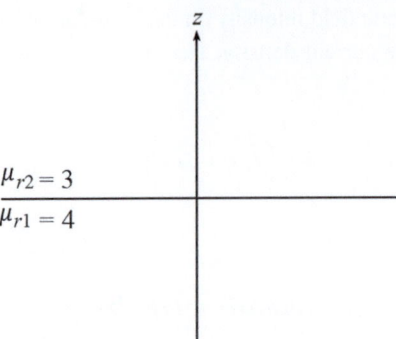

Fig. 30.2 Studying the boundary conditions for the magnetic flux density

30.4. As we know, bound surface current density can be calculated as follows.

$$\vec{J}_{bs} = \vec{M} \times \hat{n}$$

Therefore, the bound surface current densities in the first and second regions are as follows.

$$\vec{J}_{bs1} = \vec{M_1} \times (-\hat{n})$$

$$\vec{J}_{bs2} = \vec{M_2} \times \hat{n}$$

Now, the total bound surface current densities can be calculated as follows.

$$\vec{J}_{bs} = \vec{J}_{bs1} + \vec{J}_{bs2}$$

$$\Rightarrow \vec{J}_{bs} = \vec{M_2} \times \hat{n} - \vec{M_1} \times \hat{n}$$

$$\Rightarrow \vec{J}_{bs} = \hat{n} \times \vec{M_1} - \hat{n} \times \vec{M_2}$$

Choice (1) is the answer (Fig. 30.3).

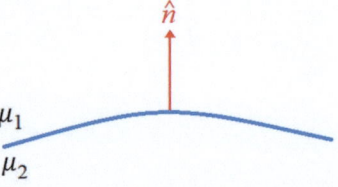

Fig. 30.3 Studying the boundary conditions for the magnetization vector

30.2 Method of Image Current in Magnetostatics

30.5. Based on the method of image current in magnetostatics, a region with infinite magnetic permeability can be replaced by the image of the current $I' = I$ at a distance $d = d'$ from the boundary of the environments. Herein, the direction of the image current is the same as the direction of the main current.

Moreover, as we know, the magnitude of magnetic force (per unit length) between two long straight wires with currents I and I' and distance x between them is calculated as follows.

$$F = \frac{\mu_0 I I'}{2\pi x}$$

Therefore, based on Fig. 30.4, the magnitude of magnetic force the per unit length of the wire can be calculated as follows.

$$F = \frac{\mu_0 I^2}{4\pi d}$$

In addition, the direction of force exerted on the wire by the region can be calculated by using the right-hand rule. Hence:

$$\vec{F} = -\frac{\mu_0 I^2}{4\pi d}\hat{z}$$

Choice (1) is the answer.

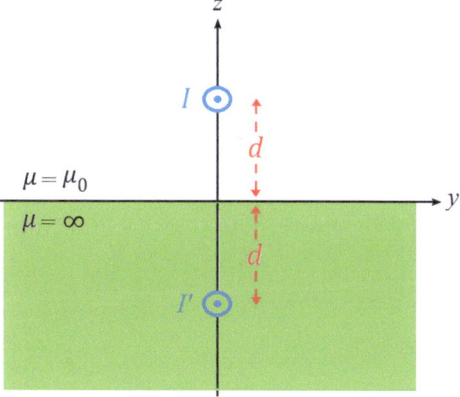

Fig. 30.4 A current-carrying wire placed in front of a large region with infinite magnetic permeability

References

1. Rahmani-Andebili, M., General Physics II – Practice Problems, Methods, and Solutions, Springer Nature, 2025.
2. Rahmani-Andebili, M., General Physics I – Practice Problems, Methods, and Solutions, Springer Nature, 2025.
3. Rahmani-Andebili, M., Mathematics of Engineering and Science – Practice Problems, Methods, and Solutions, Springer Nature, 2024.
4. Rahmani-Andebili, M., Differential Equations – Practice Problems, Methods, and Solutions, Springer Nature, 2022.
5. Rahmani-Andebili, M., Calculus III – Practice Problems, Methods, and Solutions, Springer Nature, 2023.
6. Rahmani-Andebili, M., Calculus II – Practice Problems, Methods, and Solutions, Springer Nature, 2023.
7. Rahmani-Andebili, M., Calculus I (2nd Ed.) – Practice Problems, Methods, and Solutions, Springer Nature, 2023.
8. Rahmani-Andebili, M., Precalculus (2nd Ed.) – Practice Problems, Methods, and Solutions, Springer Nature, 2024.

31

Electromagnetic Induction: Part A

Abstract

In this chapter, the basic and advanced problems of electromagnetic induction, that includes self-inductance and mutual inductance concepts, are studied. Herein, different types of problems and exercises are presented that are categorized as follows.

- *Problems with detailed solution*: They have been designed to teach students the subjects in detail. Moreover, they have been categorized in different levels based on their difficulty levels (easy, normal, and hard) and calculation amounts (small, normal, and large).
- *Partially solved exercises*: They have been designed to encourage students to practice problems while guiding them through the problem-solving procedure and hinting the required formulas.
- *Exercises with final answer*: They have been designed to encourage students to practice more by themselves while hinting them by the final answer as well as to help instructors to give tests or quizzes.

Problem

31.1. Which one of the following choices is correct regarding the relations between the induced voltage (*emf*), magnetic vector potential ($\vec{A}$), magnetic flux density ($\vec{B}$), and magnetic flux (φ) [1–8].

(a) $emf = -\dfrac{d}{dt} \int_S \left(\nabla \times \vec{A} \right) \cdot \vec{dS}$

(b) $emf = -\dfrac{d}{dt} \oint_C \vec{A} \times \vec{dl}$

(c) $\vec{B} = \nabla \times \vec{A}$

(d) $emf = -\dfrac{d\varphi}{dt}$

Difficulty level ○ Easy ● Normal ○ Hard
Calculation amount ● Small ○ Normal ○ Large

1) Only the relation (*d*) is correct.
2) Only the relations (*a*) and (*d*) are correct.
3) Only the relations (*b*), (*c*), and (*d*) are correct.
4) All the relations are correct.

Problem

31.2. In Fig. 31.1, the current-carrying wire and rectangular loop are on the same plane. Calculate the mutual inductance between the wire and loop.

Difficulty level — ○ Easy ○ Normal ● Hard
Calculation amount — ○ Small ● Normal ○ Large

1) $M_{12} = \dfrac{\mu_0 b}{2\pi} \ln\left(\dfrac{d+a}{d}\right)$

2) $M_{12} = \dfrac{\mu_0 a}{2\pi} \ln\left(\dfrac{d+a}{d}\right)$

3) $M_{12} = -\dfrac{\mu_0 b}{2\pi} \ln\left(\dfrac{d+a}{d}\right)$

4) $M_{12} = -\dfrac{\mu_0 a}{2\pi} \ln\left(\dfrac{d+a}{d}\right)$

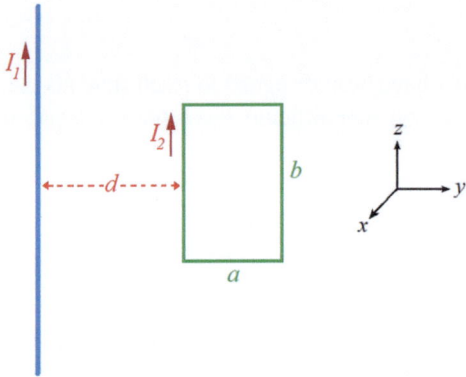

Fig. 31.1 A current-carrying wire and rectangular loop placed on the same plane

Problem

31.3. In Fig. 31.2, the current-carrying wire and rectangular loop are on the same plane. Calculate the amount of work needed to rotate the loop from position (a) to (b).

Difficulty level — ○ Easy ○ Normal ● Hard
Calculation amount — ○ Small ● Normal ○ Large

1) $W = \dfrac{\mu_0 I_1 I_2}{2\pi} \left[b\ln\left(\dfrac{d+b}{d}\right) + a\ln\left(\dfrac{d+a}{d}\right) \right]$

2) $W = \dfrac{\mu_0 I_1 I_2}{2\pi} \left[a\ln\left(\dfrac{d+b}{d}\right) + b\ln\left(\dfrac{d+a}{d}\right) \right]$

3) $W = \dfrac{\mu_0 I_1 I_2}{2\pi} \left[a\ln\left(\dfrac{d+b}{d}\right) - b\ln\left(\dfrac{d+a}{d}\right) \right]$

4) $W = \dfrac{\mu_0 I_1 I_2}{2\pi} \left[b\ln\left(\dfrac{d+b}{d}\right) - a\ln\left(\dfrac{d+a}{d}\right) \right]$

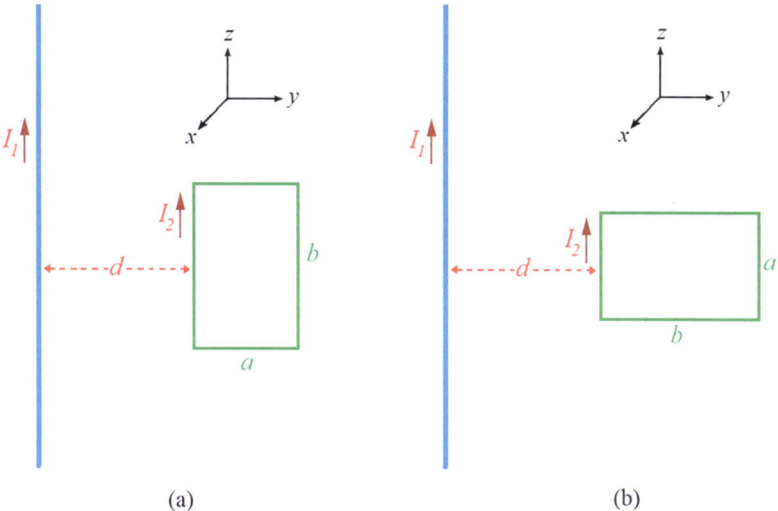

Fig. 31.2 (a) A current-carrying wire and rectangular loop placed on the same plane. (b) Rotating the loop by 90°

Problem

31.4. In Fig. 31.3, the current-carrying wire and semicircle loop are on the same plane. Calculate the mutual inductance between the wire and loop.

Difficulty level   ○ Easy   ○ Normal   ● Hard
Calculation amount   ○ Small   ● Normal   ○ Large

1) $\dfrac{\mu_0}{\pi} \displaystyle\int_a^{2a} \sqrt{\dfrac{a-y}{a+y}}\, dy$

2) $\dfrac{\mu_0}{\pi} \displaystyle\int_0^{2a} \sqrt{\dfrac{a-y}{a+y}}\, dy$

3) $\dfrac{\mu_0}{\pi} \displaystyle\int_0^{a} \sqrt{\dfrac{a+y}{a-y}}\, dy$

4) $\dfrac{\mu_0}{\pi} \displaystyle\int_0^{a} \sqrt{\dfrac{a-y}{a+y}}\, dy$

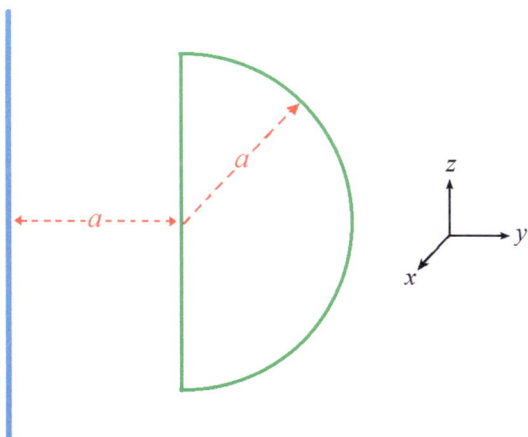

Fig. 31.3 A current-carrying wire and semicircle loop placed on the same plane

References

1. Rahmani-Andebili, M., General Physics II – Practice Problems, Methods, and Solutions, Springer Nature, 2025.
2. Rahmani-Andebili, M., General Physics I – Practice Problems, Methods, and Solutions, Springer Nature, 2025.
3. Rahmani-Andebili, M., Mathematics of Engineering and Science – Practice Problems, Methods, and Solutions, Springer Nature, 2024.
4. Rahmani-Andebili, M., Differential Equations – Practice Problems, Methods, and Solutions, Springer Nature, 2022.
5. Rahmani-Andebili, M., Calculus III – Practice Problems, Methods, and Solutions, Springer Nature, 2023.
6. Rahmani-Andebili, M., Calculus II – Practice Problems, Methods, and Solutions, Springer Nature, 2023.
7. Rahmani-Andebili, M., Calculus I (2nd Ed.) – Practice Problems, Methods, and Solutions, Springer Nature, 2023.
8. Rahmani-Andebili, M., Precalculus (2nd Ed.) – Practice Problems, Methods, and Solutions, Springer Nature, 2024.

Electromagnetic induction: Part B

Abstract

In this chapter, the problems of the thirty first chapter are fully solved, in detail, step-by-step, and with different methods.

32.1. The relation between magnetic flux density and magnetic vector potential is as follows [1–8].

$$\vec{B} = \nabla \times \vec{A} \qquad (32.1)$$

Hence, (c) is correct.

The induced voltage (*emf*) is the time derivative of magnetic flux. In other words:

$$emf = -\frac{d\varphi}{dt} \qquad (32.2)$$

Hence, (d) is correct.

The relation between magnetic flux and magnetic flux density is as follows.

$$\varphi = \int_S \vec{B} \cdot \vec{dS} \qquad (32.3)$$

Solving (32.1), (32.2) and (32.3):

$$emf = -\frac{d}{dt} \int_S \left(\nabla \times \vec{A} \right) \cdot \vec{dS} \qquad (32.4)$$

Hence, (a) is correct.

In addition, the Stokes's theorem is as follows.

$$\int_S \left(\nabla \times \vec{F} \right) \cdot \vec{dS} = \oint_C \vec{F} \cdot \vec{dl} \qquad (32.5)$$

Solving (32.4) and (32.5):

$$emf = -\frac{d}{dt}\int_C \vec{A}\cdot\vec{dl} \tag{32.6}$$

Hence, (b) is correct.

Therefore, all the relations are correct. Choice (4) is the answer.

32.2. The mutual inductance between the wire and loop can be calculated as follows.

$$M_{12} = \frac{\varphi_{12}}{I_1}$$

where, φ_{12} is the mutual magnetic flux that the wire generates and passes from the loop.

Moreover, as we know, the magnetic flux density at a distance r from a long straight wire in free space can be calculated as follows. Herein, the direction of the magnetic field can be calculated based on the right-hand rule.

$$B = \frac{\mu_0 I}{2\pi r}$$

Therefore, we can write:

$$\vec{B_1} = \frac{\mu_0 I_1}{2\pi r}\hat{\varphi}$$

The mutual magnetic flux can be calculated as follows.

$$\varphi_{12} = \int_S \vec{B_1}\cdot\vec{dS}$$

$$\Rightarrow \varphi_{12} = \int_{z=0}^{z=b}\int_{y=d}^{y=d+a} \frac{\mu_0 I_1}{2\pi y}\hat{\varphi}\cdot dydz\hat{\varphi}$$

$$\Rightarrow \varphi_{12} = \frac{\mu_0 I_1}{2\pi}\int_{z=0}^{z=b}\int_{y=d}^{y=d+a} \frac{1}{y}dydz$$

$$\Rightarrow \varphi_{12} = \frac{\mu_0 I_1}{2\pi}\Big[\ln y\Big]_d^{d+a}\Big[z\Big]_0^b$$

$$\Rightarrow \varphi_{12} = \frac{\mu_0 I_1}{2\pi}(\ln(d+a) - \ln d)(b-0)$$

$$\Rightarrow \varphi_{12} = \frac{\mu_0 I_1 b}{2\pi}\ln\left(\frac{d+a}{d}\right)$$

Therefore:

$$M_{12} = \frac{\mu_0 b}{2\pi}\ln\left(\frac{d+a}{d}\right)$$

Choice (1) is the answer (Fig. 32.1).

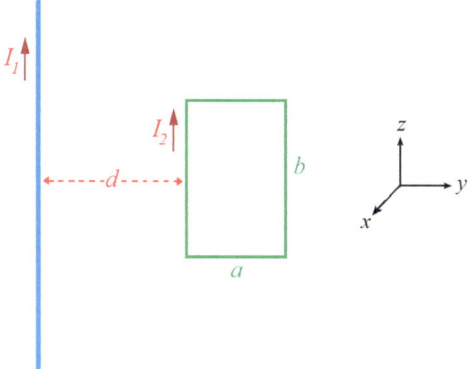

Fig. 32.1 A current-carrying wire and rectangular loop placed on the same plane

> **Notes**
> In this problem, the relations below have been used.
> $$\int \frac{dx}{x} = \ln x$$
> $$\ln a - \ln b = \ln\left(\frac{a}{b}\right)$$

32.3. The amount of work needed to rotate the loop is equal to the difference between the magnetic energy stored in the system in in the two positions. In other words:

$$W = \Delta U = U' - U$$

where,

$$U = \frac{1}{2}L_{11}I_1^2 + \frac{1}{2}L_{22}I_2^2 + M_{12}I_1I_2$$

$$U' = \frac{1}{2}L_{11}I_1^2 + \frac{1}{2}L_{22}I_2^2 + M'_{12}I_1I_2$$

Herein, L_{11}, L_{22}, and M_{12} are the self-inductance of the wire, self-inductance of the loop, and mutual inductance between the wire and loop in the first position. Moreover, M'_{12} is the mutual inductance between the wire and loop in the second position.

Thus:

$$W = I_1I_2\left(M'_{12} - M_{12}\right)$$

As can be noticed, only the mutual inductances in the two positions need to be calculated.

From the solution of the previous problem, we have:

$$M_{12} = \frac{\mu_0 b}{2\pi} \ln\left(\frac{d+a}{d}\right)$$

Likewise, for the mutual inductance in the second position, we can write as follows.

$$\varphi'_{12} = \int_S \overrightarrow{B_1} \cdot \overrightarrow{dS}$$

$$\Rightarrow \varphi'_{12} = \int_{z=0}^{z=a} \int_{y=d}^{y=d+b} \frac{\mu_0 I_1}{2\pi y} \widehat{\varphi} \cdot dy dz \widehat{\varphi}$$

$$\Rightarrow \varphi'_{12} = \frac{\mu_0 I_1}{2\pi} \int_{z=0}^{z=a} \int_{y=d}^{y=d+b} \frac{1}{y} dy dz$$

$$\Rightarrow \varphi'_{12} = \frac{\mu_0 I_1}{2\pi} \left[\ln y\right]_d^{d+b} \left[z\right]_0^a$$

$$\Rightarrow \varphi'_{12} = \frac{\mu_0 I_1}{2\pi} (\ln(d+b) - \ln d)(a - 0)$$

$$\Rightarrow \varphi'_{12} = \frac{\mu_0 I_1 a}{2\pi} \ln\left(\frac{d+b}{d}\right)$$

$$\Rightarrow M'_{12} = \frac{\mu_0 a}{2\pi} \ln\left(\frac{d+b}{d}\right)$$

Hence, the amount of work needed to rotate the loop is as follows.

$$W = I_1 I_2 \left[\frac{\mu_0 a}{2\pi} \ln\left(\frac{d+b}{d}\right) - \frac{\mu_0 b}{2\pi} \ln\left(\frac{d+a}{d}\right)\right]$$

$$\Rightarrow W = \frac{\mu_0 I_1 I_2}{2\pi} \left[a \ln\left(\frac{d+b}{d}\right) - b \ln\left(\frac{d+a}{d}\right)\right]$$

Choice (3) is the answer (Fig. 32.2).

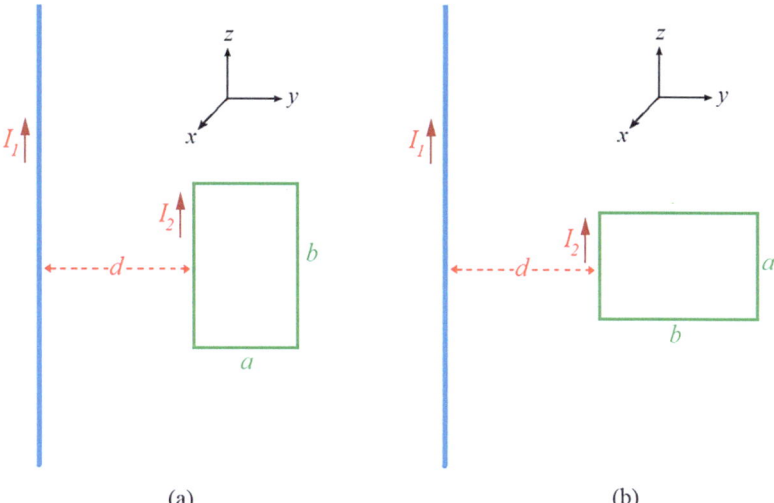

Fig. 32.2 (a) A current-carrying wire and rectangular loop placed on the same plane. (b) Rotating the loop by 90°

Notes
In this problem, the relations below have been used.

$$\int \frac{dx}{x} = \ln x$$

$$\ln a - \ln b = \ln\left(\frac{a}{b}\right)$$

32.4. As we know, the mutual inductance between the wire and loop can be calculated as follows.

$$M_{12} = \frac{\varphi_{12}}{I_1}$$

Herein, φ_{12} is the mutual magnetic flux that the wire generates and passes from the loop.

Moreover, as we know, the magnetic flux density at a distance r from a long straight wire in free space can be calculated as follows. Herein, the direction of the magnetic field can be calculated based on the right-hand rule.

$$B = \frac{\mu_0 I}{2\pi r}$$

Therefore, we can write:

$$\vec{B_1} = \frac{\mu_0 I_1}{2\pi r}\hat{\varphi}$$

The mutual magnetic flux can be calculated by using Fig. 32.3 as follows.

$$\varphi_{12} = \int_S \vec{B_1} \cdot \vec{dS}$$

$$\Rightarrow \varphi_{12} = \int_0^a \frac{\mu_0 I_1}{2\pi(a+y)}\hat{\varphi} \cdot 2z dy \hat{\varphi}$$

$$\Rightarrow \varphi_{12} = \int_0^a \frac{\mu_0 I_1}{2\pi(a+y)} 2z dy$$

Herein, there is a relation between the variables y and z that can be expressed by using Pythagorean rule as follows.

$$a^2 = z^2 + y^2 \Rightarrow z = \sqrt{a^2 - y^2}$$

Thus:

$$\Rightarrow \varphi_{12} = \int_0^a \frac{\mu_0 I_1}{2\pi(a+y)} 2\sqrt{a^2 - y^2} dy$$

$$\Rightarrow \varphi_{12} = \frac{\mu_0 I_1}{\pi} \int_0^a \frac{\sqrt{a^2 - y^2}}{a+y} dy$$

$$\Rightarrow \varphi_{12} = \frac{\mu_0 I_1}{\pi} \int_0^a \frac{\sqrt{(a+y)(a-y)}}{a+y} dy$$

$$\Rightarrow \varphi_{12} = \frac{\mu_0 I_1}{\pi} \int_0^a \frac{\sqrt{a-y}}{\sqrt{a+y}} dy$$

Therefore:

$$M_{12} = \frac{\mu_0}{\pi} \int_0^a \sqrt{\frac{a-y}{a+y}} dy$$

Choice (4) is the answer.

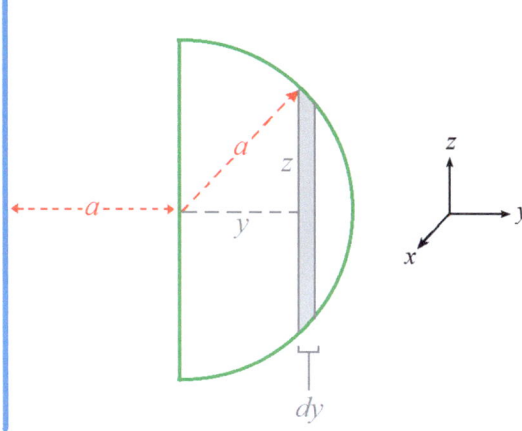

Fig. 32.3 A current-carrying wire and semicircle loop placed on the same plane

References

1. Rahmani-Andebili, M., General Physics II – Practice Problems, Methods, and Solutions, Springer Nature, 2025.
2. Rahmani-Andebili, M., General Physics I – Practice Problems, Methods, and Solutions, Springer Nature, 2025.
3. Rahmani-Andebili, M., Mathematics of Engineering and Science – Practice Problems, Methods, and Solutions, Springer Nature, 2024.
4. Rahmani-Andebili, M., Differential Equations – Practice Problems, Methods, and Solutions, Springer Nature, 2022.
5. Rahmani-Andebili, M., Calculus III – Practice Problems, Methods, and Solutions, Springer Nature, 2023.
6. Rahmani-Andebili, M., Calculus II – Practice Problems, Methods, and Solutions, Springer Nature, 2023.
7. Rahmani-Andebili, M., Calculus I (2nd Ed.) – Practice Problems, Methods, and Solutions, Springer Nature, 2023.
8. Rahmani-Andebili, M., Precalculus (2nd Ed.) – Practice Problems, Methods, and Solutions, Springer Nature, 2024.

Index

A

Ampere's circuital law, 423–427, 429–438, 453, 454, 456
Angle, 3, 4, 30, 254, 266, 387, 388, 398–402, 416–419

B

Boundary conditions, 253–282, 341–353, 355–380, 459–462, 465–469
Bound surface current density, 448, 449, 455, 457, 462, 468

C

Cartesian coordinate system, 6–7, 25–27, 33–36, 58, 59, 61–63, 67–69, 73–74, 78–80, 83–84, 88–89, 94–95, 98–99, 106–110, 116, 123, 135–139, 189–191, 205–206, 253–256, 265–269, 392
Coaxial cable, 257, 259, 271, 275, 349, 368, 369, 447, 448, 454
Continuous charge distribution, 238–241, 247–251
Conversion, 19–25, 54–58
Cross products, 4–6, 8, 14, 32, 33, 37, 46
Curl, 61–81, 83–101, 440, 444
Cylindrical capacitors, 287–290, 302–311
Cylindrical coordinate system, 7–13, 25, 26, 36–45, 58, 63–65, 69–70, 75–77, 79–80, 84–86, 90, 95–96, 100, 111–121, 139–162, 191–195, 207–216, 256–260, 270–276
Cylindrical resistors, 347–351, 363–372
Cylindrical surface, 8–11, 40–42, 145–148, 153–155, 258, 272, 274
Cylindrical volume, 12, 13, 44

D

Dielectric, 253–282, 287, 289, 291, 295
Dielectric constants, 255
Differential area vectors, 7, 8, 14, 35, 39, 47
Differential length vectors, 6, 8, 14, 34, 35, 38, 39, 46, 47
Differential volume, 7, 12, 18, 36, 43, 52
Discrete charge distribution, 233–238, 243–247
Distance vector, 26–27, 59
Divergences, 61–81, 83–101, 185
Divergence theorem, 66, 87
Dot products, 2–4, 30–31

E

Electric fields, 103–187, 189, 205, 208, 211, 217, 219, 221, 226, 247, 249, 250, 253–282, 289, 293, 301, 310, 322, 427, 438
Electric flux, 103–186, 253, 254, 256, 261, 265–267, 269, 342, 355, 356
Electric potential, 189–203, 205–231, 235, 244–246, 285, 287, 290, 298, 302, 306, 327, 329, 334, 337, 338, 345, 353, 360, 371
Electric potential energy, 233–241, 243–251, 290, 312, 338
Electrical conductivities, 341, 344–346, 348, 349, 351, 353, 358–362, 367, 368, 370, 372
Electromagnetic force, 409–422
Electromagnetic induction, 471–473, 475–480
Electromagnetic torque, 413–414, 421–422

F

Field streamlines, 106, 107
Flat capacitors, 283–287, 295–300
Flat resistors, 343–347, 357–364
Forces, 129, 184, 237, 285, 299, 300, 325, 326, 328, 329, 331–333, 335, 337, 409–412, 415–420, 462, 469
Free charge, 253, 255, 259, 273
Free space permittivity, 131, 132, 288–290, 292, 293, 303, 309, 310, 316, 323

G

Gaussian surfaces, 136, 137, 142–144, 146, 149–151, 153–156, 159, 161–164, 166, 169–174, 176, 177, 181–184, 211, 212, 214, 215, 219, 221–227, 229, 230, 248, 249, 270, 271, 276–279
Gauss's law, 131, 132, 136, 137, 142–158, 160–184, 205, 270, 271, 276–279, 310, 322
Gradient, 61–81, 83–101
Grounded conductors, 325–327, 331–334

I

Induced voltage, 471, 475
Inner product, 2–4, 30–31
Isolated conductors, 328–330, 335–339

L

Laplacian, 61–81, 83–101
Linear charge density, 109–114, 138, 141, 190, 192, 193, 205, 207, 208
Linear current, 383–390, 395–403

M

Magnetic dipole, 413, 414, 421
Magnetic energy, 423–427, 436–438, 477
Magnetic field, 383–393, 395–410, 413–419, 422, 427, 459–462, 465–468, 476, 479
Magnetic field intensities, 383–390, 392, 395–403, 413, 423, 424, 429, 431, 434, 436, 458–460, 465–467
Magnetic flux, 383–393, 395–408, 439, 443, 471, 475, 476, 479, 480
Magnetic flux densities, 390, 391, 403–408, 414, 416–419, 421, 424, 426, 433, 439, 440, 443, 444, 448, 460–462, 465, 466, 468, 471, 475, 476, 479
Magnetic material, 433, 448, 449, 453, 455–458
Magnetic permeability, 444, 449, 450, 457, 463, 469
Magnetic vector potential, 439–441, 443–444, 471, 475

Magnetization, 447–450, 453–458, 462, 468
Magnitude, 2, 4, 29, 31, 92, 105, 106, 108, 111, 129, 134, 137, 256, 260, 261, 271, 279, 301, 384–388, 390, 395, 396, 398–402, 409, 416–419, 424, 427, 450, 458, 469
Method of image charge, 325–339
Method of image current in magnetostatics, 459–462, 465–469
Mutual inductances, 472, 473, 476–479

N
Nonmagnetic material, 424, 449, 450, 457
Nonuniform, 120, 128, 161, 182, 189, 260, 286, 346, 347, 349–351, 353, 354, 364, 368–370, 378, 379, 381, 426, 435, 448, 449, 455, 456
Nonuniform dielectric, 292, 293, 318, 320

P
Perpendicular, 4, 31
Polarization surface charge density, 257, 258, 262
Polarization vector, 257, 258, 262, 263
Polarization volume charge density, 258, 259, 263, 264, 273, 275, 281, 282
Polarized charges, 259, 274, 275
Position vector, 25–26, 58–59

R
Rectangular, 6, 472, 473, 477, 479
Relative permittivity, 255, 286
Right-hand rule, 33, 37, 46, 384, 388, 395–397, 400–402, 414, 416–421, 444, 453, 469, 476, 479

S
Scalar functions, 61, 62, 64, 65, 78–81, 83, 84, 86
Self-inductance, 477
Solenoid, 441, 444, 445, 450, 458

Spherical capacitors, 283–293, 295–323
Spherical coordinate systems, 1–27, 29–59, 64–66, 70–72, 76–78, 80–81, 86–87, 91–92, 96–98, 100–101, 121–130, 133, 134, 163–187, 195–203, 216–231, 260–264, 276–282
Spherical resistors, 341–381
Spherical surface, 14–17, 48, 50, 51, 144, 152, 164–168, 172, 174–178, 180, 262, 280
Spherical volume, 18, 19, 53
Stokes's theorem, 73, 94, 443, 475
Surface charge densities, 107–109, 115, 121, 129, 137, 138, 198, 238, 258, 262, 265, 267, 269, 272, 280, 287, 289, 291, 341, 342, 355, 356
Surface current, 390–392, 403–405, 408, 459, 460
Surface current density, 345, 362, 390, 392, 403, 405–407, 459–461, 465, 467

T
Torque, 409–422

U
Uniform, 104, 116, 123, 124, 148, 168, 205, 241, 257, 259, 261, 262, 264, 280, 286, 301, 346, 362–364, 374, 390–392, 403–408, 423, 424, 427, 430, 437, 444, 448, 453, 457, 458
Unit vectors, 1–2, 6, 8, 14, 29, 33, 34, 37, 38, 46

V
Vector product, 4–6, 32
Volume charge densities, 104, 115, 116, 120, 122–124, 128, 144, 148, 152, 156, 158, 161, 164, 168, 172, 177, 180, 182, 199, 239, 241, 260
Volume current, 392, 406, 407
Volume current density, 342, 353, 380, 392, 426, 434

Z
Zenith angle, 133, 134

MIX
Papier aus verantwortungsvollen Quellen
Paper from responsible sources
FSC® C105338

If you have any concerns about our products,
you can contact us on
ProductSafety@springernature.com

In case Publisher is established outside the EU,
the EU authorized representative is:
**Springer Nature Customer Service Center GmbH
Europaplatz 3, 69115 Heidelberg, Germany**

Printed by Libri Plureos GmbH
in Hamburg, Germany